Current Topics in Microbiology and Immunology

Volume 371

For further volumes:
http://www.springer.com/series/82

Bryan R. Cullen
Editor

Intrinsic Immunity

Responsible Series Editor: Peter K. Vogt

Springer

Editor
Prof. Dr. Bryan R. Cullen
Department of Molecular Genetics
and Microbiology and
Center for Virology
Duke University Medical Center
Durham, NC
USA

ISSN 0070-217X
ISBN 978-3-642-37764-8 ISBN 978-3-642-37765-5 (eBook)
DOI 10.1007/978-3-642-37765-5
Springer Heidelberg New York Dordrecht London

Library of Congress Control Number: 2013938747

Printed on acid-free paper

Springer is part of Springer Science+Business Media (www.springer.com)

Preface

The immune responses that can limit or prevent disease induction by viruses are historically divided into innate and adaptive immune responses. Adaptive immunity refers to the selection and rapid expansion of T cell and B cell clones that have rearranged their T cell receptor and antibody genes, respectively, in ways that allow them to effectively recognize and neutralize most viruses or virus-infected cells. In contrast, innate immunity refers to cellular receptors that recognize Pathogen-Associated Molecular Patterns (PAMPs) and then activate signaling pathways leading to the production of interferons. This in turn induces the expression of a large number of Interferon-Stimulated Genes (ISGs), in both the infected cell and adjacent, as yet uninfected cells, many of which are able to alter the cellular environment in ways that inhibit viral replication. Cellular factors involved in PAMP recognition, so-called Pattern Recognition Receptors (PRRs) include the Toll-Like Receptors (TLRs) as well as cytoplasmic PRRs such as the Retinoic acid Inducible Gene-I (RIG-I) protein family, as well as the NOD-like receptors.

In addition to innate and adaptive immune responses, both of which are only activated after pathogen infection, many cells also constitutively express antiviral factors that can act as potent inhibitors of viral replication. These factors are now generally grouped together as mediators of intrinsic antiviral immunity. While several intrinsic immune factors, including APOBEC3G, TRIM5α, Tetherin, and SAMHD1, were initially discovered by researchers studying the replication of retroviruses, particularly HIV-1, it is now clear that some of these proteins, especially APOBEC3G and Tetherin, are in fact capable of inhibiting a wide range of viral species. In addition, it is becoming increasingly apparent that cells also express intrinsic factors that can limit the replication of other, non-retroviral species, including for example the inhibition of large DNA viruses by DNA damage response proteins. Finally, especially in invertebrate animals and plants, RNA-mediated intrinsic immunity can also play a key role in limiting viral replication and pathogenesis and even in vertebrates, microRNAs can play a major role in either restricting or, in some cases, facilitating viral replication. Because intrinsic immunity is, at least over short time periods, a set of fixed host antiviral mechanisms, it is not surprising that viruses have evolved numerous mechanisms to overcome both protein and RNA-mediated intrinsic immunity in their normal host species. However, as discussed in the chapter by W. E. Johnson, cellular intrinsic immune factors are also capable of evolution in ways that can circumvent viral countermeasures and this "rapid adversarial co-evolution" is clearly important in defining the host range of viruses.

This volume brings together nine papers reviewing different aspects of antiviral intrinsic immunity from scientists who have made major contributions to this area of research. I believe this field is an important one from several perspectives, including not only the potential design of antiviral drugs but also achieving a better understanding of the coevolution of viral pathogens and their hosts. I hope the reader will find these contributions as interesting as I did.

Bryan R. Cullen

Contents

The APOBEC3 Family of Retroelement Restriction Factors

Eric W. Refsland and Reuben S. Harris

Abstract The ability to regulate and even target mutagenesis is an extremely valuable cellular asset. Enzyme-catalyzed DNA cytosine deamination is a molecular strategy employed by vertebrates to promote antibody diversity and defend against foreign nucleic acids. Ten years ago, a family of cellular enzymes was first described with several proving capable of deaminating DNA and inhibiting HIV-1 replication. Ensuing studies on the apolipoprotein B mRNA-editing enzyme catalytic polypeptide-like 3 (APOBEC3) restriction factors have uncovered a broad-spectrum innate defense network that suppresses the replication of numerous endogenous and exogenous DNA-based parasites. Although many viruses possess equally elaborate counter-defense mechanisms, the APOBEC3 enzymes offer a tantalizing possibility of leveraging innate immunity to fend off viral infection. Here, we focus on mechanisms of retroelement restriction by the APOBEC3 family of restriction enzymes, and we consider the therapeutic benefits, as well as the possible pathological consequences, of arming cells with active DNA deaminases.

Contents

E. W. Refsland · R. S. Harris (✉)
Department of Biochemistry, Molecular Biology, and Biophysics, Institute for Molecular Virology, Center for Genome Engineering, and Masonic Cancer Center,

University of Minnesota, 321 Church Street S.E., 6-155 Jackson Hall, Minneapolis, MN 55455, USA
e-mail: rsh@umn.edu

B. R. Cullen (ed.), *Intrinsic Immunity*, Current Topics in Microbiology and Immunology 371, DOI: 10.1007/978-3-642-37765-5_1,

1 DNA Deaminase Evolution

Central to nucleic acid metabolism is the near-ubiquitous process of enzymatic deamination of adenine and cytosine bases, individually or in the context of larger nucleic acid constituents (Conticello et al. 2005; Grosjean 2009). For instance, in most species, the wobble base in several of the tRNA anti-codons is frequently changed by deamination of adenosine to inosine (A-to-I), which can then base pair with cytosine and thereby increase the flexibility and decoding capacity of the tRNA anti-codon (Gerber and Keller 1999). These enzymes belong to the adenosine deaminase acting on tRNA (ADAT) family. Related proteins in most metazoans from nematodes and flies to humans catalyze A-to-I editing of a variety of RNA targets (Kim et al. 1994; Nishikura 2010). These enzymes are called, appropriately, adenosine deaminases acting on RNA (ADAR). Editing events that occur in the coding region of an mRNA can result in amino acid substitutions in the resulting protein. However, the majority of editing events occur in non-coding regions of mRNA or in non-coding RNAs, and these A-to-I editing events can alter RNA secondary structure, stability, function, and/or capacity to be bound by regulatory RNAs such as siRNAs (Morse et al. 2002; Levanon et al. 2004; Agranat et al. 2008; Li et al. 2009).

Cytosine to uracil (C-to-U) deamination is almost as ancient as A-to-I editing (Conticello et al. 2007b; Grosjean 2009). The pyrimidine salvage pathways of most organisms use cytidine deaminase (CDA) to produce the essential RNA and DNA building blocks of uridine directly and thymidine after additional enzymatic steps (Zrenner et al. 2006). However, at some point near the root of the vertebrate tree, polynucleotide cytosine deaminases emerged, with the lamprey CDA being a present-day example (Rogozin et al. 2007). This enzyme is thought to underpin a unique form of adaptive immunity in which the DNA segments that encode arrays of highly diverse leucine-rich repeats are assembled into mature variable lymphocyte receptor genes by a recombination-mediated process. It is thought that an ancestor of the present-day lamprey enzyme served as the original substrate for expansion of the polynucleotide cytosine deaminase gene family during vertebrate evolution (Fig. 1a) (Rogozin et al. 2007; Conticello 2008). The result in most vertebrates alive today is a much larger repertoire of polynucleotide C-to-U editing enzymes that execute diverse biological functions from lipid metabolism to adaptive and innate immunity (Figs. 1 and 2).

All vertebrate polynucleotide cytosine deaminases belong to the so-called 'APOBEC' family. The defining feature of this family is a conserved His-X-Glu-X_{25-31}-Pro-Cys-X_{2-4}-Cys zinc (Z)-coordinating motif, which is strictly required for deaminase activity (where X can be a variety of amino acids) (Wedekind et al. 2003; Harris and Liddament 2004; Conticello et al. 2005; LaRue et al. 2009). As described in more detail below, key residues within this motif position zinc at the active site of the enzyme (Fig. 1c). The protein sequences within these motifs enable phylogenetic groupings into three subfamilies: APOBEC1, AID, and the APOBEC3s. The APOBEC3s can be further subdivided into three subgroups: Z1, Z2, and Z3 (Fig. 1a) (Conticello 2008; LaRue et al. 2009). Importantly, the number and organization of the A3 Z-domains can vary dramatically from branch

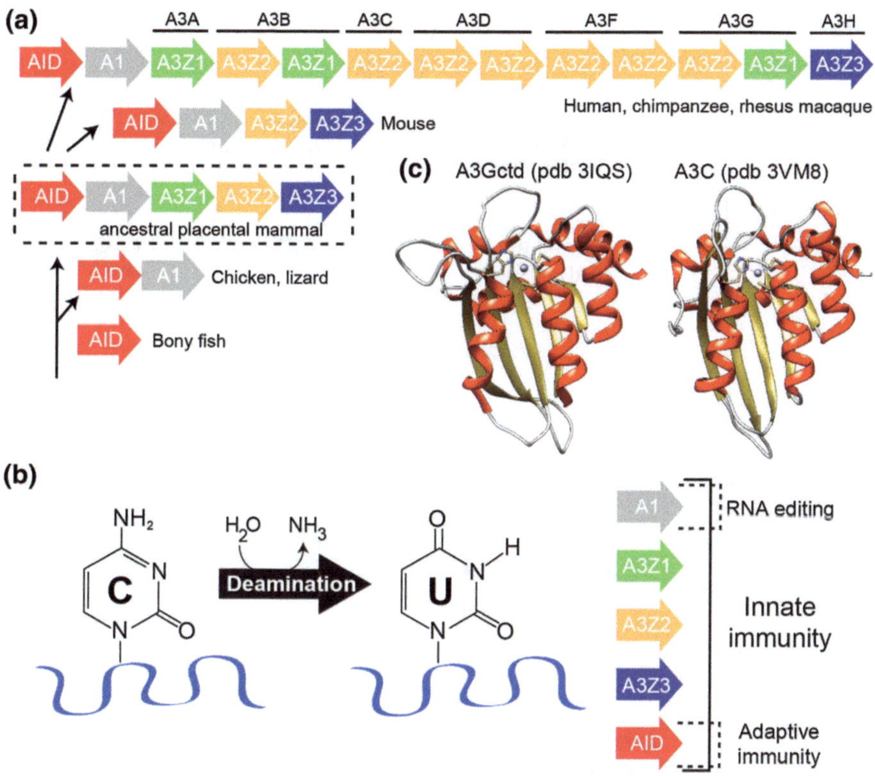

Fig. 1 Evolution and structure–function of APOBEC cytosine deaminases. **a** Expansion of the modern primate APOBEC3 locus encoding seven APOBEC3 genes with eleven zinc-coordinating (Z) domains (reprinted with permission from Lackey et al. 2012). **b** Deamination of C-to-U plays a central role in innate immunity. **c** Three-dimensional structures of the APOBEC3G C-terminal domain and APOBEC3C. A zinc ion (purple) is shown coordinated in both proteins by one histidine and two cysteine residues in the α2-β3-α3 core

to branch throughout the mammalian portion of the vertebrate phylogenetic tree (e.g., human versus mouse loci depicted in Fig. 1a).

Apolipoprotein B mRNA-editing catalytic subunit 1 (APOBEC1) has provided the namesake to the larger family. It was discovered as an enzyme that catalyzes the deamination of a specific cytosine within the APOB mRNA (Figs. 1b and 2a) (Teng et al. 1993). This produces a premature translation stop codon and a smaller secondary gene product. These two APOB proteins (APOB100 and APOB48) differentially regulate the secretion of lipoproteins from the liver (Chan 1992). Many mammalian APOBEC1 enzymes also possess DNA C-to-U deaminase activity (Harris et al. 2002; Ikeda et al. 2008; Petit et al. 2009; Ikeda et al. 2011). Taken together with the fact that earlier vertebrate lineages, such as the one represented by birds and lizards, lack an APOB-like gene, it is probable that the DNA-editing function preceded involvement in RNA editing (Severi et al. 2011).

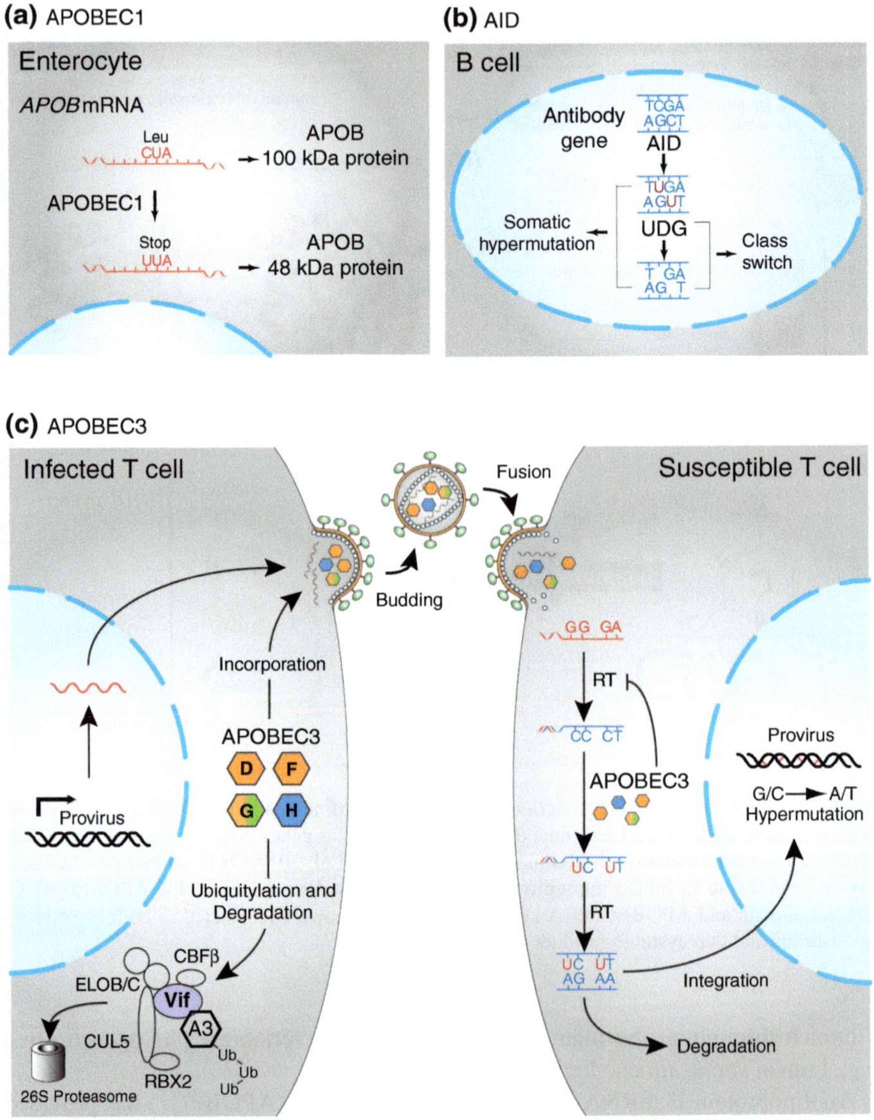

Fig. 2 The physiological functions of the APOBEC family. **a** RNA editing by APOBEC1 generates a truncated APOB protein. **b** AID activity underpins the processes of somatic hypermutation and class switch recombination in B cell germinal centers. **c** Four APOBEC3 proteins restrict HIV through cytosine deamination in the absence of Vif (reprinted with permission from Hultquist et al. 2011)

The most conserved DNA cytosine deaminase in vertebrates is activation-induced deaminase (AID; gene name *AICDA*) (Fig. 1a). AID has a central role in adaptive immunity by seeding somatic hypermutation, gene conversion, and class switch recombination with its DNA deaminase activity (Figs. 1b and 2b) [(Muramatsu et al. 1999, 2000; Di Noia and Neuberger 2002; Petersen-Mahrt et al. 2002); reviewed by (Longerich et al. 2006; Di Noia and Neuberger 2007; Conticello 2008)]. Interestingly, the genes that encode APOBEC1 and AID are positioned adjacent to each other in the genomes of most vertebrates (an inversion has placed the human gene farther away on the same chromosome). This suggests that an ancestral AID gene duplicated and diverged to produce APOBEC1 (Fig. 1a). It is likely that duplication of an ancestral AID/APOBEC1 locus produced the genetic seeds for the mammal-exclusive APOBEC3 subfamily (Fig. 1a) (Jarmuz et al. 2002; Harris and Liddament 2004).

In humans, the seven APOBEC3 proteins are encoded by a tandemly arranged gene cluster (Fig. 1a) (Jarmuz et al. 2002). These present-day genes are the products of continual evolution, in which an ancestral cluster of three Z-domains is predicted to have undergone a minimum of eight duplication events over the past 100 million years to produce the locus found in most primates (LaRue et al. 2008; Münk et al. 2012). These domains are either expressed singly or one enzyme may consist of two Z-domains (LaRue et al. 2009). In contrast, the ancestral APOBEC3 locus experienced a deletion in the rodent lineage of one of the ancestral Z-domains, leading to the present-day two domain loci, which encodes a single protein quite distinct from any of the primate enzymes (Fig. 1a).

One possible explanation for why some mammalian lineages, like primates, have many APOBEC3s, while other lineages, such as rodents, have few is that these enzymes have overlapping innate immune functions to protect the host from a variety of parasitic elements (e.g., in HIV-1 restriction, Fig. 2c; mechanism elaborated in Sect. 3, below). Because multiple distinct innate immune mechanisms serve to suppress the spread of such parasitic elements, it is reasonable to postulate that some mammals will be fortified at the APOBEC3 locus and weaker at other loci, with each mammalian lineage being distinct. For instance, primates encode a single TRIM5α protein, whereas mice have the capacity to encode a total of eight TRIM5α-like proteins (Sawyer et al. 2007; Tareen et al. 2009; Chap. 13 in Lever et al. 2010). It is likely that each species' present-day innate immune fortifications were independently shaped by past pathogenic pressures, which one can only speculate may have been the ancestors of present-day viruses and transposable elements.

2 Biochemical and Structural Insights

Zinc-dependent deaminases, such as the APOBECs, catalyze the conversion of C-to-U in polynucleotide substrates (Fig. 1b). This reaction requires the activation of water by a zinc ion coordinated by the enzyme (Fig. 1c). A glutamic acid in the active site of the enzyme protonates N3, priming the nucleophilic attack on the C4 position of the pyrimidine ring, followed by the removal and subsequent protonation of an amino group (NH_2) that results in the release of ammonia (NH_3) and uracil as products. This conversion can theoretically occur within both RNA and single-stranded DNA substrates. However, apart from APOBEC1, which has both RNA and DNA-editing activities, AID and the APOBEC3s have proven specific to DNA substrates in vitro and in vivo.

The extent of amino acid homology to APOBEC1 originally suggested that the APOBEC3 enzymes might be a family of RNA-editing proteins (Jarmuz et al. 2002). Three lines of evidence, however, demonstrated that this view was incorrect and established the APOBEC3 enzymes as single-stranded DNA cytosine deaminases. First, APOBEC3 has a high degree of homology to AID, and experiments in *E. coli* demonstrated AID, APOBEC3C, and APOBEC3G are capable of inducing high levels of mutation in an antibiotic resistance gene (Harris et al. 2002; Petersen-Mahrt et al. 2002). This was clearly due to DNA editing because mutation levels rose synergistically in a bacterial strain deficient for uracil DNA glycosylase (UDG), an enzyme that initiates base excision repair by recognizing and removing uracil exclusively from DNA (Lindahl 2000; Di Noia and Neuberger 2002; Harris et al. 2002). Second, unambiguous evidence for DNA versus RNA editing comes from head-to-head biochemical studies using recombinant enzymes. AID and APOBEC3G have a strong preference for single-stranded DNA substrates, with no detectable RNA-editing activity (Bransteitter et al. 2003; Iwatani et al. 2006). Third, a strong preference for single-stranded DNA substrates is also evident in sequencing studies of retroviruses produced in the presence of a given APOBEC3 protein, such as APOBEC3G (Fig. 2c) (Harris et al. 2003; Lecossier et al. 2003; Mangeat et al. 2003; Zhang et al. 2003). In this experimental system, each APOBEC3 protein presumably has a chance to deaminate viral genomic RNA cytosines before the reverse transcription process converts it to a single-stranded cDNA intermediate and then to the double-stranded DNA required for integration. However, the most common APOBEC3-dependent mutations detected in integrated viral DNA that has survived this process are genomic strand G-to-A mutations, entirely attributable to cDNA minus strand C-to-U deamination events. Genomic strand C-to-T editing events possibly due to RNA editing are rarely detected. Importantly, APOBEC3 DNA-editing activity is required to explain previously reported G-to-A mutation biases in HIV-1 substrates in vivo (Vartanian et al. 1994; Janini et al. 2001).

The solved structures of bacterial and yeast cytidine and cytosine deaminases were used to inform early functional and structural studies of various APOBEC3 family members (Betts et al. 1994; Ireton et al. 2003; Ko et al. 2003; Johansson et al. 2004; Xie et al. 2004). Each of these bacterial and yeast proteins, in monomeric form, is globular with a hydrophobic β-stranded core and several surrounding α-helices. The most conserved structural feature is the active site, which is defined by a histidine and two cysteines in the yeast enzyme and three cysteines in the bacterial enzymes (Xiang et al. 1997; Ireton et al. 2003; Ko et al. 2003). In both instances, these residues are positioned similarly by alpha helices and they serve to coordinate a zinc ion in the active site, which, as described above, is essential for the deamination reaction (Fig. 1b). Although these conserved features have been useful for generating models of APOBEC3 structures, they have also been misleading because the oligomeric state of each enzyme is variable. For instance, the *E. coli* CDA is homodimeric and the yeast enzyme is homotetrameric (Betts et al. 1994; Johansson et al. 2002). This has fuelled (likely incorrect) speculation that APOBEC3 family members must also function as oligomers.

Generating high-resolution structures of APOBEC3 family members has proved challenging in large part due to insolubility at higher protein concentrations [e.g., (Iwatani et al. 2006)]. However, several NMR and crystal structures have been achieved for the APOBEC3G catalytic domain (representing Z1-type deaminases), and crystal structures were obtained recently for APOBEC3C and the APOBEC3F catalytic domains (representing Z2-type deaminases) (e.g., Fig. 1c) (Chen et al. 2008; Holden et al. 2008; Furukawa et al. 2009; Shandilya et al. 2010; Kitamura et al. 2012b; Li et al. 2012; Bohn et al. 2013).

These structures have several conserved features that provide insight into how these enzymes may function. First, these proteins are all globular with a hydrophobic core consisting of five beta strands surrounded by six alpha helices and the hallmark α2-β3-α3 zinc-coordinating motif that defines the larger cytosine deaminase superfamily. Second, β-strands 3, 4, and 5 are arranged in parallel, similar to the RNA-editing enzyme TadA (an ADAT) but different from the antiparallel arrangement found in bacterial and yeast CDAs. This parallel β3-β4-β5 organization may be a key feature that distinguishes polynucleotide from non-polynucleotide deaminases. Third, although many potential oligomeric interfaces have been captured in the crystal lattices, none have proven critical for enzymatic activity and no common themes have emerged (Furukawa et al. 2009; Shandilya et al. 2010; Kitamura et al. 2012b; Bohn et al. 2013). This is consistent with a number of other studies, indicating that oligomerization may not be essential for binding and deaminating single-stranded DNA substrates (Opi et al. 2006; Nowarski et al. 2008; Shlyakhtenko et al. 2011, 2012). However, more work on this topic is clearly needed to define the role of oligomerization in vivo, because several of the family members, including APOBEC3G, elicit such a property in living cells (Bransteitter et al. 2003; Chiu et al. 2006; Soros et al. 2007; Chen et al. 2013 in preparation). Finally, it is notable that the majority of structural and amino acid differences between APOBEC3 structures are confined to non-catalytic loop regions. Such differences likely relate to substrate targeting and possible cofactor binding, ultimately reflecting physiological function.

A significant remaining question in our understanding of APOBEC3 function is how these enzymes bind single-stranded DNA substrates. A current working model proposes a positively charged brim in the region surrounding the active site consisting of R213, R215, R313, and R320 in APOBEC3G (Chen et al. 2008; Shindo et al. 2012). These residues are predicted to position single-stranded DNA substrates in a manner that allows the target cytosine to enter the active site (Chen et al. 2008). This model also predicts that in order to access to the catalytic glutamic acid, the target C will be flipped out with respect to the phosphodiester backbone. A base-flipping mechanism is in good agreement with the structure of the adenosine deaminase TadA complexed with its RNA substrate (Losey et al. 2006).

Finally, the brim-domain model and TadA structures suggest an explanation for the different local single-stranded DNA deamination preferences among APOBEC family members (Conticello et al. 2007a; Chen et al. 2008). Unlike bacterial restriction enzymes with 4, 6, or 8 base palindromic recognition sequences, APOBEC3 family members have a notable preference for the base immediately 5′ of the target C (Harris et al. 2002, 2003; Mangeat et al. 2003; Zhang et al. 2003; Bishop et al. 2004; Liddament et al. 2004; Wiegand et al. 2004; Yu et al. 2004a, b; Zheng et al. 2004; Doehle et al. 2005a; Langlois et al. 2005; Dang et al. 2006; Aguiar et al. 2008; Harari et al. 2009; Stenglein et al. 2010). Specifically, AID prefers a 5′ purine base (5′-AC or GC), APOBEC3G a 5′ cytosine (5′-CC), and all other family members a 5′ thymine (5′-TC). Several studies have recently mapped this activity to a loop adjacent to the active site, positioned between β4 and α4 secondary structural elements (Conticello et al. 2007b; Chen et al. 2008; Holden et al. 2008; Kohli et al. 2009, 2010; Rathore et al. 2013 *in preparation*). This is most dramatically evidenced by loop grafting experiments, in which this loop in AID can be replaced by the homologous loop from APOBEC3G or APOBEC3F resulting in a complete switch of the preferred base immediately 5′ of the target cytosine (Kohli et al. 2009, 2010; Carpenter et al. 2010; Wang et al. 2010). Moreover, exchanging the same loop (or even a single amino acid) between APOBEC3A and APOBEC3G completely swaps the dinucleotide preference of these enzymes (Rathore et al. 2013 in preparation). Despite this progress, the field still anxiously awaits high-resolution structures of enzyme–substrate complexes that will more precisely define the substrate binding mechanism and advance our understanding of how these enzymes function in vivo.

3 Biological Functions

3.1 APOBEC3 Proteins in HIV Restriction

Pathogens including the retrovirus HIV-1 (hereafter HIV) must both engage and avoid numerous host factors to replicate and cause disease. Genome-wide knockdown and proteomic studies suggest that up to 10 % of human proteins either directly or indirectly impact HIV replication (Brass et al. 2008; Konig et al. 2008; Zhou et al. 2008; Yeung et al. 2009; Jäger et al. 2012a, b). The majority of these proteins are required in some capacity for virus replication (i.e., dependency factors). In contrast, a small number of these cellular proteins are dominant proteins that directly suppress virus replication (i.e., restriction factors). Restriction factor hallmarks include the capacity to potently inhibit virus replication, signatures of rapid evolution (positive selection), responsiveness to interferon, and neutralization by at least one viral counter-restriction strategy (Malim and Emerman 2008; Harris et al. 2012; Malim and Bieniasz 2012). Here, we focus on the mechanism of HIV restriction by APOBEC3 DNA cytosine deaminases, and we encourage readers to see chapters on equally interesting restriction and counter-restriction mechanisms

(TRIM5α in the Chap. 2 by A.J. Fletcher and G.J. Towers, Tetherin/BST-2 in the Chap. 3 by S.J.D. Neil, and SAMHD1 in the Chap. 4 by M. Sharkey, this volume).

Original studies showed that the viral infectivity factor (Vif) protein of HIV is required for virus replication in primary $CD4^{+}$ lymphocytes and in several common laboratory T cell lines (e.g., CEM, H9, so-called non-permissive), but it was dispensable in several others (e.g., CEM-SS, SupT1, or permissive lines) (Fisher et al. 1987; Strebel et al. 1987; Gabuzda et al. 1992). This phenotypic difference led to the cloning of APOBEC3G as one cDNA sequence expressed differentially (of many) between CEM and CEM-SS (Sheehy et al. 2002). However, APOBEC3G proved remarkable as it could convert a permissive cell line to a non-permissive phenotype (Sheehy et al. 2002). Taken together with the independent and near-simultaneous discoveries of APOBEC3G as a putative RNA-editing factor and as a DNA-editing enzyme, an editing mechanism of restriction was predicted and shortly after demonstrated (Harris et al. 2002, 2003; Lecossier et al. 2003; Mangeat et al. 2003; Zhang et al. 2003).

Over five hundred papers have now been published on APOBEC3G and the related APOBEC3 proteins that have culminated in a current Trojan horse-like model for HIV restriction as shown in Fig. 2c. To be effective as an HIV restriction factor, APOBEC3G must first be expressed in the cytoplasmic compartment of an infected virus-producing cell (Mangeat et al. 2003). Second, cytoplasmic APOBEC3G is thought to interact with a Gag ribonucleoprotein complex, and this interaction is required for APOBEC3G packaging (Alce and Popik 2004; Luo et al. 2004; Schafer et al. 2004; Svarovskaia et al. 2004; Khan et al. 2005; Burnett and Spearman 2007; Bogerd and Cullen 2008). This interaction can be disrupted by RNase treatment and is therefore thought to involve a bridging RNA between APOBEC3G and the nucleocapsid region of Gag (Cen et al. 2004; Svarovskaia et al. 2004; Iwatani et al. 2007; Bogerd and Cullen 2008). The identity of the bridging RNA is still an active area of investigation; while several reports indicate a role for the Alu-like 7SL RNA, a role for viral genomic RNA has not been excluded (Khan et al. 2005, 2007; Tian et al. 2007; Wang et al. 2007; Bogerd and Cullen 2008). Third, by an ill-defined mechanism, packaged APOBEC3G must breach the nucleocapsid core of the viral particle. This is a genetically (but not mechanistically) defined step, as various chimeric constructs have the capacity to be packaged into viral particles, yet do not enter the core or restrict (Haché et al. 2005; Martin et al. 2011; Song et al. 2012).

Once an APOBEC3G-loaded viral core is deposited into a target cell, reverse transcription proceeds and a susceptible single-stranded cDNA intermediate is generated. Here, APOBEC3G deaminates C-to-U (Harris et al. 2003; Mangeat et al. 2003; Zhang et al. 2003; Yu et al. 2004b). The uracilated cDNA is either subjected to degradation or templates the second-strand synthesis, preceding integration into the target cell's genome (Kaiser and Emerman 2006; Mbisa et al. 2007; Yang et al. 2007). The uracils in the cDNA strand template the insertion of adenines in the nascent plus strand and thereby immortalize G-to-A mutations that limit subsequent rounds of viral replication (Harris et al. 2003; Lecossier et al. 2003; Mangeat et al. 2003; Zhang et al. 2003).

Human T cell lines that express near-physiological levels of an APOBEC3G catalytic mutant do not suppress the replication of a Vif-deficient virus, indicating that the predominant mechanism of HIV restriction depends upon deaminase activity (Bishop et al. 2004; Miyagi et al. 2007; Schumacher et al. 2008; Browne et al. 2009). However, a component of APOBEC3G's capacity to restrict HIV may be deaminase independent (Newman et al. 2005; Bishop et al. 2006, 2008; Iwatani et al. 2006; Opi et al. 2006; Iwatani et al. 2007; Mbisa et al. 2007). The most convincing of these studies has shown that APOBEC3G is capable of binding viral genomic RNA and sterically hindering the processivity of reverse transcriptase (Bishop et al. 2006, 2008). Nevertheless, the importance of deaminase-independent mechanisms, at least for APOBEC3G and HIV, is questionable given the aforementioned results at physiological expression levels.

Soon after the initial discovery and functional characterization of APOBEC3G, attention turned toward its six, most closely related family members. Proviral sequences isolated from HIV-infected individuals exhibit two distinct patterns of G-to-A mutation consistent with APOBEC3-mediated deamination, both 5′-GG-to-AG and 5′-GA-to-AA (Fitzgibbon et al. 1993; Janini et al. 2001; Caride et al. 2002; Kieffer et al. 2005; Pace et al. 2006; Gandhi et al. 2008; Land et al. 2008). As mentioned previously, APOBEC3G has a demonstrated preference for deaminating 5′-CC dinucleotides on the viral minus strand resulting in 5′-GG-to-AG mutations and is therefore the most likely source of this mutational pattern (Harris et al. 2003; Mangeat et al. 2003; Zhang et al. 2003; Liddament et al. 2004; Yu et al. 2004b). However, high levels of proviral 5′-GA-to-AA mutation in patients' proviral sequences suggested the involvement of at least one additional family member. Unfortunately, all six other APOBEC3 family members prefer to deaminate a 5′-TC, therefore excluding the possibility of identifying an additional APOBEC3 source based solely on mutational preference [e.g., (Bishop et al. 2004); reviewed in (Albin and Harris 2010)].

In cell-based model systems with forced cDNA expression, all seven APOBEC3s have reported activity [e.g., (Bishop et al. 2004; Liddament et al. 2004; Wiegand et al. 2004; Yu et al. 2004a; Zheng et al. 2004; Doehle et al. 2005a; Rose et al. 2005; Dang et al. 2006, 2008; Goila-Gaur et al. 2007; OhAinle et al. 2008); reviewed in (Albin and Harris 2010)]. However, cellular expression patterns, ability to encapsidate into viral particles, and restriction in relevant T cell models indicate not all the APOBEC3s function in vivo as HIV restriction factors. For example, in primary $CD4^+$ T lymphocytes, a major target of HIV in vivo, six APOBEC3s are expressed at appreciable levels, but APOBEC3A mRNA is undetectable (Koning et al. 2009; Refsland et al. 2010). In addition, only four: APOBEC3D, APOBEC3F, APOBEC3G, and APOBEC3H (but not APOBEC3A, APOBEC3B, and APOBEC3C) package into viral-like particles and inhibit viral replication when stably expressed in human T cell lines (Hultquist et al. 2011). Finally, endogenous APOBEC3D and APOBEC3F combine to explain the 5′-GA-to-AA mutation pattern observed in the non-permissive T cell line CEM2n (Refsland et al. 2012). Of note, CEM2n does not express appreciable levels of APOBEC3H mRNA; therefore, the endogenous contribution of this protein could not be evaluated. Seven variants of APOBEC3H have been reported, and cell-based studies indicate only a subset of these haplotypes are stable at the protein level and capable of HIV restriction (Dang et al. 2008; OhAinle et al. 2008; Harari et al. 2009; Wang et al. 2011). The significance of this natural variation in human populations and the relevance of it to HIV restriction require further investigation. Taken together, while all the APOBEC3s can be compelled to deaminate a single-stranded viral DNA in vitro or in heterologous cell lines by gross overexpression, only four APOBEC3s—APOBEC3D, APOBEC3F, APOBEC3G, and APOBEC3H—appear to have the capacity to function as HIV restriction factors in T lymphocytes.

All successful viruses have evolved sophisticated immune suppression and/or evasion mechanisms to neutralize the cellular defense systems they must face. One of the best-studied APOBEC3 counter-restriction mechanisms is orchestrated by the HIV Vif. This protein is small (23 kDa), highly basic, and strictly required for pathogenesis in vivo as well as for virus replication ex vivo in monocytes, macrophages, primary $CD4^+$ T lymphocytes, and non-permissive T cell lines (Kan et al. 1986; Lee et al. 1986; Sodroski et al. 1986; Fisher et al. 1987; Strebel et al. 1987; Gabuzda et al. 1992; von Schwedler et al. 1993). Vif recruits a multiprotein E3 ubiquitin ligase complex consisting of CUL5/NEDD8, ELOB, ELOC, RBX2, and CBFβ to mediate proteasomal degradation of the cellular APOBEC3 restriction factors (Fig. 2c) (Conticello et al. 2003; Marin et al. 2003; Sheehy et al. 2003; Yu et al. 2003; Mehle et al. 2004; Jäger et al. 2012b). In addition to this primary role, it has been suggested that Vif can relieve APOBEC3G-mediated restriction by alternative mechanisms such as directly blocking deaminase activity, preventing encapsidation, sequestering the protein in catalytically inactive conformations, or impeding APOBEC3G translation (Mariani et al. 2003; Stopak et al. 2003; Kao et al. 2004; Santa-Marta et al. 2005; Opi et al. 2007; Goila-Gaur et al. 2008; Britan-Rosich et al. 2011). The extent to which these mechanisms function to counteract APOBEC3 during a productive infection warrants further investigation but is likely to be secondary to the proteasomal degradation mechanism.

3.2 APOBEC3 Proteins in General Innate Immune Defense

Exogenous retroviruses are rare in humans, consistent with the idea that the multifaceted APOBEC3 defense provides robust protection against this type of pathogen. HIV is one exception to this rule. It is successful, at least in part, due to Vif and degradation of the cellular APOBEC3s. However, evidence for a long history of positive selection acting on the human APOBEC3 locus suggests this family has been defending the genomic integrity of its host's cells long before HIV was transmitted into the human population (Sawyer et al. 2004; Zhang and Webb 2004; Sanville et al. 2010). Given the significant presence of endogenous retroelements in the genomes of mammals (nearly 50 %), suppression of these elements may represent the subfamily's true *raison d'être* and the primary source of the positive selection that has shaped the complex present-day locus and the diverse functions of these innate immunity factors.

Endogenous retroelements, including those containing long terminal repeats (LTR) like endogenous retroviruses, as well as non-LTR elements like long interspersed nuclear elements (LINEs) and short interspersed nuclear elements (SINEs), may have provided the evolutionary pressure necessary for the maintained expansion of the APOBEC3 locus in primates. Additionally, the differences in retrotransposition frequency between rodents and primates could be attributable to possessing an arsenal of seven APOBEC3 proteins versus only a single one (Maksakova et al. 2006; Stenglein and Harris 2006). In support of this hypothesis, human APOBEC3s have demonstrated activity on LTR retrotransposons from mice and yeast (Dutko et al. 2005; Esnault et al. 2005, 2006; Schumacher et al. 2005; Bogerd et al. 2006a; Chen et al. 2006; Lee and Bieniasz 2007; Jern and Coffin 2008; Lee et al. 2008). Additionally, some endogenous retroviruses exhibit the characteristic scars from APOBEC3F and APOBEC3G activity in their genomes (Anwar et al. 2013 in press). Non-LTR elements, including LINE1 and Alu, are also restricted by human APOBEC3s but, in contrast to the HIV restriction mechanism described above, this mechanism appears entirely deamination independent (Bogerd et al. 2006b; Chiu et al. 2006; Muckenfuss et al. 2006; Stenglein and Harris 2006; Carmi et al. 2011).

The generation of a single-stranded DNA intermediate in the life cycle of a parasitic element may render it susceptible to APOBEC3 restriction. In addition to HIV, other viruses including simian immunodeficiency virus, murine leukemia virus, foamy virus, porcine endogenous retrovirus, human T cell leukemia virus, and hepatitis B virus have all been reported to be susceptible to APOBEC3-mediated editing of their genomes (Harris et al. 2003; Mangeat et al. 2003; Mariani et al. 2003; Kobayashi et al. 2004; Turelli et al. 2004; Yu et al. 2004a; Doehle et al. 2005b; Lochelt et al. 2005; Russell et al. 2005; Suspène et al. 2005; Abudu et al. 2006; Delebecque et al. 2006; Jonsson et al. 2007). Whether or not the APOBEC3-mediated restriction of all of these viral pathogens is part of a natural innate immune response or a by-product of the particular model systems used awaits further investigation.

Endogenous retroelements and their mobility are believed to have played a central role early in shaping the human genome during speciation (Kazazian 2004; Carmi et al. 2011). However, this process has its associated costs. Ultimately, cells have devised strategies to defend and preserve genomic integrity by curbing the movement of these genetic elements. The evolution of the APOBEC3 family has likely played a prominent role in this defense and in diversifying the retroelements to make them more useful for the host species [e.g., (Carmi et al. 2011)].

4 Pathological Consequences of DNA Deamination

Although DNA cytosine deamination has obvious benefits for individual cells and the organism as a whole, this process may have considerable pathological consequences, most notably cancer. For instance, overexpression of APOBEC1 in the liver of mice has been shown to cause hepatocellular carcinoma and liver dysplasia (Yamanaka et al. 1995). Likewise, AID transgenesis also causes cancer, and it has been implicated in initiating the chromosomal translocations responsible for some lymphocyte neoplasias including the hallmark c-myc/IgH rearrangement in Burkitt's lymphoma [(Okazaki et al. 2003; Ramiro et al. 2004; Unniraman et al. 2004); reviewed by (Perez-Duran et al. 2007)]. AID may also promote resistance to the chemotherapeutic drug imatinib (Klemm et al. 2009). However, the overall impact of APOBEC1 and AID on human cancer is questionable because their expression is largely limited to tissues associated with their biological functions, APOBEC1 in the enterocytes of the small intestine and AID in B cells (Fig. 3a) (Powell et al. 1987; Teng et al. 1993; Muramatsu et al. 1999, 2000).

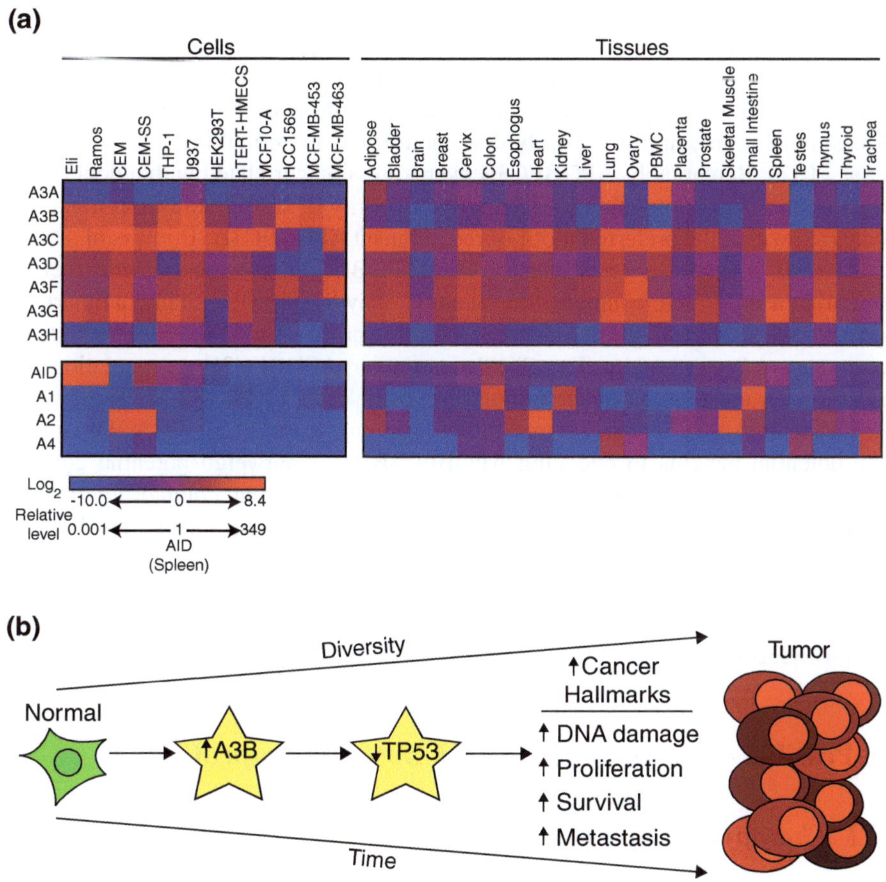

Fig. 3 APOBEC expression and model of carcinogenesis. **a** APOBEC family mRNA expression in the indicated cell lines and normal human tissues. Data are relative to levels of AID in the spleen (reprinted with permission from Burns et al. 2013). **b** Model depicting the effect of APOBEC3B overexpression over time

In contrast, most APOBEC3s have much broader expression ranges that span most tissues in the human body (Fig. 3a) (Koning et al. 2009; Refsland et al. 2010; Burns et al. 2013). Broad expression profiles, potent DNA deaminase activity, and C-to-T transition biases in tumor genome sequences strongly suggested that one or more of the APOBEC3 proteins may be a source of mutation in different cancers (Harris et al. 2002; Nik-Zainal et al. 2012; Roberts et al. 2012; Burns et al. 2013). APOBEC3B became a leading candidate as it uniquely and constitutively localizes to the nucleus by inheriting a nuclear import mechanism from AID (Bogerd et al. 2006b; Bonvin et al. 2006; Muckenfuss et al. 2006; Stenglein and Harris 2006; Kinomoto et al. 2007; Stenglein et al. 2008; Hultquist et al. 2011; Pak et al. 2011; Lackey et al. 2012, 2013). Recently, APOBEC3B was found overexpressed in some laboratory breast cancer cell lines, but not in available control cell lines (Fig. 3a; for example, compare HCC1569, MCF-MB-453, and MCF-MB-463 to telomerase immortalized human mammary epithelial cells, hTERT HMECS) (Burns et al. 2013). APOBEC3B up-regulation was shown to be responsible for elevated genomic uracil levels and increased mutation rates in breast cancer cell lines (Burns et al. 2013). APOBEC3B up-regulation was similarly robust in the majority of human breast tumors, in contrast to barely detectable levels in normal breast tissue (Burns et al. 2013). Remarkably, APOBEC3B overexpression correlated with a doubling in the tumor genomic mutation loads, and the majority of C-to-T transition mutations occurred within the preferred motif of recombinant APOBEC3B (Burns et al. 2013).

Overall, a model is emerging for how APOBEC3B provides genetic fuel for tumorigenesis, which, coupled with selection, may help explain many hallmarks of cancer such as increased DNA damage, elevated proliferation, decreased apoptosis, and massive heterogeneity (Fig. 3b). In particular, APOBEC3B up-regulation correlates with inactivation of the tumor suppressor gene *TP53*, which strongly suggests that it may be an early tumor-initiating event (Burns et al. 2013). Obviously, the potential benefits to encoding APOBEC3B must outweigh potential costs of carcinogenesis. An attractive explanation for this apparent conundrum may be that its innate immune function is important early in life and for the health of the species, for instance, in germ cells or early development (Bogerd et al. 2006b; Wissing et al. 2011), whereas the toll of cancer is not imposed in most instances until after the reproductive years. In any event, much more work is now justified on APOBEC3B and its role in breast and, potentially, other human cancers.

5 Possible Avenues to APOBEC3-Based Therapeutics

5.1 Therapy by Hypermutation

If left unimpeded by Vif, APOBEC3 proteins such as APOBEC3G can convert up to 10 % of viral cDNA cytosines into uracils in a single round of virus replication [e.g., (Harris et al. 2003; Yu et al. 2004b)]. The resulting massive levels of G-to-A mutations effectively ruin the genetic potential of the retrovirus in a process called lethal mutagenesis (Loeb et al. 1999; Haché et al. 2006). Moreover, the preferred context of APOBEC3G deamination events often results in the conversion of the tryptophan codon TGG into a premature stop codon TAG, which is more detrimental to the virus than a simple amino acid change. Overall, physiological levels of APOBEC3 proteins largely, if not fully, suppress the replication of Vif-deficient HIV.

This remarkable potency therefore raises the prospect of developing drugs to leverage the APOBEC3 restriction mechanism against HIV (Fig. 4a). Direct inhibition of Vif is certainly one strategy, but such an approach is destined to be susceptible to problems imposed by extensive natural HIV variation and the rapid evolution of drug resistance. A more appealing alternative may be to develop a drug toward one of the more genetically stable cellular proteins recruited to the APOBEC3 degradation complex. In particular, Vif requires at least four heterologous protein–protein interactions to successfully counteract the APOBEC3 proteins. The first strategy, of course, is targeting the direct interaction with the APOBEC3 proteins themselves, which may occur through conserved structural motifs (Albin et al. 2010; Kitamura et al. 2012a). Second is targeting the recently discovered interaction with CBFβ, which is essential for Vif stability and function (Jäger et al. 2012b; Zhang et al. 2012). Third and fourth are targeting the distinct Vif interaction motifs in ELOC (the SLQ motif) or CUL5, which are also essential for activity of the APOBEC3 Ub ligase complex (Marin et al. 2003; Yu et al. 2003). Finally, it may be possible to target an upstream component of the proteasomal pathway involved in APOBEC3 degradation [e.g., (Kim et al. 2013)].

However, despite these clear opportunities for drug development, progress has been relatively underwhelming. Proof-of-concept experiments have been achieved through cell-based screens for preservation of APOBEC3G-GFP fluorescence in the presence of HIV Vif (Nathans et al. 2008). The lead molecule from these studies, RN18, however, has only been subject to modest additional development (Ali et al. 2012). An additional concern holding back development of RN18 is the major challenge of identifying the molecular target, which is essential for structural studies and rational improvements through medicinal chemistry.

Independent lead compounds have also been identified based on predicted fit into the Vif-binding pocket in ELOC, which would effectively outcompete the SLQ motif of Vif (Huang et al. 2013a; b). However, the best of these indolizine-type compounds has a modest IC50 value of 11 μM. This will likely need improvements in potency and solubility before efficacy can be achieved in live cell studies. In any event, much more work in this area is needed, including larger-scale cellular screens, biochemical screens, and computational screens, followed by a comprehensive set of secondary and tertiary screens to narrow in on the most effective drug candidates.

5.2 Therapy by Hypomutation

Vif efficiently counteracts the HIV-relevant APOBEC3 repertoire (Conticello et al. 2003; Kao et al. 2003; Marin et al. 2003; Sheehy et al. 2003; Stopak et al. 2003; Hultquist et al. 2011). Nonetheless, evidence of APOBEC3 activity is found in the

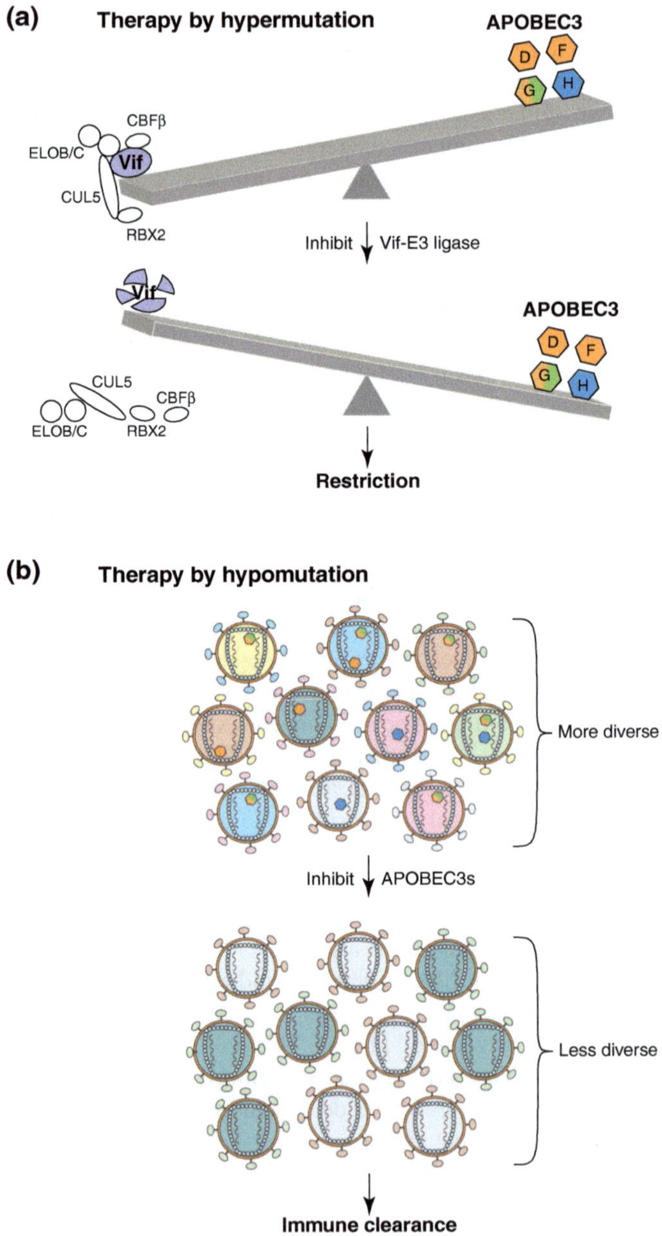

Fig. 4 Potential therapies harnessing APOBEC3 innate immunity. **a** Inhibition of the Vif–E3 ligase complex could result in virus restriction. **b** Inhibiting the deaminase activity of APOBEC3 proteins could deprive HIV of a source of variation and allow immune clearance

proviral sequences derived from infected patients [e.g., (Janini et al. 2001)]. This raises the possibilities that either Vif cannot completely neutralize all of the cytoplasmic APOBEC3 before encapsidation or HIV regulates its high mutation rate, in part, through controlled degradation of the APOBEC3 proteins with Vif acting as a molecular rheostat optimizing the levels of cytosine deamination necessary for immune evasion and potentially drug resistance (Haché et al. 2006; Harris 2008).

A counterintuitive but potentially more effective strategy to decrease the pathogenesis of HIV may be to inhibit the enzymatic activity of the APOBEC3s, thereby eliminating a potential source of variation for the virus (Fig. 4b) (Harris 2008; Li et al. 2012; Olson et al. 2013). Toward this end, high-throughput screens have resulted in proof-of-concept experiments that APOBEC3G activity can be blocked with small molecules (Li et al. 2012; Olson et al. 2013). However, much additional work is still necessary to improve the solubility, potency, and bioavailability of APOBEC3 inhibitors, and efficacy will need to be demonstrated in cell-based studies before critical animal experiments can be done. Nevertheless, a hypomutation strategy is attractive because decreasing the HIV mutation rate has the potential to limit the diversity of the viral population and render it susceptible to normal immune clearance mechanisms (Fig. 4b). Indeed, the adaptive T cell- and B cell-mediated immune responses manage to keep the virus in check initially in most patients, and it is tempting to speculate that even a slight tip in favor of host immunity may enable complete virus clearance (i.e., analogous to most other viral infections). Given the fact that existing anti-retroviral drugs are numerous and largely effective, a big challenge facing the field is the development of a curative therapy. It remains possible that promoting either virus hypermutation or hypomutation could be part of such a cure.

Acknowledgments We thank J. Hultquist and A. Land for comments, and B. Leonard for Fig.3b. We apologize to colleagues whose work could not be cited due to space limitations. Support was provided by grants from the National Institutes of Health R01-AI064046 and P01-GM091743 to RSH and F31-DA033186 to EWR.

References

Abudu A, Takaori-Kondo A, Izumi T et al (2006) Murine retrovirus escapes from murine APOBEC3 via two distinct novel mechanisms. Curr Biol 16:1565–1570

Agranat L, Raitskin O, Sperling J, Sperling R (2008) The editing enzyme ADAR1 and the mRNA surveillance protein hUpf1 interact in the cell nucleus. Proc Natl Acad Sci USA 105:5028–5033

Aguiar RS, Lovsin N, Tanuri A, Peterlin BM (2008) Vpr.A3A chimera inhibits HIV replication. J Biol Chem 283:2518–2525

Albin JS, Harris RS (2010) Interactions of host APOBEC3 restriction factors with HIV-1 in vivo: implications for therapeutics. Expert Rev Mol Med 12:e4

Albin JS, LaRue RS, Weaver JA et al (2010) A single amino acid in human APOBEC3F alters susceptibility to HIV-1 Vif. J Biol Chem 285:40785–40792

Alce TM, Popik W (2004) APOBEC3G is incorporated into virus-like particles by a direct interaction with HIV-1 Gag nucleocapsid protein. J Biol Chem 279:34083–34086

Ali A, Wang J, Nathans RS, Cao H, Sharova N, Stevenson M, Rana TM (2012) Synthesis and structure-activity relationship studies of HIV-1 virion infectivity factor (Vif) inhibitors that block viral replication. ChemMedChem 7:1217–1229

Anwar F, Davenport MP, Ebrahimi D (2013) Footprint of APOBEC3 on the genome of retrotransposons. J Virol (in press)

Betts SD, Hachigian TM, Pichersky E, Yocum CF (1994) Reconstitution of the spinach oxygen-evolving complex with recombinant *Arabidopsis* manganese-stabilizing protein. Plant Mol Biol 26:117–130

Bishop KN, Holmes RK, Malim MH (2006) Antiviral potency of APOBEC proteins does not correlate with cytidine deamination. J Virol 80:8450–8458

Bishop KN, Holmes RK, Sheehy AM, Davidson NO, Cho SJ, Malim MH (2004) Cytidine deamination of retroviral DNA by diverse APOBEC proteins. Curr Biol 14:1392–1396

Bishop KN, Verma M, Kim EY, Wolinsky SM, Malim MH (2008) APOBEC3G inhibits elongation of HIV-1 reverse transcripts. PLoS Pathog 4:e1000231

Bogerd HP, Cullen BR (2008) Single-stranded RNA facilitates nucleocapsid: APOBEC3G complex formation. RNA 14:1228–1236

Bogerd HP, Wiegand HL, Doehle BP, Lueders KK, Cullen BR (2006a) APOBEC3A and APOBEC3B are potent inhibitors of LTR-retrotransposon function in human cells. Nucleic Acids Res 34:89–95

Bogerd HP, Wiegand HL, Hulme AE, Garcia-Perez JL, O'Shea KS, Moran JV, Cullen BR (2006b) Cellular inhibitors of long interspersed element 1 and Alu retrotransposition. Proc Natl Acad Sci U S A 103:8780–8785

Bohn M-F, Shandilya SMD, Albin JS, Kouno T, Anderson BD, McDougle RM, Carpenter MA, Rathore A, Evans L, Davis AN, Zhang J, Lu Y, Somasundaran M, Matsuo H, Harris RS & Schiffer CA (2013) Crystal structure of the HIV-1 Vif-binding, catalytically active domain of the DNA cytosine deaminase APOBEC3F. Structure, in press.

Bonvin M, Achermann F, Greeve I et al (2006) Interferon-inducible expression of APOBEC3 editing enzymes in human hepatocytes and inhibition of hepatitis B virus replication. Hepatology 43:1364–1374

Bransteitter R, Pham P, Scharff MD, Goodman MF (2003) Activation-induced cytidine deaminase deaminates deoxycytidine on single-stranded DNA but requires the action of RNase. Proc Natl Acad Sci USA 100:4102–4107

Brass AL, Dykxhoorn DM, Benita Y et al (2008) Identification of host proteins required for HIV infection through a functional genomic screen. Science 319:921–926

Britan-Rosich E, Nowarski R, Kotler M (2011) Multifaceted counter-APOBEC3G mechanisms employed by HIV-1 Vif. J Mol Biol 410:1065–1076

Browne EP, Allers C, Landau NR (2009) Restriction of HIV-1 by APOBEC3G is cytidine deaminase-dependent. Virology 387:313–321

Burnett A, Spearman P (2007) APOBEC3G multimers are recruited to the plasma membrane for packaging into human immunodeficiency virus type 1 virus-like particles in an RNA-dependent process requiring the NC basic linker. J Virol 81:5000–5013

Burns MB, Lackey L, Carpenter MA et al (2013) APOBEC3B is an enzymatic source of mutation in breast cancer. Nature 494:366–370

Caride E, Brindeiro RM, Kallas EG, de Sa CA, Eyer-Silva WA, Machado E, Tanuri A (2002) Sexual transmission of HIV-1 isolate showing G→A hypermutation. J Clin Virol 23:179–189

Carmi S, Church GM, Levanon EY (2011) Large-scale DNA editing of retrotransposons accelerates mammalian genome evolution. Nat Commun 2:519

Carpenter MA, Rajagurubandara E, Wijesinghe P, Bhagwat AS (2010) Determinants of sequence-specificity within human AID and APOBEC3G. DNA Repair 9:579–587

Cen S, Guo F, Niu M, Saadatmand J, Deflassieux J, Kleiman L (2004) The interaction between HIV-1 Gag and APOBEC3G. J Biol Chem 279:33177–33184

Chan L (1992) Apolipoprotein B, the major protein component of triglyceride-rich and low density lipoproteins. JBiol Chem 267:25621–25624

Chen H, Lilley CE, Yu Q et al (2006) APOBEC3A is a potent inhibitor of adeno-associated virus and retrotransposons. Curr Biol 16:480–485

Chen KM, Harjes E, Gross PJ et al (2008) Structure of the DNA deaminase domain of the HIV-1 restriction factor APOBEC3G. Nature 452:116–119

Chen YJ, Li M, Carpenter MA, McDougle RM, Macdonald PJ, Harris RS, Mueller JD (2013) APOBEC3 multimerization correlates with HIV-1 restriction activity in living cells. J Mol Biol (in preparation)

Chiu YL, Witkowska HE, Hall SC et al (2006) High-molecular-mass APOBEC3G complexes restrict Alu retrotransposition. Proc Natl Acad Sci USA 103:15588–15593

Conticello SG (2008) The AID/APOBEC family of nucleic acid mutators. Genome Biol 9:229

Conticello SG, Harris RS, Neuberger MS (2003) The Vif protein of HIV triggers degradation of the human antiretroviral DNA deaminase APOBEC3G. Curr Biol 13:2009–2013

Conticello SG, Langlois MA, Neuberger MS (2007a) Insights into DNA deaminases. Nat Struct Mol Biol 14:7–9

Conticello SG, Langlois MA, Yang Z, Neuberger MS (2007b) DNA deamination in immunity: AID in the context of its APOBEC relatives. Adv Immunol 94:37–73

Conticello SG, Thomas CJ, Petersen-Mahrt SK, Neuberger MS (2005) Evolution of the AID/APOBEC family of polynucleotide (deoxy) cytidine deaminases. Mol Biol Evol 22:367–377

Dang Y, Siew LM, Wang X, Han Y, Lampen R, Zheng YH (2008) Human cytidine deaminase APOBEC3H restricts HIV-1 replication. J Biol Chem 283:11606–11614

Dang Y, Wang X, Esselman WJ, Zheng YH (2006) Identification of APOBEC3DE as another antiretroviral factor from the human APOBEC family. J Virol 80:10522–10533

Delebecque F, Suspene R, Calattini S et al (2006) Restriction of foamy viruses by APOBEC cytidine deaminases. J Virol 80:605–614

Di Noia J, Neuberger MS (2002) Altering the pathway of immunoglobulin hypermutation by inhibiting uracil-DNA glycosylase. Nature 419:43–48

Di Noia JM, Neuberger MS (2007) Molecular mechanisms of antibody somatic hypermutation. Annu Rev Biochem 76:1–22

Doehle BP, Schäfer A, Cullen BR (2005a) Human APOBEC3B is a potent inhibitor of HIV-1 infectivity and is resistant to HIV-1 Vif. Virology 339:281–288

Doehle BP, Schafer A, Wiegand HL, Bogerd HP, Cullen BR (2005b) Differential sensitivity of murine leukemia virus to APOBEC3-mediated inhibition is governed by virion exclusion. J Virol 79:8201–8207

Dutko JA, Schafer A, Kenny AE, Cullen BR, Curcio MJ (2005) Inhibition of a yeast LTR retrotransposon by human APOBEC3 cytidine deaminases. Curr Biol 15:661–666

Esnault C, Heidmann O, Delebecque F et al (2005) APOBEC3G cytidine deaminase inhibits retrotransposition of endogenous retroviruses. Nature 433:430–433

Esnault C, Millet J, Schwartz O, Heidmann T (2006) Dual inhibitory effects of APOBEC family proteins on retrotransposition of mammalian endogenous retroviruses. Nucleic Acids Res 34:1522–1531

Fisher AG, Ensoli B, Ivanoff L et al (1987) The sor gene of HIV-1 is required for efficient virus transmission in vitro. Science 237:888–893

Fitzgibbon JE, Mazar S, Dubin DT (1993) A new type of G→A hypermutation affecting human immunodeficiency virus. AIDS Res Hum Retroviruses 9:833–838

Furukawa A, Nagata T, Matsugami A et al (2009) Structure, interaction and real-time monitoring of the enzymatic reaction of wild-type APOBEC3G. EMBO j 28:440–451

Gabuzda DH, Lawrence K, Langhoff E, Terwilliger E, Dorfman T, Haseltine WA, Sodroski J (1992) Role of vif in replication of human immunodeficiency virus type 1 in CD4+ T lymphocytes. J Virol 66:6489–6495

Gandhi SK, Siliciano JD, Bailey JR, Siliciano RF, Blankson JN (2008) Role of APOBEC3G/F-mediated hypermutation in the control of human immunodeficiency virus type 1 in elite suppressors. J Virol 82:3125–3130

Gerber AP, Keller W (1999) An adenosine deaminase that generates inosine at the wobble position of tRNAs. Science 286:1146–1149
Goila-Gaur R, Khan MA, Miyagi E, Kao S, Opi S, Takeuchi H, Strebel K (2008) HIV-1 Vif promotes the formation of high molecular mass APOBEC3G complexes. Virology 372:136–146
Goila-Gaur R, Khan MA, Miyagi E, Kao S, Strebel K (2007) Targeting APOBEC3A to the viral nucleoprotein complex confers antiviral activity. Retrovirology 4:61
Grosjean H (2009) DNA and RNA modification enzymes: structure, mechanism, function and evolution. Landes Bioscience, Austin
Haché G, Liddament MT, Harris RS (2005) The retroviral hypermutation specificity of APOBEC3F and APOBEC3G is governed by the C-terminal DNA cytosine deaminase domain. Journal Bio Chem 280:10920–10924
Haché G, Mansky LM, Harris RS (2006) Human APOBEC3 proteins, retrovirus restriction, and HIV drug resistance. AIDS Rev 8:148–157
Harari A, Ooms M, Mulder LC, Simon V (2009) Polymorphisms and splice variants influence the antiretroviral activity of human APOBEC3H. J Virol 83:295–303
Harris RS (2008) Enhancing immunity to HIV through APOBEC. Nat Biotechnol 26:1089–1090
Harris RS, Bishop KN, Sheehy AM et al (2003) DNA deamination mediates innate immunity to retroviral infection. Cell 113:803–809
Harris RS, Hultquist JF, Evans DT (2012) The restriction factors of human immunodeficiency virus. J Bio Chem 287:40875–40883
Harris RS, Liddament MT (2004) Retroviral restriction by APOBEC proteins. Nat Rev Immunol 4:868–877
Harris RS, Petersen-Mahrt SK, Neuberger MS (2002) RNA editing enzyme APOBEC1 and some of its homologs can act as DNA mutators. Mol Cell 10:1247–1253
Holden LG, Prochnow C, Chang YP et al (2008) Crystal structure of the anti-viral APOBEC3G catalytic domain and functional implications. Nature 456:121–124
Huang W, Zuo T, Jin H, Liu Z, Yang Z, Yu X, Zhang L (2013a) Design, synthesis and biological evaluation of indolizine derivatives as HIV-1 VIF-ElonginC interaction inhibitors. Mol Diversity doi:10.1007/s11030-013-9424-3
Huang W, Zuo T, Luo X, et al. (2013b) Indolizine derivatives as HIV-1 VIF-ElonginC interaction inhibitors. Chem Biol Drug Des doi:10.1111/cbdd.12119
Hultquist JF, Lengyel JA, Refsland EW, LaRue RS, Lackey L, Brown WL, Harris RS (2011) Human and rhesus APOBEC3D, APOBEC3F, APOBEC3G, and APOBEC3H demonstrate a conserved capacity to restrict Vif-deficient HIV-1. J Virol 85:11220–11234
Ikeda T, Abd El Galil KH, Tokunaga K et al (2011) Intrinsic restriction activity by apolipoprotein B mRNA editing enzyme APOBEC1 against the mobility of autonomous retrotransposons. Nucleic Acids Res 39:5538–5554
Ikeda T, Ohsugi T, Kimura T, Matsushita S, Maeda Y, Harada S, Koito A (2008) The antiretroviral potency of APOBEC1 deaminase from small animal species. Nucleic Acids Res 36:6859–6871
Ireton GC, Black ME, Stoddard BL (2003) The 1.14 A crystal structure of yeast cytosine deaminase: evolution of nucleotide salvage enzymes and implications for genetic chemotherapy. Structure 11:961–972
Iwatani Y, Chan DS, Wang F et al (2007) Deaminase-independent inhibition of HIV-1 reverse transcription by APOBEC3G. Nucleic Acids Res 35:7096–7108
Iwatani Y, Takeuchi H, Strebel K, Levin JG (2006) Biochemical activities of highly purified, catalytically active human APOBEC3G: correlation with antiviral effect. J Virol 80:5992–6002
Jäger S, Cimermancic P, Gulbahce N et al (2012a) Global landscape of HIV-human protein complexes. Nature 481:365–370
Jäger S, Kim DY, Hultquist JF et al (2012b) Vif hijacks CBF-beta to degrade APOBEC3G and promote HIV-1 infection. Nature 481:371–375
Janini M, Rogers M, Birx DR, McCutchan FE (2001) Human immunodeficiency virus type 1 DNA sequences genetically damaged by hypermutation are often abundant in patient peripheral blood mononuclear cells and may be generated during near-simultaneous infection and activation of CD4(+) T cells. J Virol 75:7973–7986

Jarmuz A, Chester A, Bayliss J, Gisbourne J, Dunham I, Scott J, Navaratnam N (2002) An anthropoid-specific locus of orphan C to U RNA-editing enzymes on chromosome 22. Genomics 79:285–296

Jern P, Coffin JM (2008) Effects of retroviruses on host genome function. Annu Rev Genet 42:709–732

Johansson E, Mejlhede N, Neuhard J, Larsen S (2002) Crystal structure of the tetrameric cytidine deaminase from Bacillus subtilis at 2.0 A resolution. Biochemistry 41:2563–2570

Johansson E, Neuhard J, Willemoes M, Larsen S (2004) Structural, kinetic, and mutational studies of the zinc ion environment in tetrameric cytidine deaminase. Biochemistry 43:6020–6029

Jonsson SR, LaRue RS, Stenglein MD, Fahrenkrug SC, Andresdottir V, Harris RS (2007) The restriction of zoonotic PERV transmission by human APOBEC3G. PLoS ONE 2:e893

Kaiser SM, Emerman M (2006) Uracil DNA glycosylase is dispensable for human immunodeficiency virus type 1 replication and does not contribute to the antiviral effects of the cytidine deaminase Apobec3G. J Virol 80:875–882

Kan NC, Franchini G, Wong-Staal F, DuBois GC, Robey WG, Lautenberger JA, Papas TS (1986) Identification of HTLV-III/LAV sor gene product and detection of antibodies in human sera. Science 231:1553–1555

Kao S, Khan MA, Miyagi E, Plishka R, Buckler-White A, Strebel K (2003) The human immunodeficiency virus type 1 Vif protein reduces intracellular expression and inhibits packaging of APOBEC3G (CEM15), a cellular inhibitor of virus infectivity. J Virol 77:11398–11407

Kao S, Miyagi E, Khan MA, Takeuchi H, Opi S, Goila-Gaur R, Strebel K (2004) Production of infectious human immunodeficiency virus type 1 does not require depletion of APOBEC3G from virus-producing cells. Retrovirology 1:27

Kazazian HH Jr (2004) Mobile elements: drivers of genome evolution. Science 303:1626–1632

Khan MA, Goila-Gaur R, Opi S, Miyagi E, Takeuchi H, Kao S, Strebel K (2007) Analysis of the contribution of cellular and viral RNA to the packaging of APOBEC3G into HIV-1 virions. Retrovirology 4:48

Khan MA, Kao S, Miyagi E et al (2005) Viral RNA is required for the association of APOBEC3G with human immunodeficiency virus type 1 nucleoprotein complexes. J Virol 79:5870–5874

Kieffer TL, Kwon P, Nettles RE, Han Y, Ray SC, Siliciano RF (2005) G→A hypermutation in protease and reverse transcriptase regions of human immunodeficiency virus type 1 residing in resting CD4+ T cells in vivo. J Virol 79:1975–1980

Kim DY, Kwon E, Hartley PD, Crosby DC, Mann S, Krogan NJ, Gross JD (2013) CBFbeta stabilizes HIV Vif to counteract APOBEC3 at the expense of RUNX1 Target Gene Expression. Molecular Cell Target Gene Expression.Molecular Cell 49:632–44

Kim U, Wang Y, Sanford T, Zeng Y, Nishikura K (1994) Molecular cloning of cDNA for double-stranded RNA adenosine deaminase, a candidate enzyme for nuclear RNA editing. Proc Natl Acad Sci USA 91:11457–11461

Kinomoto M, Kanno T, Shimura M et al (2007) All APOBEC3 family proteins differentially inhibit LINE-1 retrotransposition. Nucleic Acids Res 35:2955–2964

Kitamura S, Ode H, Nakashima M et al (2012a) The APOBEC3C crystal structure and the interface for HIV-1 Vif binding. Nat Struct Mol Biol 19:1005–1010

Kitamura S, Ode H, Nakashima M et al (2012b) The APOBEC3C crystal structure and the interface for HIV-1 Vif binding. Nat Struct Mol Biol 19:1005–1010

Klemm L, Duy C, Iacobucci I et al (2009) The B cell mutator AID promotes B lymphoid blast crisis and drug resistance in chronic myeloid leukemia. Cancer Cell 16:232–245

Ko TP, Lin JJ, Hu CY, Hsu YH, Wang AH, Liaw SH (2003) Crystal structure of yeast cytosine deaminase. Insights into enzyme mechanism and evolution. J Bio Chem 278:19111–19117

Kobayashi M, Takaori-Kondo A, Shindo K, Abudu A, Fukunaga K, Uchiyama T (2004) APOBEC3G targets specific virus species. J Virol 78:8238–8244

Kohli RM, Abrams SR, Gajula KS, Maul RW, Gearhart PJ, Stivers JT (2009) A portable hot spot recognition loop transfers sequence preferences from APOBEC family members to activation-induced cytidine deaminase. J Bio Chem 284:22898–22904

Kohli RM, Maul RW, Guminski AF et al (2010) Local sequence targeting in the AID/APOBEC family differentially impacts retroviral restriction and antibody diversification. J Bio Chem 285:40956–40964
Konig R, Zhou Y, Elleder D et al (2008) Global analysis of host-pathogen interactions that regulate early-stage HIV-1 replication. Cell 135:49–60
Koning FA, Newman EN, Kim EY, Kunstman KJ, Wolinsky SM, Malim MH (2009) Defining APOBEC3 expression patterns in human tissues and hematopoietic cell subsets. J Virol 83:9474–9485
Lackey L, Demorest ZL, Land AM, Hultquist JF, Brown WL, Harris RS (2012) APOBEC3B and AID have similar nuclear import mechanisms. J Mol Biol 419:301–314
Lackey L, Law EK, Brown WL, Harris RS (2013) Subcellular localization of the APOBEC3 proteins during mitosis and implications for genomic DNA deamination. Cell Cycle 12:762–772
Land AM, Ball TB, Luo M et al (2008) Human immunodeficiency virus (HIV) type 1 proviral hypermutation correlates with CD4 count in HIV-infected women from Kenya. J Virol 82:8172–8182
Langlois MA, Beale RC, Conticello SG, Neuberger MS (2005) Mutational comparison of the single-domained APOBEC3C and double-domained APOBEC3F/G anti-retroviral cytidine deaminases provides insight into their DNA target site specificities. Nucleic Acids Res 33:1913–1923
LaRue RS, Andrésdóttir V, Blanchard Y et al (2009) Guidelines for naming nonprimate APOBEC3 genes and proteins. J Virol 83:494–497
LaRue RS, Jonsson SR, Silverstein KA et al (2008) The artiodactyl APOBEC3 innate immune repertoire shows evidence for a multi-functional domain organization that existed in the ancestor of placental mammals. BMC Mol Biol 9:104
Lecossier D, Bouchonnet F, Clavel F, Hance AJ (2003) Hypermutation of HIV-1 DNA in the absence of the Vif protein. Science 300:1112
Lee TH, Coligan JE, Allan JS, McLane MF, Groopman JE, Essex M (1986) A new HTLV-III/LAV protein encoded by a gene found in cytopathic retroviruses. Science 231:1546–1549
Lee YN, Bieniasz PD (2007) Reconstitution of an infectious human endogenous retrovirus. PLoS Pathog 3:e10
Lee YN, Malim MH, Bieniasz PD (2008) Hypermutation of an ancient human retrovirus by APOBEC3G. J Virol 82:8762–8770
Levanon EY, Eisenberg E, Yelin R et al (2004) Systematic identification of abundant A-to-I editing sites in the human transcriptome. Nat Biotechnol 22:1001–1005
Lever AML, Jeang K-T, Berkhout B (2010) Recent advances in human retroviruses: principles of replication and pathogenesis: advances in retroviral research. World Scientific, Singapore, Hackensack
Li JB, Levanon EY, Yoon JK et al (2009) Genome-wide identification of human RNA editing sites by parallel DNA capturing and sequencing. Science 324:1210–1213
Li M, Shandilya SM, Carpenter MA et al (2012) First-in-class small molecule inhibitors of the single-strand DNA cytosine deaminase APOBEC3G. ACS Chem Biol 7:506–517
Liddament MT, Brown WL, Schumacher AJ, Harris RS (2004) APOBEC3F properties and hypermutation preferences indicate activity against HIV-1 in vivo. Curr Biol 14:1385–1391
Lindahl T (2000) Suppression of spontaneous mutagenesis in human cells by DNA base excision-repair. Mutat Res 462:129–135
Lochelt M, Romen F, Bastone P et al (2005) The antiretroviral activity of APOBEC3 is inhibited by the foamy virus accessory Bet protein. Proc Natl Acad Sci USA 102:7982–7987
Loeb LA, Essigmann JM, Kazazi F, Zhang J, Rose KD, Mullins JI (1999) Lethal mutagenesis of HIV with mutagenic nucleoside analogs. Proc Natl Acad Sci USA 96:1492–1497
Longerich S, Basu U, Alt F, Storb U (2006) AID in somatic hypermutation and class switch recombination. Curr Opin Immunol 18:164–174
Losey HC, Ruthenburg AJ, Verdine GL (2006) Crystal structure of Staphylococcus aureus tRNA adenosine deaminase TadA in complex with RNA. Nat Struct Mol Biol 13:153–159

Luo K, Liu B, Xiao Z, Yu Y, Yu X, Gorelick R, Yu XF (2004) Amino-terminal region of the human immunodeficiency virus type 1 nucleocapsid is required for human APOBEC3G packaging. J Virol 78:11841–11852

Maksakova IA, Romanish MT, Gagnier L, Dunn CA, van de Lagemaat LN, Mager DL (2006) Retroviral elements and their hosts: insertional mutagenesis in the mouse germ line. PLoS Genet 2:e2

Malim MH, Bieniasz PD (2012) HIV restriction factors and mechanisms of evasion. Cold Spring Harb Perspect Med 2:a006940

Malim MH, Emerman M (2008) HIV-1 accessory proteins–ensuring viral survival in a hostile environment. Cell Host Microbe 3:388–398

Mangeat B, Turelli P, Caron G, Friedli M, Perrin L, Trono D (2003) Broad antiretroviral defence by human APOBEC3G through lethal editing of nascent reverse transcripts. Nature 424:99–103

Mariani R, Chen D, Schrofelbauer B et al (2003) Species-specific exclusion of APOBEC3G from HIV-1 virions by Vif. Cell 114:21–31

Marin M, Rose KM, Kozak SL, Kabat D (2003) HIV-1 Vif protein binds the editing enzyme APOBEC3G and induces its degradation. Nat Med 9:1398–1403

Martin KL, Johnson M, D'Aquila RT (2011) APOBEC3G complexes decrease human immunodeficiency virus type 1 production. J Virol 85:9314–9326

Mbisa JL, Barr R, Thomas JA et al (2007) Human immunodeficiency virus type 1 cDNAs produced in the presence of APOBEC3G exhibit defects in plus-strand DNA transfer and integration. J Virol 81:7099–7110

Mehle A, Strack B, Ancuta P, Zhang C, McPike M, Gabuzda D (2004) Vif overcomes the innate antiviral activity of APOBEC3G by promoting its degradation in the ubiquitin-proteasome pathway. J Biol Chem 279:7792–7798

Miyagi E, Opi S, Takeuchi H, Khan M, Goila-Gaur R, Kao S, Strebel K (2007) Enzymatically active APOBEC3G is required for efficient inhibition of human immunodeficiency virus type 1. J Virol 81:13346–13353

Morse DP, Aruscavage PJ, Bass BL (2002) RNA hairpins in noncoding regions of human brain and Caenorhabditis elegans mRNA are edited by adenosine deaminases that act on RNA. Proc Natl Acad Sci USA 99:7906–7911

Muckenfuss H, Hamdorf M, Held U et al (2006) APOBEC3 proteins inhibit human LINE-1 retrotransposition. J Bio Chem 281:22161–22172

Münk C, Willemsen A, Bravo IG (2012) An ancient history of gene duplications, fusions and losses in the evolution of APOBEC3 mutators in mammals. BMC Evol Biol 12:71

Muramatsu M, Kinoshita K, Fagarasan S, Yamada S, Shinkai Y, Honjo T (2000) Class switch recombination and hypermutation require activation-induced cytidine deaminase (AID), a potential RNA editing enzyme. Cell 102:553–563

Muramatsu M, Sankaranand VS, Anant S, Sugai M, Kinoshita K, Davidson NO, Honjo T (1999) Specific expression of activation-induced cytidine deaminase (AID), a novel member of the RNA-editing deaminase family in germinal center B cells. J Biol Chem 274:18470–18476

Nathans R, Cao H, Sharova N et al (2008) Small-molecule inhibition of HIV-1 Vif. Nat Biotechnol 26:1187–1192

Newman EN, Holmes RK, Craig HM, Klein KC, Lingappa JR, Malim MH, Sheehy AM (2005) Antiviral function of APOBEC3G can be dissociated from cytidine deaminase activity. Curr Biol 15:166–170

Nik-Zainal S, Alexandrov LB, Wedge DC et al (2012) Mutational processes molding the genomes of 21 breast cancers. Cell 149:979–993

Nishikura K (2010) Functions and regulation of RNA editing by ADAR deaminases. Annu Rev Biochem 79:321–349

Nowarski R, Britan-Rosich E, Shiloach T, Kotler M (2008) Hypermutation by intersegmental transfer of APOBEC3G cytidine deaminase. Nat Struct Mol Biol 15:1059–1066

OhAinle M, Kerns JA, Li MM, Malik HS, Emerman M (2008) Antiretroelement activity of APOBEC3H was lost twice in recent human evolution. Cell Host Microbe 4:249–259

Okazaki IM, Hiai H, Kakazu N, Yamada S, Muramatsu M, Kinoshita K, Honjo T (2003) Constitutive expression of AID leads to tumorigenesis. J Exp Med 197:1173–1181

Olson ME, Li M, Harris RS, Harki DA (2013) Small-molecule APOBEC3G DNA cytosine deaminase inhibitors based on a 4-amino-1,2,4-triazole-3-thiol scaffold. ChemMedChem 8:112–117

Opi S, Kao S, Goila-Gaur R, Khan MA, Miyagi E, Takeuchi H, Strebel K (2007) Human immunodeficiency virus type 1 Vif inhibits packaging and antiviral activity of a degradation-resistant APOBEC3G variant. J Virol 81:8236–8246

Opi S, Takeuchi H, Kao S et al (2006) Monomeric APOBEC3G is catalytically active and has antiviral activity. J Virol 80:4673–4682

Pace C, Keller J, Nolan D, James I, Gaudieri S, Moore C, Mallal S (2006) Population level analysis of human immunodeficiency virus type 1 hypermutation and its relationship with APOBEC3G and vif genetic variation. J Virol 80:9259–9269

Pak V, Heidecker G, Pathak VK, Derse D (2011) The role of amino-terminal sequences in cellular localization and antiviral activity of APOBEC3B. J Virol 85:8538–8547

Perez-Duran P, de Yebenes VG, Ramiro AR (2007) Oncogenic events triggered by AID, the adverse effect of antibody diversification. Carcinogenesis 28:2427–2433

Petersen-Mahrt SK, Harris RS, Neuberger MS (2002) AID mutates *E. coli* suggesting a DNA deamination mechanism for antibody diversification. Nature 418:99–103

Petit V, Guetard D, Renard M, Keriel A, Sitbon M, Wain-Hobson S, Vartanian JP (2009) Murine APOBEC1 is a powerful mutator of retroviral and cellular RNA in vitro and in vivo. J Mol Biol 385:65–78

Powell LM, Wallis SC, Pease RJ, Edwards YH, Knott TJ, Scott J (1987) A novel form of tissue-specific RNA processing produces apolipoprotein-B48 in intestine. Cell 50:831–840

Ramiro AR, Jankovic M, Eisenreich T et al (2004) AID is required for c-myc/IgH chromosome translocations in vivo. Cell 118:431–438

Rathore A, Carpenter MA, Demir O, Ikeda T, Li M, Shaban N, Anokhin D, Law EK, Brown WL, Amaro R & Harris RS (2013) A single amino acid switches the intrinsic dinucleotide DNA cytosine deamination preference of APOBEC3G. Nucleic Acids Research, in preparation

Refsland EW, Hultquist JF, Harris RS (2012) Endogenous origins of HIV-1 G-to-A hypermutation and restriction in the nonpermissive T Cell Line CEM2n. PLoS Pathog 8:e1002800

Refsland EW, Stenglein MD, Shindo K, Albin JS, Brown WL, Harris RS (2010) Quantitative profiling of the full APOBEC3 mRNA repertoire in lymphocytes and tissues: implications for HIV-1 restriction. Nucleic Acids Res 38:4274–4284

Roberts SA, Sterling J, Thompson C et al (2012) Clustered mutations in yeast and in human cancers can arise from damaged long single-strand DNA regions. Mol Cell 46:424–435

Rogozin IB, Iyer LM, Liang L et al (2007) Evolution and diversification of lamprey antigen receptors: evidence for involvement of an AID-APOBEC family cytosine deaminase. Nat Immunol 8:647–656

Rose KM, Marin M, Kozak SL, Kabat D (2005) Regulated production and anti-HIV type 1 activities of cytidine deaminases APOBEC3B, 3F, and 3G. AIDS Res Hum Retroviruses 21:611–619

Russell RA, Wiegand HL, Moore MD, Schafer A, McClure MO, Cullen BR (2005) Foamy virus Bet proteins function as novel inhibitors of the APOBEC3 family of innate antiretroviral defense factors. J Virol 79:8724–8731

Santa-Marta M, da Silva FA, Fonseca AM, Goncalves J (2005) HIV-1 Vif can directly inhibit apolipoprotein B mRNA-editing enzyme catalytic polypeptide-like 3G-mediated cytidine deamination by using a single amino acid interaction and without protein degradation. J Bio Chem 280:8765–8775

Sanville B, Dolan MA, Wollenberg K et al (2010) Adaptive evolution of Mus Apobec3 includes retroviral insertion and positive selection at two clusters of residues flanking the substrate groove. PLoS Pathog 6:e1000974

Sawyer SL, Emerman M, Malik HS (2004) Ancient adaptive evolution of the primate antiviral DNA-editing enzyme APOBEC3G. PLoS Biol 2:E275

Sawyer SL, Emerman M, Malik HS (2007) Discordant evolution of the adjacent antiretroviral genes TRIM22 and TRIM5 in mammals. PLoS Pathog 3:e197

Schafer A, Bogerd HP, Cullen BR (2004) Specific packaging of APOBEC3G into HIV-1 virions is mediated by the nucleocapsid domain of the gag polyprotein precursor. Virology 328:163–168

Schumacher AJ, Haché G, MacDuff DA, Brown WL, Harris RS (2008) The DNA deaminase activity of human APOBEC3G is required for Ty1, MusD, and human immunodeficiency virus type 1 restriction. J Virol 82:2652–2660

Schumacher AJ, Nissley DV, Harris RS (2005) APOBEC3G hypermutates genomic DNA and inhibits Ty1 retrotransposition in yeast. Proc Natl Acad Sci U S A 102:9854–9859

Severi F, Chicca A, Conticello SG (2011) Analysis of reptilian APOBEC1 suggests that RNA editing may not be its ancestral function. Mol Biol Evol 28:1125–1129

Shandilya SM, Nalam MN, Nalivaika EA et al (2010) Crystal structure of the APOBEC3G catalytic domain reveals potential oligomerization interfaces. Structure 18:28–38

Sheehy AM, Gaddis NC, Choi JD, Malim MH (2002) Isolation of a human gene that inhibits HIV-1 infection and is suppressed by the viral Vif protein. Nature 418:646–650

Sheehy AM, Gaddis NC, Malim MH (2003) The antiretroviral enzyme APOBEC3G is degraded by the proteasome in response to HIV-1 Vif. Nat Med 9:1404–1407

Shindo K, Li M, Gross PJ et al (2012) A comparison of two single stranded DNA binding models by mutational analysis of APOBEC3G. Biology 1:260–276

Shlyakhtenko LS, Lushnikov AY, Li M, Lackey L, Harris RS, Lyubchenko YL (2011) Atomic force microscopy studies provide direct evidence for dimerization of the HIV restriction factor APOBEC3G. J Bio Chem 286:3387–3395

Shlyakhtenko LS, Lushnikov AY, Miyagi A, Li M, Harris RS, Lyubchenko YL (2012) Nanoscale structure and dynamics of ABOBEC3G complexes with single-stranded DNA. Biochemistry 51:6432–6440

Sodroski J, Goh WC, Rosen C, Tartar A, Portetelle D, Burny A, Haseltine W (1986) Replicative and cytopathic potential of HTLV-III/LAV with sor gene deletions. Science 231:1549–1553

Song C, Sutton L, Johnson ME, D'Aquila RT, Donahue JP (2012) Signals in APOBEC3F N-terminal and C-terminal deaminase domains each contribute to encapsidation in HIV-1 virions and are both required for HIV-1 restriction. J Bio Chem 287:16965–16974

Soros VB, Yonemoto W, Greene WC (2007) Newly synthesized APOBEC3G is incorporated into HIV virions, inhibited by HIV RNA, and subsequently activated by RNase H. PLoS Pathog 3:e15

Stenglein MD, Burns MB, Li M, Lengyel J, Harris RS (2010) APOBEC3 proteins mediate the clearance of foreign DNA from human cells. Nat Struct Mol Biol 17:222–229

Stenglein MD, Harris RS (2006) APOBEC3B and APOBEC3F inhibit L1 retrotransposition by a DNA deamination-independent mechanism. J Bio Chem 281:16837–16841

Stenglein MD, Matsuo H, Harris RS (2008) Two regions within the amino-terminal half of APOBEC3G cooperate to determine cytoplasmic localization. J Virol 82:9591–9599

Stopak K, de Noronha C, Yonemoto W, Greene WC (2003) HIV-1 Vif blocks the antiviral activity of APOBEC3G by impairing both its translation and intracellular stability. Mol Cell 12:591–601

Strebel K, Daugherty D, Clouse K, Cohen D, Folks T, Martin MA (1987) The HIV 'A' (sor) gene product is essential for virus infectivity. Nature 328:728–730

Suspène R, Guetard D, Henry M, Sommer P, Wain-Hobson S, Vartanian JP (2005) Extensive editing of both hepatitis B virus DNA strands by APOBEC3 cytidine deaminases in vitro and in vivo. Proc Natl Acad Sci USA 102:8321–8326

Svarovskaia ES, Xu H, Mbisa JL et al (2004) Human apolipoprotein B mRNA-editing enzyme-catalytic polypeptide-like 3G (APOBEC3G) is incorporated into HIV-1 virions through interactions with viral and nonviral RNAs. J Bio Chem 279:35822–35828

Tareen SU, Sawyer SL, Malik HS, Emerman M (2009) An expanded clade of rodent Trim5 genes. Virology 385:473–483

Teng B, Burant CF, Davidson NO (1993) Molecular cloning of an apolipoprotein B messenger RNA editing protein. Science 260:1816–1819

Tian C, Wang T, Zhang W, Yu XF (2007) Virion packaging determinants and reverse transcription of SRP RNA in HIV-1 particles. Nucleic Acids Res 35:7288–7302
Turelli P, Mangeat B, Jost S, Vianin S, Trono D (2004) Inhibition of hepatitis B virus replication by APOBEC3G. Science 303:1829
Unniraman S, Zhou S, Schatz DG (2004) Identification of an AID-independent pathway for chromosomal translocations between the Igh switch region and Myc. Nat Immunol 5:1117–1123
Vartanian JP, Meyerhans A, Sala M, Wain-Hobson S (1994) G→A hypermutation of the human immunodeficiency virus type 1 genome: evidence for dCTP pool imbalance during reverse transcription. Proc Natl Acad Sci USA 91:3092–3096
von Schwedler U, Song J, Aiken C, Trono D (1993) Vif is crucial for human immunodeficiency virus type 1 proviral DNA synthesis in infected cells. J Virol 67:4945–4955
Wang M, Rada C, Neuberger MS (2010) Altering the spectrum of immunoglobulin V gene somatic hypermutation by modifying the active site of AID. J Exp Med 207:141–153
Wang T, Tian C, Zhang W et al (2007) 7SL RNA mediates virion packaging of the antiviral cytidine deaminase APOBEC3G. J Virol 81:13112–13124
Wang X, Abudu A, Son S, Dang Y, Venta PJ, Zheng YH (2011) Analysis of human APOBEC3H haplotypes and anti-human immunodeficiency virus type 1 activity. J Virol 85:3142–3152
Wedekind JE, Dance GS, Sowden MP, Smith HC (2003) Messenger RNA editing in mammals: new members of the APOBEC family seeking roles in the family business. Trends Genet 19:207–216
Wiegand HL, Doehle BP, Bogerd HP, Cullen BR (2004) A second human antiretroviral factor, APOBEC3F, is suppressed by the HIV-1 and HIV-2 Vif proteins. EMBO J 23:2451–2458
Wissing S, Montano M, Garcia-Perez JL, Moran JV, Greene WC (2011) Endogenous APOBEC3B restricts LINE-1 retrotransposition in transformed cells and human embryonic stem cells. J Biol Chem 286:36427–36437
Xiang S, Short SA, Wolfenden R, Carter CW Jr (1997) The structure of the cytidine deaminase-product complex provides evidence for efficient proton transfer and ground-state destabilization. Biochemistry 36:4768–4774
Xie K, Sowden MP, Dance GS, Torelli AT, Smith HC, Wedekind JE (2004) The structure of a yeast RNA-editing deaminase provides insight into the fold and function of activation-induced deaminase and APOBEC-1. Proc Natl Acad Sci USA 101:8114–8119
Yamanaka S, Balestra ME, Ferrell LD et al (1995) Apolipoprotein B mRNA-editing protein induces hepatocellular carcinoma and dysplasia in transgenic animals. Proc Natl Acad Sci USA 92:8483–8487
Yang B, Chen K, Zhang C, Huang S, Zhang H (2007) Virion-associated uracil DNA glycosylase-2 and apurinic/apyrimidinic endonuclease are involved in the degradation of APOBEC3G-edited nascent HIV-1 DNA. J Bio Chem 282:11667–11675
Yeung ML, Houzet L, Yedavalli VS, Jeang KT (2009) A genome-wide short hairpin RNA screening of jurkat T-cells for human proteins contributing to productive HIV-1 replication. J Biol Chem 284:19463–19473
Yu Q, Chen D, König R, Mariani R, Unutmaz D, Landau NR (2004a) APOBEC3B and APOBEC3C are potent inhibitors of simian immunodeficiency virus replication. J Biol Chem 279:53379–53386
Yu Q, Konig R, Pillai S et al (2004b) Single-strand specificity of APOBEC3G accounts for minus-strand deamination of the HIV genome. Nat Struct Mol Biol 11:435–442
Yu X, Yu Y, Liu B, Luo K, Kong W, Mao P, Yu XF (2003) Induction of APOBEC3G ubiquitination and degradation by an HIV-1 Vif-Cul5-SCF complex. Science 302:1056–1060
Zhang H, Yang B, Pomerantz RJ, Zhang C, Arunachalam SC, Gao L (2003) The cytidine deaminase CEM15 induces hypermutation in newly synthesized HIV-1 DNA. Nature 424:94–98
Zhang J, Webb DM (2004) Rapid evolution of primate antiviral enzyme APOBEC3G. Hum Mol Genet 13:1785–1791
Zhang W, Du J, Evans SL, Yu Y, Yu XF (2012) T-cell differentiation factor CBF-beta regulates HIV-1 Vif-mediated evasion of host restriction. Nature 481:376–379

Zheng YH, Irwin D, Kurosu T, Tokunaga K, Sata T, Peterlin BM (2004) Human APOBEC3F is another host factor that blocks human immunodeficiency virus type 1 replication. J Virol 78:6073–6076

Zhou H, Xu M, Huang Q et al (2008) Genome-scale RNAi screen for host factors required for HIV replication. Cell Host Microbe 4:495–504

Zrenner R, Stitt M, Sonnewald U, Boldt R (2006) Pyrimidine and purine biosynthesis and degradation in plants. Annu Rev Plant Biol 57:805–836

Inhibition of Retroviral Replication by Members of the TRIM Protein Family

Adam J. Fletcher and Greg J. Towers

Abstract The TRIM protein family is emerging as a central component of mammalian antiviral innate immunity. Beginning with the identification of TRIM5α as a mammalian post-entry restriction factor against retroviruses, to the repeated observation that many TRIMs ubiquitinate and regulate signaling pathways, the past decade has witnessed an intense research effort to understand how TRIM proteins influence immunity. The list of viral families targeted directly or indirectly by TRIM proteins has grown to include adenoviruses, hepadnaviruses, picornaviruses, flaviviruses, orthomyxoviruses, paramyxoviruses, herpesviruses, rhabdoviruses and arenaviruses. We have come to appreciate how, through intense bouts of positive selection, some TRIM genes have been honed into species-specific restriction factors. Similarly, in the case of TRIMCyp, we are beginning to understand how viruses too have mutated to evade restriction, suggesting that TRIM and viruses have coevolved for millions of years of primate evolution. Recently, TRIM5α returned to the limelight when it was shown to trigger the expression of antiviral genes upon recognition of an incoming virus, a paradigm shift that demonstrated that restriction factors make excellent pathogen sensors. However, it remains unclear how many of ~100 human TRIM genes are antiviral, despite the expression of many of these genes being upregulated by interferon and upon viral infection. TRIM proteins do not conform to one type of antiviral mechanism, reflecting the diversity of viruses they target. Moreover, the cofactors of restriction remain largely enigmatic. The control of retroviral replication remains an important medical subject and provides a useful backdrop for reviewing how TRIM proteins act to repress viral replication.

A. J. Fletcher (✉) · G. J. Towers (✉)
MRC Centre for Medical Molecular Virology, Division of Infection and Immunity, University College London, 90 Gower St, London WC1E 6BT, UK
e-mail: adam.fletcher@ucl.ac.uk

G. J. Towers
e-mail: g.towers@ucl.ac.uk

B. R. Cullen (ed.), *Intrinsic Immunity*, Current Topics in Microbiology and Immunology 371, DOI: 10.1007/978-3-642-37765-5_2,

Contents

1 Introduction

Tripartite motif-containing (TRIM) proteins are an ancient family of E3 ligases. They are characterized by a conserved architecture of two or three zinc-binding folds (one RING domain, one or two B-Box domains), a predicted coiled coil (collectively an RBCC), and usually a C-terminal domain that recruits specific interaction partners (Nisole et al. 2005). TRIM genes predate the emergence of metazoans (Marin 2012), appear sporadically in invertebrates (Sardiello et al. 2008), and experienced substantial species-specific expansion in vertebrates (Reddy et al. 1991; Meroni and Diez-Roux 2005; Boudinot et al. 2011). They are extremely numerous in the higher vertebrates; humans encode nearly 100 TRIM genes, several of which are so young that they are absent in other ape species (Han et al. 2011).

Several lines of evidence suggest that a large proportion of TRIM proteins have roles in innate immunity. One, TRIM genes evolve by segmental duplication events followed by selection-driven adaption. Viruses themselves are candidate sources of selective pressure, as a panel of TRIMs has been shown to inhibit the replication of particular viruses (Mallery et al. 2010; Stremlau et al. 2004; Wolf and Goff 2007; Everett et al. 2006; Taylor et al. 2011; Poole et al. 2009; Kajaste-Rudnitski et al. 2011). In turn, TRIM proteins can influence the evolution of their viral adversaries, as exemplified by TRIM5α and TRIMCyp, which are immersed in genetic conflicts with retroviruses (Ylinen et al. 2010; Price et al. 2009; Newman et al. 2006). Two, several TRIMs have been shown to influence the flux through innate signaling pathways, by ligating ubiquitin to component signal transducers (McNab et al. 2011; Kawai and Akira 2011; Tsuchida et al. 2010). Often, this activity is triggered by viral infection. Perhaps related to this, some TRIM proteins can potentiate immune signaling pathways when transfected into reporter assays, an effect comparable to known signaling transducers like MAVS/IPS-1 or TRAF6 (Uchil et al. 2013; Pertel et al. 2011; Yu et al. 2011; Versteeg et al. 2013). Three, expression of many TRIM genes is acutely sensitive to both interferon (Carthagena et al. 2009; Ozato et al. 2008; Asaoka et al. 2005) and viral infection. This is true for TRIMs from fish (van der Aa et al. 2009) to mammals (Rajsbaum et al. 2008). In one study, half of all human TRIM transcripts were sensitive to regulation by interferons (Carthagena et al. 2009). Four, TRIM genes can be polymorphic, which can have the effect of broadening the antiviral repertoire within a species (Newman et al. 2006; Wilson et al. 2008a; Kirmaier et al. 2010). Five, the most ancient of TRIM genes, represented by human TRIM37, is related to the TRAF family of ubiquitin E3 ligases (Marin 2012), which regulate innate signal transduction through their ubiquitin ligase activity. Human TRAFs bear an N-terminal RING and C-terminal MATH domain, like TRIM37, and these flank tandem zinc fingers. It is becoming apparent that TRIMs and TRAFs have some comparable roles in innate immune signaling (Xia et al. 2009; Gack et al. 2007; Pertel et al. 2011; Tsuchida et al. 2010). And six, although not evidence in itself, the TRIM family experiences rapid gene gain and loss, a phenotype indicative of adaptation that is often exhibited by immunity-related genes (Barreiro and Quintana-Murci 2010; Han et al. 2011).

In this chapter, we focus our attention on the antiretroviral properties of the TRIM family members TRIM5, TRIM22/Staf50, and TRIM28/Kap-1/TIF1β, as the best characterized TRIM proteins with antiretroviral activity. We review recent data suggesting a role for the TRIM28-related protein TRIM24 in retroviral restriction. We discuss how the biology of these proteins relates to their antiviral mechanism and comment on the diversity of retroviruses they target. Finally, we review the use of screens as a means to uncover novel antiretroviral activity in the TRIM family.

2 TRIM5α and TRIMCyp Bind Retroviral Capsids to Accelerate Their Degradation and Stimulate Innate Signaling

2.1 Combining Two Activities

TRIM5α has two known functions in the cell. The first is as a defensive retroviral-capsid-binding protein that can inhibit or restrict retroviral infection (Stremlau et al. 2004; Keckesova et al. 2004; Kirmaier et al. 2010). The second is an E3 ubiquitin ligase that promotes NFκB and AP-1 transcription factor signaling (Pertel et al. 2011; Tareen and Emerman 2011). These two activities are mechanistically integrated in that E3 activity potentiates restriction of retroviral infectivity

(Wu et al. 2006) while binding retroviral capsids with high avidity enhances innate signaling (Pertel et al. 2011). In this manner, TRIM5α mounts a proteasome-dependent block to viral DNA synthesis while simultaneously acting as a pattern recognition receptor (PRR) for the retroviral capsid. In this section, we discuss some of the salient and more recent findings in TRIM5α/TRIMCyp biology. For a discussion on the evolution of these proteins, we direct the reader to a recent review by Daugherty and Malik (Daugherty and Malik 2012).

2.2 The C-Terminal PRYSPRY has Surface Variable Loops that Determine Restriction Specificity

Human *TRIM5* belongs to a *TRIM* cluster on chromosome 11p15, alongside its paralogues *TRIM6*, *TRIM22*, *TRIM34*, and *TRIM21*, which arose by gene duplication events after the divergence of placental and marsupial mammals ~180 million years ago (Sawyer et al. 2007). In humans, TRIM5α is the longest of five putative isoforms encoded by the *TRIM5* gene and is unique in bearing a PRYSPRY domain, alternately called a B30.2, at its C terminus (Reymond et al. 2001; Stremlau et al. 2004). The PRYSPRY, an amalgam of PRY and SPRY subdomains, is a distorted β-sandwich, two layers of antiparallel β-sheets connected by non-structured loops. In TRIM5α, the PRYSPRY is the point of contact with the viral capsid, and it bears the telltale signs of antagonistic host–virus evolution: The variable loops are hot spots of positive selection through primate evolution and are poorly conserved in sequence and length between primates and harbor determinants of antiviral specificity. Being so mobile, these loops have eluded the efforts of crystallographers until only recently, and we now have a model of the rhesus macaque TRIM5α PRYSPRY, derived by superposition of a 1.55 Å crystal structure (solved by substituting the most dynamic variable loop (v1) with a two-amino acid linker) with NMR relaxation parameters of the variable loops (Biris et al. 2012). The TRIM5α PRYSPRY is similar to the few other PRYSPRY structures available, including that from the *Drosophila* SPRY-SOCS box family protein Gustavus (Woo et al. 2006a) and human TRIM21 (James et al. 2007). As previously predicted, the major differences between TRIM5 species variants cluster on one surface, comprising the longest variable loops. Intriguingly, there is striking congruence between the positions of residues in this surface that influence capsid recognition in TRIM5α, antibody binding in TRIM21, and hereditary disease when mutated in Pyrin/TRIM20 and MID1/TRIM18 (Woo et al. 2006b).

2.3 TRIM5α Recognizes Multiple Capsid Epitopes to Maintain Broad Specificity

Collectively, TRIM5α proteins restrict a surprising diversity of retroviruses, bearing capsids that are only distantly related, at least at the level of primary sequence. For example, the tamarin monkey TRIM5α restricts both spumaviruses and the betaretrovirus Mason-Pfizer monkey virus (MPMV) (Diehl et al. 2008; Yap et al. 2008); the ring-tailed lemur TRIM5α restricts feline, equine and primate lentiviruses as well as a murine gammaretrovirus (Rahm et al. 2011) human TRIM5α restricts an equine lentivirus and a murine gammaretrovirus (Stremlau et al. 2004; Keckesova et al. 2004). It transpires that different retroviral capsid proteins (CA), despite having distinct primary sequence, comprise subunits of similar helical tertiary structures, which assemble into hexamers or pentamers within symmetrical lattices (Ganser et al. 2003; Li et al. 2000; Ganser-Pornillos et al. 2004; Mortuza et al. 2008). Binding of TRIM5α to monomeric or dimeric CA is difficult to detect by conventional biochemical or yeast two-hybrid techniques, suggesting low-affinity interactions (Nisole et al. 2004; Sebastian and Luban 2005). However, mature retroviral cores, comprising complexed CA monomers, isolated from virions with gentle detergents, are bound by immobilized TRIM5α, supporting the hypothesis that TRIM5α recognizes complex multivalent epitopes in the capsid structure (Pornillos et al. 2011; Ganser-Pornillos et al. 2007; Welker et al. 2000; Sebastian and Luban 2005). Furthermore, capsid-nucleocapsid (CANC) tubes, which are built from the same hexagonal lattice as the natural HIV-1 core and are often used as structural surrogates of mature capsids, are bound by TRIM5α in vitro (Black and Aiken 2010; Li et al. 2000; Stremlau et al. 2006; Ganser et al. 1999). Within cells, TRIM5α restriction can be saturated by mature, stable retroviral cores but not by fragments of Gag (Towers et al. 2003; Cowan et al. 2002; Sebastian and Luban 2005; Forshey et al. 2005; Shi and Aiken 2006; Dodding et al. 2005), further supporting the hypothesis that TRIM5α recruitment to virus requires the intact CA complex structure. Recently, direct binding between recombinant proteins has also been observed. Recombinant rhesus macaque TRIM5α, the RING substituted for the TRIM21 RING (TRIM5-21R) to improve solubility, decorates HIV-1 CANC tubes when viewed by electron microscopy (EM). This TRIM5 chimera also sediments with capsids derived from membrane-stripped EIAV, despite HIV-1 and EIAV CAs sharing only ~25 % sequence identity (Langelier et al. 2008; Kar et al. 2008).

This model of multiple low-affinity interactions between TRIM5α and capsids requires multiple points of contact in the CA lattice (Ganser-Pornillos et al. 2011; Li et al. 2000). Supporting this hypothesis, two remarkably similar clefts exist in the N-terminal domains of both murine leukemia virus (MLV) and HIV-1 CA and are sampled by PRYSPRY variable loops (Fig. 1). One cleft, formed by CA α-helices 4–6 in MLV and 4–7 in HIV-1, contains determinants of viral tropism (Ohkura et al. 2011; Maillard et al. 2011; Maillard et al. 2007; Owens et al. 2004). For example, MLV CA residue 110 is a determinant for N- or B-tropism, and residues 82 and 117 toggle B-MLV sensitivity to human TRIM5α variants (Kozak and Chakraborti 1996; Maillard et al. 2007; Towers et al. 2000). HIV-1 CA mutated at residues E128 and R132 (to alanines) are less infectious in human cells but more infectious in Old World primate cells (Hatziioannou et al. 2004). Similarly, mutation of HIV-1 CA residues 83, 120, and 122—residues analogous to those in MLV that determine sensitivity to human TRIM5α—can potentiate HIV-1 restriction by human TRIM5α (Maillard et al. 2007, 2011). The second cleft, between the MLV N-terminal

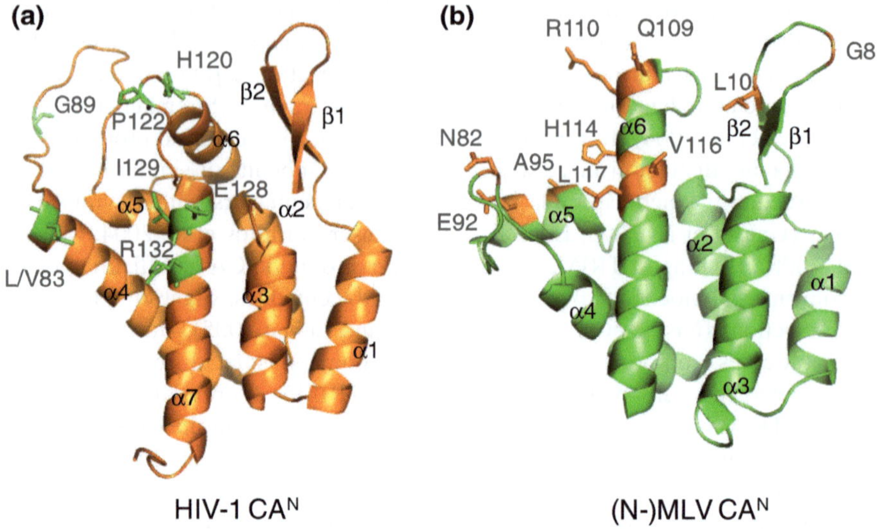

Fig. 1 Residues in N-terminal domains of HIV-1 and MLV CA proteins that influence sensitivity to TRIM5α occupy two surface-exposed clefts. N-terminal CA domains (CA^N) from HIV-1 (**a**) (PDB 1GWP) and N-MLV (**b**) (PDB 1U7 K). Residues that impact on TRIM5α restriction are labeled *green* (HIV-1) or *orange* (N-MLV). Both CA^N are viewed from the same orientation, with the β-sheet-α-helix 6 cleft (HIV-1 and MLV) and the α-helix 4–7 (HIV-1) and α-helix 4–6 (N-MLV) clefts visible. Based on studies in (Ohkura et al. 2011) and (Maillard et al. 2011)

β-strands and α-helix 6, and HIV-1 β1 and α-helix 6, is critical for MLV restriction by rhesus TRIM5α; the L10 W mutation obscures the cleft, prevents rhesus TRIM5α binding to N-MLV CA, and allows the virus to escape restriction (Ohkura et al. 2011). Overall, the combinatorial involvement of several regions of the PRYSPRY and CA in restriction fit a model of multivalent interaction.

2.4 TRIMCyp is Also a Versatile Capsid Recognition Domain

Examination of TRIM5 variants in primate species revealed that on several occasions, the TRIM5α PRYSPRY has been replaced by a cyclophilin A (CypA) cDNA by retrotransposition. The fact that this event has occurred independently on more than one occasion is evidenced by differences in the position of the CypA cDNA, as well as differences in the splicing events that give rise to the TRIMCyp chimeric proteins in each case (Wilson et al. 2008b; Sayah et al. 2004; Nisole et al. 2004; Virgen et al. 2008; Lin and Emerman 2008; Newman et al. 2008; Brennan et al. 2008; Malfavon-Borja et al. 2013; Liao et al. 2007). It transpires that, like a PRYSPRY, the CypA domain is also a very versatile recognition domain for retroviral CA proteins and is able to switch capsid specificity through single amino acid substitutions. Whereas *Aotus* (Owl monkey) TRIMCyp recognizes HIV-1 and FIV

(Diaz-Griffero et al. 2006b), as does the parental CypA protein (Lin and Emerman 2006), *Macaca. mulatta* (rhesus macaque) TRIMCyp restricts HIV-2 and HIV-1 O-groups but not HIV-1 M-group viruses (Price et al. 2009; Virgen et al. 2008; Wilson et al. 2008b). Thus, these TRIMCyps differentiate between viral lineages (HIV-1 and HIV-2) and independent HIV-1 zoonoses (HIV-1 M- and O-groups).

The switch in restriction specificity from M-group HIV-1 to most isolates of O-group HIV-1 and HIV-2 in *M. mulatta* TRIMCyp is due to 2 amino acid changes in the Cyp domain with respect to the parental CypA sequence. Substitutions D66N and R69H, alter Cyp conformation and physical properties allowing conformational diversity not present in the parental CypA molecule (Price et al. 2009; Ylinen et al. 2010; Caines et al. 2012). Intriguingly, R69H expands antiviral specificity by allowing an open binding site conformation that allows restriction of HIV-2 but maintains the ability to hold the closed binding site conformation that can recruit and restrict HIV-1 (Fig. 2a). The second substitution in rhesus TRIMCyp D66 N focuses specificity away from HIV-1 M-group while maintaining binding and restriction of HIV-1 O-group and HIV-2 capsids, by maintaining conformational flexibility (Caines et al. 2012). The natural germline CypA is fixed in the closed binding site conformation and binds HIV-2 CA poorly. The significant difference between HIV-1 and HIV-2 capsids in regard to TRIMCyp specificity is HIV-2 bearing a cyclophilin-binding loop that is shorter by one residue (Price et al. 2009). Furthermore, the sensitivity of naturally occurring HIV-1 isolates to *M. mulatta* TRIMCyp can be predicted based on the residue at position CA88 of the cyclophilin-binding loop. O-group isolates bearing methionine, valine, or isoleucine at CA position 88 are sensitive to *M. mulatta* TRIMCyp, whereas those bearing an alanine at this position in CA are not. Alanine is found in only 7 of 66 O-group sequences but almost all M-group sequences and these viruses remain insensitive to *M. mulatta* TRIMCyp restriction (Price et al. 2009).

Interestingly, like *Aotus* TRIMCyp, the *M. fasicularis* (cynomolgus macaque) TRIMCyp restricts HIV-1, FIV, and SIVtan but not HIV-2 (Ylinen et al. 2010). *M. fasicularis* TRIMCyp bears the specificity enhancing R69H substitution in the CypA domain, which is expected to broaden restriction to HIV-2. However, *M. fasicularis* TRIMCyp also bears a second change, in this case E143K, that focuses specificity away from HIV-2 while maintaining restriction of HIV-1 and SIVtan (Ylinen et al. 2010) (Fig. 2a). Considering the substitutions that have occurred in the TRIMCyp molecules in the context of the macaque phylogeny suggests a complex evolutionary history (Fig. 2b). Note that R69H, which expands specificity, is present in all of the macaque TRIMCyps, whereas D66N, which narrows specificity toward HIV-2, is present in *M. nemestrina* (pigtail macaque) and *M. mulatta* but not *M. fasicularis*. This is despite *M. fascicularis* being more closely related to *M. mulatta* than *M. nemestrina* and is suggestive of waves of infection that selected for alternate antiviral specificities during macaque evolution. Structural studies of the Cyp domain also illustrate the plasticity of this molecule and its ability to adapt to recruiting different lentiviral capsids. For example, although K143 is distant from the CypA active site, mutation to glutamate causes a conformational change that is propagated to the binding site, altering CA recognition (Fig. 2c) (Ylinen et al. 2010).

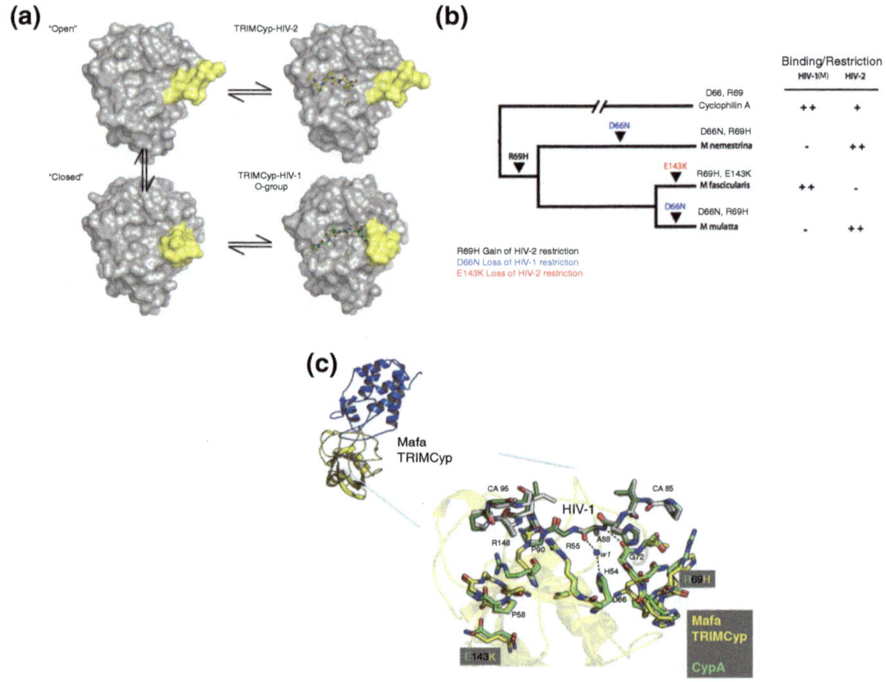

Fig. 2 Evolution of TRIMCyp in Asian macaques. **a** Mutations D66 N and R69H convey the property of conformational diversity onto the Asian monkey TRIMCyp cyclophilin domain (Caines et al. 2012). Thus, the molecule exists between "open" and "closed" conformations. This diversity allows this particular TRIMCyp to restrict both HIV-2 via recruitment of HIV-2 capsid to an "open" conformation cyclophilin and HIV-1 O-group capsid via a "closed" conformation. The HIV-2 capsid derived binding peptide is shown in *yellow*, and the HIV-1 O-group capsid derived peptide is shown in *green*. The mobile cyclophilin active site loop encompassing residues 66 and 69 is shown in *yellow*. **b** Dendrogram illustrating the macaque phylogeny with the restriction specificity of the TRIMCyp species variants shown on the right. The changes to Cyp domain sequence that have occurred are shown on the branches. The specificity switches suggest waves of selective forces from pathogenic viruses during macaque evolution (Ylinen et al. 2010). **c** X-ray crystal structures of HIV-1 CA in complex with Cyp molecules illustrate the plasticity of Cyp recruitment to capsid (Ylinen et al. 2010). HIV-1 CA (*dark green*) in complex with wild-type cyclophilin A (*light green*) superposed on the structure of HIV-1 CA (*gray*) and the *M. fasicularis* TRIMCyp Cyp domain (*yellow*) that bears substitutions R69H and E143 K. Mutation to lysine attracts the peptide oxygen at position P58. This propagates changes to other Cyp active site residues, including R148 that is coordinated by the P58 peptide oxygen in CypA structure. The altered position of P58 releases the side chain of R148, allowing it to interact with capsid residues downstream of residue A88. The structure around Cyp residue 69 is not significantly changed by the R69H change in the crystal structure

Despite the absence of any known extant lentiviruses in Asian macaques, these observations imply that historic lentiviruses may have provided selection pressures to drive diversification of the CypA domain of macaque TRIMCyps. Indeed, fossil records demonstrate the ancestors of Asian monkeys migrated across Europe ~5.5 million years ago, and lentiviruses are known to have been circulating in European mammals at least 12 million years ago, according to endogenized lentiviral proviruses found in rabbits and hares (Katzourakis et al. 2007; Keckesova et al. 2009; Li et al. 2009).

Most recently, a third example of a TRIMCyp fusion has been described in Old World primates (TRIMCypA3), although in this case the gene has decayed, perhaps due to a lack of recent selective advantage (Malfavon-Borja et al. 2013). Reconstruction of the ancient TRIMCyp suggested that it too had broad antiviral activity able to restrict HIV-1 and HIV-2 lineage viruses. However, the protein subsequently evolved to target unknown viruses, presumably those that circulated at this time, and restriction of the viruses we know today was lost. Why lentiviruses retain cyclophilin binding in the face of persistent restriction and TRIMCyp evolution remains the subject of some speculation. Cyclophilin A is certainly a cofactor for HIV-1 replication and probably FIV too (Franke et al. 1994; Thali et al. 1994; Lin and Emerman 2006). The interaction with cyclophilins is certainly ancient, as evidenced by the demonstration of cyclophilin A binding by the extinct endogenous lentiviruses RELIK and PSIV (Goldstone et al. 2010). It is possible that cyclophilin recruitment is preserved through necessity to recruit another cyclophilin, for example the nuclear pore–associated cyclophilin Nup358, which acts as a cofactor for HIV-1 and possibly other lentiviruses (Ocwieja et al. 2011; Schaller et al. 2011).

2.5 TRIM5α Causes Structural Damage to Capsids

An important goal in the TRIM5 field is to understand the details of the restriction mechanism and the fate of restricted virions. The best supported model suggests that TRIM5α binding to capsids causes their untimely disassembly (Zhao et al. 2011; Stremlau et al. 2006). Capsid disassembly has been measured by the so-called fate-of-capsid (FOC) assay, which correlates sensitivity to TRIM5α restriction with loss of particulate, pelletable capsid. For example, centrifugation of SIVmac-, N-MLV- or HIV-1-infected cell lysates expressing restrictive squirrel monkey, human, or rhesus TRIM5α, respectively, demonstrated a decrease in capsid sedimentation compared to non-restrictive cells (Stremlau et al. 2006; Perron et al. 2007). The loss of CA sedimentation has been interpreted as showing that TRIM5α restriction causes CA disassembly. This assay was validated with the observations that PRYSPRY deletion or HIV-1 capsid mutation H87Q that reduces restriction, restores the particulate capsid that is found in the pellet. The FOC assay has also been useful in demonstrating the importance of the TRIM5α RBCC motif in restriction. Deletion of the RING or B-Box2 prevents loss of pelletable N-MLV capsids, even though these two domains are not necessary for binding to CANC arrays in vitro (Perron et al. 2007; Zhao et al. 2011). This suggests that to mediate uncoating in cells, in addition to capsid binding, TRIM5α requires the capacity to multimerize and recruit the proteasome.

Another assay supporting a role for premature uncoating in TRIM5 restriction involves incubation of CANC tubes with cell lysates expressing TRIM5α or TRIMCyp, followed by EM. Generally, CANC tubes are shortened or damaged under such conditions (Black and Aiken 2010). Rhesus TRIM5-21R, or another recombinant TRIM5α mutant comprising simply the CC to PRYSPRY, have also been shown to alter the appearance of both CANC tubes and isolated HIV-1 cores, respectively (Zhao et al. 2011; Langelier et al. 2008). Specifically, the TRIM5 CC-PRYSPRY truncation released linear fragments of CA rather than soluble monomers (Zhao et al. 2011). Using this truncation mutant, an elegant approach to identify the regions of HIV-1 CA targeted by TRIM5α was devised, whereby hexamers in the form of CANC tubes were cross-linked by the introduction of cysteines at inter- or intra-hexamer interfaces. In this system, only the oxidation of cysteines positioned at the inter-hexamer interface prevented CC-PRYSPRY-induced CANC damage, suggesting that TRIM5α causes capsid disassembly by separating adjacent hexamers. In these experiments, a RING or B-Box2 is not required, presumably because the *in vitro* reactions use high protein concentrations that obviate the requirement for RING–B-Box2-mediated multimerization and any proteasome recruitment that the TRIM5N-terminus may normally orchestrate within cells.

2.6 TRIM5α Ubiquitination

If the first stage of TRIM5α restriction involves formation of a TRIM5α–capsid complex that decreases capsid integrity and prevents nuclear entry, the second stage is defined by two observations. One, it precludes viral reverse transcription (Hofmann et al. 1999; Towers et al. 2000), and two, it is overcome by proteasome inhibition (Wu et al. 2006). There is little evidence for capsid degradation during restriction (Chatterji et al. 2006; Stremlau et al. 2006; Yueh et al. 2006; Perron et al. 2007), and neither ubiquitin nor the RING domain is required for the structural damage caused to CANC tubes by TRIM5α in vitro (Zhao et al. 2011; Black and Aiken 2010). Rather, the generally accepted model of TRIM5α restriction is that TRIM5α bound to capsids recruits proteasomes that potentiate capsid disassembly (Stremlau et al. 2006; Perron et al. 2007). In agreement, proteasome inhibition in the FOC assay inhibits HIV-1 uncoating by TRIM5α/TRIMCyp and restores pelletable capsid (Diaz-Griffero et al. 2007). Indeed, TRIM5α can bind to proteasome components, for example the regulatory protein RPT1/PSMC2 (Lukic et al. 2011). The fate of the non-infectious TRIM5α–capsid complex is uncertain in the absence of the second proteasome-dependent stage of restriction. Although the virus has sufficient time and resources to undergo reverse transcription and is not rapidly uncoated, it is prevented from entering the nucleus and forming DNA circles, a surrogate marker for nuclear entry (Wu et al. 2006; Shank and Varmus 1978).

2.6.1 The Role of the RING

The molecular details of the second stage of restriction remain poorly defined. Specifically, how does TRIM5α recruit the proteasome? A role for TRIM5α ubiquitination has been established in innate immune signaling (see below) but not during restriction. TRIM5α is evidently an E3 ligase capable of autoubiquitination in vitro, and it accumulates in a ubiquitinated form under proteasome inhibition (Xu et al. 2003; Yamauchi et al. 2008; Pertel et al. 2011; Diaz-Griffero et al. 2006a). Additionally, for a protein of 57 kDa, TRIM5α has an unusually short half-life (less than an hour), indicating that it is subject to rapid turnover (Diaz-Griffero et al. 2006a; Javanbakht et al. 2005). Indeed, proteasome inhibitors simultaneously lengthen TRIM5α half-life and rescue viral reverse transcription, suggesting that rapid TRIM5α turnover might be sufficient to levy the block to reverse transcription. As an example, *Aotus* TRIMCyp has a short half-life and restricts before reverse transcription, whereas the artificial factor Fv1Cyp has a long half-life and blocks after DNA synthesis (Schaller et al. 2007; Yap et al. 2007). Following this argument, it is surprising that in several cases, RING deletions abrogate TRIM5α restriction completely rather than phenocopy proteasome inhibition. Rhesus TRIM5αΔRING or the C15A/C18A RING mutants have threefold longer half-lives and lose the majority of their ability to restrict HIV-1 and N-MLV, respectively (Yap et al. 2007; Stremlau et al. 2004; Diaz-Griffero et al. 2007). Furthermore, human TRIM5αΔRING loses all but a twofold restriction against N-MLV infection (Perez-Caballero et al. 2005a; Perron et al. 2007). *Aotus* TRIMCypΔRING persists for ten times its usual half-life and loses 90 % of its restriction against HIV-1 (Diaz-Griffero et al. 2006a, b, 2007; Perez-Caballero et al. 2005b).

A more comprehensive mutagenesis of the rhesus TRIM5α RING, in either the putative E2-binding site or RING–RING dimerization interface, can inhibit autoubiquitination in vitro and completely inhibit restriction of HIV-1 and EIAV, suggesting that ubiquitination is required for viral restriction (Lienlaf et al. 2011). The same observation is made upon Ubc13/Uev1a depletion, the E2 heterodimer used by TRIM5α to synthesis ubiquitin chains (see below) (Pertel et al. 2011). These observations suggest that the nature of the TRIM5α–capsid complex is dependent upon ubiquitination, although the RING mutants are no less able to bind CANC tubes in vitro (Lienlaf et al. 2011). Therefore, ΔRING or cysteine mutants are inconsistent with proteasome inhibition and might behave aberrantly through misfolding, mislocalization, or dominant-negative effects and are perhaps not the best tool for dissecting TRIM5α ubiquitination. This might explain the observation that rhesus TRIM5αΔRING is unable to hetero-oligomerize wild-type TRIM5α when both proteins are over-expressed, leading to a proposed role for the RING domain in TRIM5α oligomerization (Li et al. 2011). Interestingly, other rhesus TRIM5α RING mutants have been described that retain potent restriction of HIV-1 but have lost the ability to block viral reverse transcription (Roa et al. 2012). Substitutions at residue Tyr63 in the rhesus protein, which faces into the putative E2-binding cleft, bind CANC tubes in vitro but restrict after reverse transcription and fail to dismantle HIV-1 capsids in the FOC assay, phenocopying proteasome inhibition. These mutants are likely to be informative for the dissection of the mechanism behind proteasome recruitment.

2.6.2 TRIM5 Signaling

A recent development in the field has been the suggestion that TRIM5α constitutes a critical component of an innate signaling pathway. LPS stimulation of TLR4 generates an antiviral state that inhibits the infectivity of diverse viruses including Newcastle disease virus and vesicular stomatitis virus, and TRIM5α appears to contribute to this process (Pertel et al. 2011). In this regard, TRIM5α ubiquitination has been dissected in detail and shown to involve the synthesis of lysine 63 (K63)-linked polyubiquitin using the E2 heterodimer Ubc13/Uev1a. In vitro ubiquitination reactions suggest that these ubiquitin chains are unanchored, a recently described phenomenon that is proposed to mediate TRAF6 signaling (Xia et al. 2009). In this model, unanchored K63-ubiquitin chains promote the autophosphorylation activity of transforming growth factor-β-activated kinase 1 (TAK1), resulting in IKK activation and NFκB nuclear translocation. Notably, recognition of the retroviral lattice potentiates TRIM5α ubiquitination in vitro and innate signaling in cells, broadening our appreciation of what defines a pattern recognition receptor (PRR), or indeed, a pathogen-associated molecular pattern (PAMP) which, in this case, may be the CA lattice itself. This observation parallels another intriguing discovery that TRIM5-21R dimers themselves assemble into hexameric arrays, multimerizing via the B-Box2 (Ganser-Pornillos et al. 2011). Importantly, two-dimensional hexagonal CANC crystals accelerate TRIM5-21R polymerization, further suggesting TRIM5α might dynamically respond to its target in cells and form a hexameric cage around the incoming virion (Diaz-Griffero 2011). In this regard, recognition of capsid has been shown to accelerate TRIM5α turnover (Rold and Aiken 2008), suggesting that recruitment to virions facilitates TRIM5α enzymatic activity and turnover through virus-induced multimerization.

2.7 TRIM5α Cytoplasmic Bodies as Components of Restriction

Like many TRIM proteins, TRIM5α forms highly dynamic cytoplasmic bodies, the function of which is not clear (Reymond et al. 2001; Campbell et al. 2007). However, their molecular composition is slowly becoming known. For example, there is mounting evidence that TRIM5α cytoplasmic bodies are lipid-rich structures. Rhesus TRIM5α immunoprecipitates flotillin-1 and co-localizes by immunofluorescence (IF) with caveolin-1, both of which are components of lipid microdomains (Ohmine et al. 2010). Cholesterol depletion disrupts cytoplasmic bodies, and lipid starvation increases HIV-1 or N-MLV infection of rhesus macaque or human cells, respectively, suggesting that lipid-rich bodies are important for the restriction mechanism (Ohmine et al. 2010). In support of this, replenishing lipids with serum restores restriction. A recent proteomic analysis of fractionated lipid microdomains from human fibroblasts identified TRIM5α

(Kim et al. 2009). Furthermore, a screen in 293T cells to identify binding partners of TRIM5α found that the factor accumulates in a detergent-insoluble fraction, potentially enriched for lipids (Hwang et al. 2010). Membranous lipid rafts are important signaling platforms (Triantafilou et al. 2011), and cytoplasmic organelles such as peroxisomes play a role in innate immune signaling (Dixit et al. 2010). Potentially, TRIM5α signaling and restriction are coordinated by lipid-rich assembly platforms, a notion which remains to be tested.

A role for TRIM5α bodies in restriction is further hinted at in the observation, demonstrated by immunofluorescence, that proteasome inhibition sequesters HIV-1 at rhesus TRIM5α bodies (Campbell et al. 2008; Danielson et al. 2012). Under these conditions, the TRIM5α bodies also stain positively for proteasomes, ubiquitin, and CA, suggesting proteasomes are recruited in order to solubilize the CA/TRIM5α complexes, which is also inferred by the FOC assay (Diaz-Griffero et al. 2006a, 2007; Campbell et al. 2008; Danielson et al. 2012; Lukic et al. 2011). Indeed, the amount of particulate capsid in the FOC assay correlates with the number of viral reverse transcripts in those cell lysates (Kim et al. 2011), demonstrating a direct relationship between the proteasome-mediated uncoating and a block to reverse transcription. Interestingly, some TRIM5α bodies are seen to form rapidly around HIV-1 in the presence of proteasome inhibitor, reaching full intensity within a minute from first appearing, suggesting TRIM5α is able to traffic to incoming virions in response to infection (Campbell et al. 2008). Indeed, TRIM5α cytoplasmic bodies are highly motile, exhibiting short saltatory or long-distance unidirectional movements at speeds of >2 μm/s (Campbell et al. 2007). They also co-localize with microtubules, as do HIV-1 cores, providing a potential mechanism for their motility and recruitment to virions (Sodeik 2002; Diaz-Griffero et al. 2006a). In the absence of proteasome inhibitors, TRIM5α, HIV-1, and proteasome co-localization is minimal, implying restriction and viral destruction ordinarily occurs rapidly via transient associations, as demonstrated by measuring the kinetics of restriction (Perron et al. 2007; Perez-Caballero et al. 2005b).

3 TRIM28 is a Transcriptional Corepressor of Endogenous and Exogenous Retroviruses

3.1 TRIM28 is a Molecular Scaffold for Heterochromatin-Inducing Enzymes

Whereas TRIM5α was investigated in the context of restriction before the discovery of its constitutive role in immune signaling, TRIM28 was first considered the requisite corepressor of Krüppel-associated box (KRAB) domain zinc finger proteins (KRAB-ZFP) (Friedman et al. 1996; Kim et al. 1996; Moosmann et al. 1996), before becoming implicated in retroviral immunity. Numbering nearly 400 members in humans and mice, KRAB-ZFPs are the largest family of transcriptional repressors in vertebrates (Emerson and Thomas 2009). Members of this superfamily bear an N-terminal KRAB domain—a conserved 75-residue repressor motif—and between 2 and 40 C-terminal C_2H_2 zinc finger motifs, which bind DNA and direct target specificity (Thiesen et al. 1991; Bellefroid et al. 1991). The KRAB domain recruits TRIM28 (also called KRAB-associated protein 1 (KAP1), TIF1-β, or KRIP-1), which provides a platform from which chromatin-remodeling enzymes gain access to nucleosomes (Bellefroid et al. 1991; Huntley et al. 2006; Urrutia 2003). This results in the introduction and spreading of heterochromatin at specific loci and transcriptional shutdown. Tethering TRIM28 directly to a promoter causes transcriptional silencing (Friedman et al. 1996; Sripathy et al. 2006), demonstrating the protein's ability to deliver repression.

The various domains of TRIM28 have been attributed specific roles (Fig. 3). TRIM28 homodimerizes and binds a KRAB domain (Peng et al. 2000) via its RBCC motif, which contains two B-Boxes. Beyond the RBCC, a hydrophobic PxVxL pentapeptide binds heterochromatin protein 1 (HP1) (Ryan et al. 1999; Nielsen et al. 1999; Le Douarin et al. 1996), an interaction essential for both nuclear retention and repression (Iyengar et al. 2011; Sripathy et al. 2006). At TRIM28's C terminus, a tandem PHD (plant homeodomain) finger and bromodomain recruit chromatin remodifiers, for example the histone 3 lysine 9 (H3K9)-specific methyltransferase SETDB1/ESET (Schultz et al. 2002) and the nucleosome-remodeling and histone deacetylation (NuRD) component CHD3/Mi2α (Schultz et al. 2001), which associates with the histone deacetylases HDAC1 and HDAC2. The mechanism by which TRIM28 recruits these enzymes is small ubiquitin-like modifier (SUMO) protein dependent. The PHD domain is a zinc-coordinating RING-like C_4HC_3 fold and binds Ubc9, the sole SUMO E2 conjugation enzyme. Consequently, the domain is an intramolecular E3 ligase for SUMO, autosumoylating the neighboring bromodomain (Mascle et al. 2007; Ivanov et al. 2007; Zeng et al. 2008). Mutation of target lysines abolishes TRIM28 repression activity (Mascle et al. 2007), interaction with SETDB1 and CHD3 (Ivanov et al. 2007), and prevents the introduction of H3K9me3 marks on target DNA (Ivanov et al. 2007), without impacting on KRAB–TRIM28 binding or TRIM28-DNA localization (Iyengar et al. 2011). SUMO interaction motifs (SIMs) in SETDB1 and CHD3 are necessary for both their binding to TRIM28 and optimal enzymatic activity (Ivanov et al. 2007). TRIM28 represents the first example of a PHD harboring such activity; indeed, very few E3 ligases for SUMO are currently known, with the protein inhibitor of activated STAT (PIAS) proteins being the best characterized to date (Shuai and Liu 2005). Some other TRIMs have been shown to possess SUMO E3 activity, for example TRIM19 (PML) (Boutell et al. 2003) and TRIMs 22, 27, and 32 (Chu and Yang 2011) .

The consequences of TRIM28 chromatin recruitment are still being unraveled. A high-H3K9me3, low-histone acetylation state as a result of SETDB1 and CHD3 activities reduces RNA polymerase II binding (Groner et al. 2010; Sripathy et al. 2006). Moreover, the HP1β protein binds to nearby H3K9me3 marks and recruits additional heterochromatin-inducing factors (Kwon and Workman 2011). This

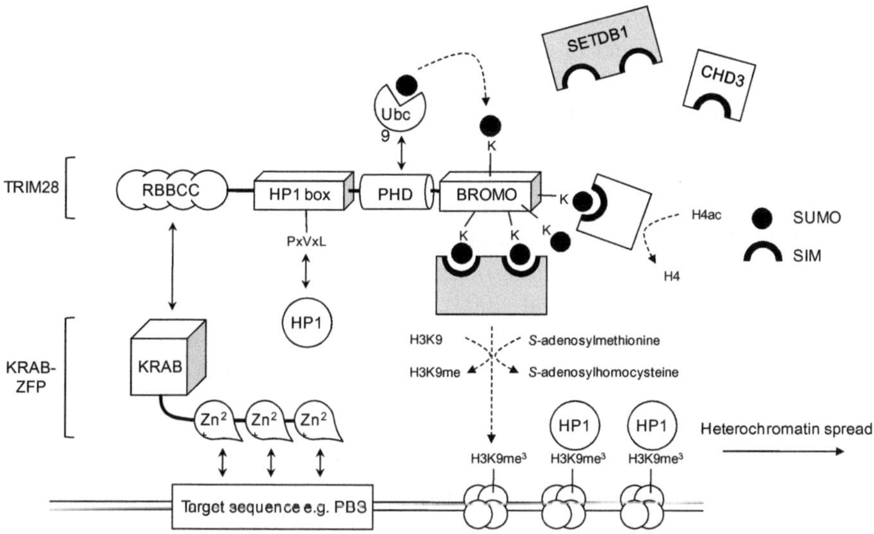

Fig. 3 Introduction of heterochromatin by TRIM28 and associated factors. The TRIM28 RBCC interacts with the KRAB domain of a KRAB-ZFP, which is targeted to a specific target sequence by DNA-binding zinc fingers. The TRIM28 plant homeodomain (PHD) is a SUMO E3 enzyme that binds the SUMO conjugation E2 enzyme Ubc9 to SUMOylate the bromodomain. Chromatin-remodifying enzymes like lysine methyltransferase SETDB1 and histone deacetylase complex NuRD bear SUMO-interacting motifs (SIM) that bind the SUMOylated bromodomain, and heterochromatin markers like H3K9me3 and deacetylated H4 are established. By binding the PxVxL pentapeptide in the TRIM28 HP1 box, HP1 proteins are recruited to H3K9me3 sites, contributing to heterochromatin propagation

could enable TRIM28-mediated silencing to spread over long distances—possibly several kilobases (kb)—from the original site of repressor binding (Groner et al. 2010). Indeed, TRIM28 is shown to bind intragenic loci many kb from promoters, including the 3′ ends of genes (Iyengar et al. 2011). Another consequence of TRIM28 recruitment is *de novo* DNA methylation. TRIM28 and the KRAB protein Zfp57 implement the DNA methylation pattern underlying gene imprinting, the phenomenon of monoallelic gene expression in plants and mammals (Quenneville et al. 2011; Li et al. 2008; Zuo et al. 2012). DNA methylation has also been observed in a conditional gene regulation system, whereby TRIM28 is artificially targeted to a particular gene in a doxycycline-dependent manner. In this system, a KRAB domain is fused to the *Escherichia coli* tetracycline repressor (tTRKRAB) and co-transduced with the target gene under the control of tetracycline operator (*tetO*) sequences (Szulc et al. 2006; Herchenroder et al. 1999). Using this system, TRIM28 delivers irreversible transcriptional silencing if induced during the first few days of mouse development, due to the establishment of repressive promoter DNA methylation (Wiznerowicz et al. 2007). However, introduction of repressive marks using the tTRKRAB system appears to be reversible when introduced into a variety of differentiated human and mouse cells, mouse ES cells, and in adult mice (Szulc et al. 2006; Groner et al. 2012), suggesting that the nature of the repression complex might vary in a cell-type- and temporal-specific manner. Indeed, TRIM28-mediated repression and subsequent DNA methylation appears to be irreversible in mouse ES cells when the target is an endogenous retrovirus (Rowe et al. 2013a).

Surprisingly, relatively few targets of TRIM28 are known, despite a prevalence of binding sites. Genome-wide chromatin immunoprecipitation followed by sequencing (ChIP-seq profiling) suggests ~7,000 TRIM28-binding sites in the human genome (O'Geen et al. 2007), yet the expression of relatively few host genes is affected by TRIM28 depletion in cell lines (Iyengar et al. 2011; Groner et al. 2010). For example, it has been postulated that enrichment of TRIM28 and subsequent heterochromatin formation at genes encoding KRAB-ZFPs protects these loci from deleterious recombination; KRAB-ZFP genes are highly similar as a result of gene duplication (Emerson and Thomas 2009; Vogel et al. 2006). However, TRIM28 depletion only weakly affects KRAB-ZFP expression levels in cell lines. Moreover, upon TRIM28 depletion in cell lines (Iyengar et al. 2011) or deletion in ES cells (Rowe et al. 2013b), some gene transcripts are downregulated, while others are upregulated, suggestive of a role for TRIM28 in more than just transcriptional repression. This ambiguity in gene targets is in part explained by the recent finding that TRIM28 underlies an ES-specific restriction of endogenous retroviruses (ERV).

3.2 TRIM28 Mediates Repression of Endogenous Retroviruses in Embryonic Stem Cells

Nearly half of human and mouse genomes are derived from LTR (~8 %)- and non-LTR (~35 %)-containing retroelements. For example, there are around 2,500 intracisternal A-type particle (IAP) elements (a so-called class II ERV) in the mouse genome (Zhang et al. 2008). Although the majority of ERVs have accumulated mutations that render them unable to retrotranspose, many ERV LTRs represent *cis*-acting regulatory elements that can influence the expression of neighboring genes and require silencing (Rowe et al. 2013b). Similarly, active retroelements must be regulated in order to avoid insertional mutagenesis. Interestingly, TRIM28-knockout mice are embryonic lethal as they fail to gastrulate (Cammas et al. 2000). Furthermore, zygotic heterozygous *Trim28* embryos, with just maternal *Trim28* deleted, die before birth due to demethylation of various promoters (Messerschmidt et al. 2012). Wild-type embryos undergo massive genome-wide demethylation after fertilization, followed by a program of selective remethylation (Feng et al. 2010). To prevent spurious gene and ERV activation during the wave of demethylation, repression is imparted by histone methylation. It was recently shown that ERV LTRs and 5′UTRs are decorated with H3K9me3, TRIM28, and SETDB1 in mouse embryonic stem (ES) cells but not in differentiated embryonic fibroblasts (MEF) (Mikkelsen et al. 2007; Matsui et al. 2010). A conditional knockout mouse revealed the importance of TRIM28 in this process, as loss of TRIM28 in mouse ES cells simultaneously removes H3K9me3 marks and SETDB1 binding from IAP promoters (Rowe et al. 2010). Similarly, H3K9me3 is removed following deletion of SETDB1 in ES cells while having no impact on TRIM28 docking at DNA (Matsui et al. 2010), confirming SETDB1 as the downstream effector of a TRIM28-mediated histone methylation of ERVs. Accordingly, deletion of either of these factors causes large increases in both class I (MLV) and class II (IAP and MusD) ERV mRNAs in mouse ES cells but not in differentiated MEFs (Matsui et al. 2010; Rowe et al. 2010). Strikingly, IAP mRNA was upregulated by ~500-fold in TRIM28-deleted embryos (Rowe et al. 2010), perhaps underlying the embryonic lethality of TRIM28-knockout mice. Indeed, TRIM28 deletion caused a proportional increase in IAV DNA, indicating that activated ERVs are competent for reverse transcription and integration (Matsui et al. 2010), a known mechanism for genome mutagenesis (Zhang et al. 2008). Furthermore, single-stranded DNA derived from reverse transcribing endogenous retroviruses has the potential to trigger pattern recognition (Stetson et al. 2008), placing TRIM28 in a critical position in the regulation of host innate immune responses.

TRIM28-mediated histone methylation and transcriptional repression occur upstream of DNA methylation, as enrichment of SETDB1 and H3K9me3 at ERVs was unaffected by triple deletion of the three mammalian DNA methyltransferases DNMT1, DNMT3a, and DNMT3b (Matsui et al. 2010). However, TRIM28 recruitment rapidly leads to *de novo* DNA methylation at ERVs, which becomes critical for their silencing later in embryogenesis (Rowe et al. 2013a). TRIM28's target within ERVs in ES cells maps to the 5′UTR, as this ~500-bp region cloned from derepressed IAP cDNAs in TRIM28-deleted ES cells, confers repression of a heterologous promoter in a position-independent manner (Rowe et al. 2010). Therefore, it would seem that a TRIM28, KRAB-ZFP repressor complex mediates an essential histone and DNA methylation of ERV 5′UTRs in mouse ES cells that sustains cell viability. It transpires that the same mechanism underlies a potent restriction of exogenous MLV transcription in the same cells.

3.3 *TRIM28 is Responsible for the Stem Cell–Specific Restriction of Murine Leukemia Viruses*

Of a panel of RNA and DNA viruses tested for their abilities to replicate in undifferentiated murine embryonal carcinoma (EC) cells, MLVs were unique in their inability to establish a productive infection (Teich et al. 1977). The restriction was shown to occur in the absence of viral proteins and was subsequent to integration. Production of viral RNA was limiting, implying a block to transcription or viral RNA stability (Teich et al. 1977; Loh et al. 1988). Importantly, the restriction was lost following cell differentiation (Teich et al. 1977). De novo DNA methylation (Stewart et al. 1982) and poor enhancer activity of the 5′ LTR (Feuer et al. 1989) were among several mechanisms initially demonstrated to contribute to the restriction in EC cells. However, another critical determinant of restriction was found to reside within the MLV 18 base pair proline transfer RNA ($tRNA^{Pro}$) primer-binding site (PBS^{Pro}) (Feuer et al. 1989; Barklis et al. 1986; Kempler et al. 1993). Substitution for a different PBS, complementary to $tRNA^{Gln}$ (PBS^{Gln}), rescued MLV from restriction (Petersen et al. 1991).

The PBS makes a particularly effective target for restriction, as it is complementary to the 3′ end of a specific cellular tRNA. The tRNA is recruited to the viral RNA to act as a primer for reverse transcription. Thus, the PBS sequence is dictated by the tRNA, making escape by mutation difficult. This so-called repressor-binding site (RBS) was shown to confer restriction in an orientation- and position-independent manner, against MLV or heterologous promoters (Petersen et al. 1991; Loh et al. 1990), suggesting that the restriction was DNA rather than RNA mediated. Furthermore, DNA-binding factors were suspected based on the observation that competitive DNA bearing the M-MLV RBS could alleviate restriction (Loh et al. 1988). Similarly, exonuclease III protection assays and electrophoretic mobility shift assays (EMSA), using an M-MLV PBS for a probe, revealed that DNA binding was enriched in EC nuclear extracts (Petersen et al. 1991; Loh et al. 1990). Selection for rare escape mutants identified a single-point mutation within the RBS—referred to as the B2 RBS—that bypassed restriction in EC cells (Barklis et al. 1986), and was not bound by RBS-specific factors in EC EMSA or DNA protection experiments (Petersen et al. 1991; Loh et al. 1990). Collectively, these findings paved the way for identification of the nuclear factor/s that bound the MLV provirus to prevent transcription. In 2007, thirty years after the phenotype was first described, TRIM28 was identified by mass spectrometry (MS) as the EC-specific component of the restriction complex (Wolf and Goff 2007). Antibodies against TRIM28 were able to supershift the PBS-binding complex in EMSAs, while cell differentiation saw a decline in TRIM28 protein expression concurrent with the loss in probe binding by EMSA, explaining the differentiation state specificity of the restriction phenotype (Wolf and Goff 2007). Depletion of TRIM28 in EC or ES cells alleviated restriction of MLV, while ChIPs revealed enrichment of TRIM28 at an MLV proviral WT PBS^{Pro}, but not a B2 PBS^{Pro}, revealing the mechanism behind the escape of the B2 mutant virus.

Given this handle on the molecular composition of the MLV restriction complex, the identity of the KRAB-ZFP protein that bound the MLV PBS was revealed to be ZFP809 (Wolf and Goff 2009). EMSA was used to isolate the complex, which was cross-linked to the PBS probe, immunoprecipitated with anti-TRIM28 antibodies, resolved by SDS-PAGE and bands identified by mass spectrometry (Wolf and Goff 2009). Of three candidate proteins, only recombinant GST-tagged ZFP809 bound the MLV PBS probe in EMSA (Wolf and Goff 2009). As with TRIM28, ZFP809 depletion from EC cells attenuates restriction, while ChIP reveals enrichment of ZFP809 at the proviral PBS^{Pro} (Wolf and Goff 2009). Furthermore, the MLV restriction phenotype of the expected specificity (i.e., targeting WT PBS^{Pro} but not PBS^{Gln}) could be introduced to 293 cells by over-expression of ZFP809, and reversed by TRIM28 depletion (Wolf and Goff 2009). As expected from the model of TRIM28–KRAB-ZFP–mediated repression, TRIM28 binding to the MLV

PBS is independent of its ability to bind HP1γ, although TRIM28 PxVxL mutants are unable to restrict MLV (Wolf et al. 2008a). Indeed, HP1γ was also enriched at the MLV PBS^{Pro} (Wolf and Goff 2007). Although we do not yet have a complete picture of TRIM28-mediated restriction at the PBS, the silencing of PBS^{Pro} in endogenous retroviruses (ERV), by *de novo* DNA methylation, was recently shown to require TRIM28 and SETDB1 (Rowe et al. 2013a), strongly suggesting TRIM28 mediated restriction mechanisms behind repression of both exogenous and endogenous retroviruses are comparable in nature.

3.4 TRIM28-Mediated Restriction of Complex Retroviruses

In a screen to identify other retroviral PBS sequences bound by the EC cell restriction complex, $PBS^{Lys\text{-}1,2}$, as used by spumavirus, maedi-visna virus and Mason-Pfizer monkey virus, was comparably bound to PBS^{Pro} in gel shift assays (Yamauchi et al. 1995), by a complex containing TRIM28 (Wolf et al. 2008b). Indeed, substitution of PBS^{Pro} in MLV for $PBS^{Lys\text{-}1,2}$ demonstrates that this PBS is also able to confer repression of viral transcription (Yamauchi et al. 1995), suggesting $PBS^{Lys\text{-}1,2}$-using retroviruses might also suffer silencing in EC cells. However, there is no cross-competition between PBS^{Pro} and $PBS^{Lys\text{-}1,2}$ probes in EMSA experiments, meaning that TRIM28 is recruited to $PBS^{Lys\text{-}1,2}$ by an alternative KRAB-ZFP protein (Yamauchi et al. 1995; Wolf et al. 2008b), the identity of which is not yet known. Human T cell leukemia virus 1 (HTLV-1) also uses $tRNA^{Pro}$ for replication and is restricted in 293 cells expressing ectopic ZFP809 (Wolf et al. 2008b); interestingly, restriction was not overcome in this setting by the HTLV-1 trans-activator Tax (Wolf et al. 2008b), suggesting TRIM28 might contribute to HTLV-1 tropism.

TRIM28 has not been shown to mediate repression of the HIV-1 LTR, although there are several circumstantial observations that suggest it might. First, ectopic expression of the KRAB-ZFP ZBRK1/ZNF350 represses HIV-1 LTR-driven luciferase activity and its depletion proportionally augments HIV-1 LTR activity in 293T cells or the T cell line MT-4 (Nishitsuji et al. 2012). A target for ZBRK1 activity was mapped to the U3 region of the HIV-1 LTR; ZBRK1 bound an LTR-containing probe in EMSA and was enriched at stably integrated LTRs as shown by ChIP (Nishitsuji et al. 2012). Furthermore, the repression activity of ZBRK1 was dependent on TRIM28, as implicated by KRAB mutations that inhibit TRIM28 binding, or depletion of TRIM28. Furthermore, although the HDAC inhibitor trichostatin A (TSA) inhibited ZBRK1-mediated repression, depletion of either SETDB1 or HP1 did not (Nishitsuji et al. 2012), suggesting that TRIM28 might mediate repression of HIV-1 through histone deacetylation and not establishment of an H3K9me3-rich heterochromatin. However, it should be noted that HDAC inhibitors are effective activators of the HIV-1 LTR (Van Lint et al. 1996), and TSA treatment reduces HP1γ expression (Bartova et al. 2005), suggesting that the effects of TSA and ZBRK1 might be antagonistic and unrelated.

Nonetheless, depletion of endogenous ZBRK1 also diminished a modest enrichment of TRIM28 at the HIV-1 LTR as shown by ChIP (Nishitsuji et al. 2012). Similarly, weak repression of the HIV-1 LTR has been demonstrated with the macrophage-specific KRAB-ZFP protein OTK18/ZNF175 (Carlson et al. 2004), although the effect was not abrogated by deletion of its two KRAB domains. Second, artificial targeting of KRAB domains from KOX1—a human KRAB-ZFP known to bind TRIM28 (Moosmann et al. 1996)—to the HIV-1 genome by fusing them to GAL4 or *tetR* DNA-binding domains inhibits HIV-1 Tat-dependent transcription in a dose-dependent manner (Herchenroder et al. 1999). In these experiments, repression was observed when KRAB binding was 6 kb from the 5′ LTR, consistent with the ability of TRIM28 to mediate long-range repression. Similarly, downregulation of HIV-1 transcription was seen with a KRAB-Tat chimera (Pengue et al. 1995), although using Tat as the DNA-targeting domain makes interpretation less straightforward. More recently, custom-designed KRAB-ZFPs that recognize endogenous HIV-1 genomic sequences were shown to levy substantial transcriptional repression in reporter assays in stimulated T cells and to restrict full-length HIV-1 replication (Reynolds et al. 2003; Segal et al. 2004), providing a proof-of-concept for the design of artificial restriction factors based on a TRIM28-recruiting KRAB domain. Third, latent HIV-1 proviruses are enriched for HP1γ and H3K9me3, and HP1γ depletion enhances both basal- and Tat-mediated transcription from an integrated LTR (du Chene et al. 2007). Fourth, HIV-1 bears a PBS complementary to cellular $tRNA^{Lys\text{-}3}$, which mediates modest restriction when substituted for PBS^{Pro} in a recombinant MLV (Yamauchi et al. 1995). A subset of MusD retroelements, possibly targeted by mouse TRIM28, also use $PBS^{Lys\text{-}3}$ (Rowe et al. 2010). Together these experiments suggest that HIV-1 proviruses might be targets of KRAB-ZFP proteins and are enriched for heterochromatin signatures. It remains to be shown conclusively whether TRIM28 specifically regulates HIV-1 transcription.

3.5 TRIM28 Restricts HIV-1 Integration by Facilitating Integrase Deacetylation

A different model for TRIM28 restriction has been proposed, whereby TRIM28 induces deacetylation of HIV-1 Integrase (IN) to decrease integration efficiency. This finding is based on the observation that HIV-1 IN is acetylated at lysines in its C-terminal domain, by histone acetyl transferases (HAT) GCN5 or p300, and this modification is required for optimal integration efficiency (Cereseto et al. 2005; Terreni et al. 2010). However, the significance of HIV-1 IN acetylation is contentious, not least because substitution of purported target lysine residues in IN appears to impact on HIV-1 replication in a manner dependent on the whether IN has been epitope tagged (Terreni et al. 2010; Topper et al. 2007). Nonetheless, a yeast two-hybrid (Y2H) screen for human lymphocyte proteins that bind a constitutively acetylated IN-HAT chimera revealed TRIM28 as an IN interaction partner

(Terreni et al. 2010; Allouch and Cereseto 2009; Allouch et al. 2011). Potentially, the TRIM28 bromodomain recognizes acetyl-lysine residues in IN, although structural studies of the tandem PHD–bromodomain suggest this binding affinity has been lost in TRIM28 (Zeng et al. 2008). Over-expression of TRIM28 decreases HIV-1 integrant numbers in single-cycle assays, while depletion of TRIM28 in primary lymphocytes increased the number of integrants by threefold (Allouch et al. 2011). Histone deacetylase 1 (HDAC1), able to deacetylate IN in vitro, was found to coimmunoprecipitate with the IN–TRIM28 complex (Allouch et al. 2011). Therefore, TRIM28 might recruit HDAC1—as it does the NuRD complex (see above)—to eliminate IN acetylation and integration efficiency. In support of this model, treatment of cells with HDAC inhibitors, or depletion of HDAC1, increases integration in restrictive cells. However, this observation is at odds with a described role for HDAC1 in HIV-1 reverse transcription (Sorin et al. 2009). Moreover, GCN5 depletion has little impact on HIV-1 integration (Terreni et al. 2010). Thus, this proposed role for TRIM28 in HIV-1 restriction requires further investigation.

3.6 *TRIM28-Related Protein TRIM24 Represses ERV Transcription in Hepatocytes*

TRIM28 is highly similar to two other TRIMs, TRIM24/TIF1α and TRIM33/TIF1γ, a TRIM subfamily sharing the RBCC–PHD–bromodomain architecture. In mouse hepatocytes, these proteins hetero-oligomerize to form complexes that antagonize transcriptional activation by the all-*trans* retinoic acid receptor (RAR) α. Genetic disruption of these TRIM complexes leads to large-scale transcriptional upregulation and ultimately hepatocellular carcinoma (HCC) (Khetchoumian et al. 2007; Herquel et al. 2011). Interestingly, it was recently shown that among transcripts elevated in TRIM24-knockout hepatocytes were several host ISGs, suggesting either that TRIM24 might repress ISG transcription directly or perhaps like TRIM28, it represses ERV transcription and subsequent reverse transcription—a known trigger of pattern recognition and autoimmunity (Stetson et al. 2008). Although the TRIM28-sensitive class II ERV IAPs were not expressed in TRIM24-knockout hepatocytes, a particular class of mouse retrotransposons called VL30 ERVs were strongly derepressed specifically in liver tissue (Herquel et al. 2013). Notably, these VL30 ERVs contain all-*trans* retinoic acid response elements in their LTRs, suggesting that TRIM24 mediates their transcriptional repression through its inhibition of RARα activity. Moreover, in a manner reminiscent of TRIM28-mediated restriction, TRIM24 binding is enriched at VL30 LTRs (Herquel et al. 2013).

Whether this example of ERV silencing by TRIM24 occurs in other species remains to be seen, and the precise mechanism of restriction requires further delineation. Nonetheless, these findings suggest that this subfamily of TRIM proteins has been employed repeatedly for retroviral restriction and the prevention of autoimmunity, which might even contribute to the oncogenic nature of TRIM24 loss. Indeed, derepressed VL30 cDNAs in TRIM24-knockout hepatocytes coincide with a modest upregulation of the ISG IRF9 (Herquel et al. 2013). In a more comprehensive screen, loss of TRIM24 function in HCC was shown to correlate with widespread upregulation of IFN signaling pathway components and ISGs (Tisserand et al. 2011)—for example *Ifit* genes, *Gbp* genes, *Herc5, Isg15, Ifi204* and weak upregulation of other known retroviral restriction factors *Samhd1* and *Mov10*—possibly exacerbated by accumulation of cytosolic retroviral cDNAs. Although TRIM24 dose not interact with KRAB domains (Abrink et al. 2001; Peng et al. 2000), its PHD and bromodomain are able to recognize histone lysines and acetyl-lysines (Tsai et al. 2010), suggesting TRIM24 likely regulates transcriptional control in a distinct manner to TRIM28. Given what we now understand about TRIM28 and TRIM24 biology, the implication is that the related protein TRIM33 might also restrict a specific set of endogenous retroviruses in a tissue-specific manner.

4 TRIM22 Represses HIV-1 LTR Transcription and is a Correlate of Viral Control in HIV-1-Infected Patients

TRIM22, also called stimulated transactivating factor of 50 kDa (Staf50), was first cloned from a library of type I IFN-upregulated cDNAs in human lymphoblastoid Daudi cells (Tissot and Mechti 1995). Like TRIM5, TRIM22 expression is strongly stimulated by type 1 interferon treatment in several cell types (Gao et al. 2009) and expression is elevated following HIV-1 infection of PBMCs (Singh et al. 2011). TRIM22 also has a PRYSPRY domain, which like TRIM5α's PRYSPRY is highly variable between species. Indeed, phylogenetic analyses reveal that TRIM22 has evolved through periodic episodes of positive selection (Sawyer et al. 2007). Such a pattern of evolution is strongly suggestive of genetic conflict with pathogens and supports the hypothesis that TRIM22 is a restriction factor. Indeed, the positions of positively selected amino acids in TRIM22, concentrated in the PRYSPRY, are strikingly similar to the amino acid positions in TRIM5α, which determine restriction specificity in primates (see TRIM5).

TRIM22 was first recognized by its sequence homology to murine Rpt-1 (Tissot and Mechti 1995; Patarca et al. 1988), also called TRIM30α, which is one of several mouse paralogues of human TRIM5 (Tareen et al. 2009). Based on this homology and in light of a modest repression by murine TRIM30α on transcription from the HIV-1 LTR (Patarca et al. 1988), human TRIM22 was shown to mediate modest restriction against the HIV-1 LTR (Tissot and Mechti 1995). Further, type I IFN-restricted infection of HIV-1 in Jurkat cells is counteracted by TRIM22 depletion. In other studies, TRIM22 was shown to underlie a difference in HIV-1 permissivity between two clones of monocytic U937 cells (Franzoso et al. 1994; Kajaste-Rudnitski et al. 2011). While various factors had been proposed to underlie the phenotype, including lysosomal serine proteases (Franzoso et al. 1994), poor fusogenic activity in the non-permissive clone (Moriuchi et al. 1997), and possibly HIV-1 cellular coreceptor CXCR4 expression (Biswas et al. 1998), the cause of restriction in the non-permissive U937 clone remained ill defined. However, the non-permissive clone became permissive to HIV-1 replication on depletion of its constitutively expressed TRIM22, while the permissive clone became non-permissive on ectopic expression of TRIM22 (Kajaste-Rudnitski et al. 2011). TRIM22 represses basal transcription from the LTR and, unlike TRIM28-mediated repression of HTLV-1 transactivation by Tax (Wolf and Goff 2009), TRIM22 cannot overcome Tat-stimulated transcription. TRIM22 repression is also independent of the HIV-1 LTR NF-κB sites, but it does moderately interfere with the ability of TNFα, or a combination of the phorbal ester PMA and ionomycin, to stimulate HIV-1 transcription, suggesting that TRIM22 might interfere with specific signaling cascades, calcium regulation, or the binding of transcription factors like AP-1, NFAT, or Sp1 to the HIV-1 LTR. Indeed, TRIM22 activity against HIV-1 might be indirect, as, again like TRIM5α, it is able to selectively induce innate signaling pathways when over-expressed (Yu et al. 2011).

In vivo support for an association between HIV-1 replication and TRIM22 expression comes from the observation that TRIM22 transcripts are elevated in HIV-1-positive patients versus controls, and their levels positively correlate with type I IFN levels (Singh et al. 2011). In the same cohort, TRIM22 levels inversely correlated with HIV-1 plasma viral load and positively correlated with CD4 + T cell count, findings that suggest the protein may be involved in control of viral replication but may simply illustrate that TRIM22 expression levels act as a marker of type I IFN levels. It is interesting that there is also evidence for TRIM22 being a transcriptional repressor at the hepatitis B virus core promoter (Gao et al. 2009), ubiquitinating the 3C protease of encephalomyocarditis virus (Eldin et al. 2009) and ubiquitinating influenza A virus nucleoprotein (Di Pietro et al. 2013).

The mechanism of TRIM22-mediated restriction of HIV-1 remains undefined. It is apparently RING independent (Kajaste-Rudnitski et al. 2011), suggesting that repression does not require ubiquitin conjugation E2 enzyme–binding activity. Unlike TRIM28, TRIM22 is not a well-characterized transcriptional corepressor, and it lacks the SUMO-regulated PHD and bromodomains that in TRIM28 recruit chromatin-remodifying enzymes (Ivanov et al. 2007) (see TRIM28). Presumably, the TRIM22 PRYSPRY forms a protein–protein interface, as it does in TRIM5α. However, TRIM22 is predominantly nuclear (Kajaste-Rudnitski et al. 2011) and localizes to nuclear bodies (Sivaramakrishnan et al. 2009), as does another nuclear antiviral TRIM protein, TRIM19/PML (Reichelt et al. 2011), although these may be different nuclear bodies. It is interesting that TRIM22′s PRYSPRY appears to regulate nuclear localization (Sivaramakrishnan et al. 2009; Herr et al. 2009; Gao et al. 2009; Yu et al. 2011), in conjunction with a bipartite NLS located between the coiled coil and PRYSPRY (Tissot and Mechti 1995). Like TRIM19, TRIM22 nuclear localization is required for its restriction activity (Reichelt et al. 2011; Gao et al. 2009). In particular, residues Val493 and Cys494 at the C terminus of human TRIM22 regulate nuclear body formation, potentially by mediating interaction with other nuclear body proteins. Whether the PRYSPRY is necessary for the observed repression of the HIV-1 LTR is an important question that remains to be answered.

Finally, TRIM22 has been reported to inhibit HIV-1 particle release in a variety of type I IFN stimulated cell lines, although only weak effects on HIV-1 replication were described (Barr et al. 2008; Ohmine et al. 2011). In fact, HIV-1 Gag downregulation from plasma membranes has been described following over-expression of a variety of TRIMs (Ohmine et al. 2011; Uchil et al. 2008), but the phenotypes are comparably weak, suggesting artifacts of transient over-expression.

5 Do Other TRIM Family Proteins Inhibit Retroviral Replication?

Of the ~100 human TRIM genes, only the four discussed have been described as having any significant impact on the life cycles of retroviruses. Do any other TRIM proteins counteract retroviruses? TRIM proteins cluster into discrete monophyletic clades (Han et al. 2011; Marin 2012), the members of which share domain organization (Ozato et al. 2008). Where one clade member is antiviral, it is tempting to speculate that other members might also have evolved immune activity. The assumption is not entirely unfounded, as TRIM5 and TRIM22, paralogues on human chromosome 11p15, have both evolved by positive selection in their PRYSPRY domains and both possess antiretroviral activities, albeit with apparently unrelated mechanisms. Moreover, TRIM21, the most distantly related TRIM in this cluster, is a restriction factor in a similar vein to TRIM5α. TRIM21 targets an early stage of adenoviral replication, although this is through recruitment of intracellular neutralizing antibodies attached to incoming viral particles rather than through recruitment of the viral capsid itself (Mallery et al. 2010).

An obvious obstacle in the search for novel antiretroviral function is having a sensitive virus and a replication assay in hand. Nonetheless, expression screens have been used to uncover retroviral restriction activities in other TRIM proteins. In one screen, 36 human TRIMs were co-transfected with MLV or HIV-1 vectors and 13 of these decreased viral titre in the supernatants of TRIM-expressing cells (Uchil et al. 2008). Human TRIM8, 15, 19, 25, 26, 28, and 35 reduced MLV titre; human TRIM15, 26, and 32 reduced HIV-1 titre. Although the restriction mechanisms have not yet been fully examined, evidence suggests restriction by these TRIM proteins might proceed via novel mechanisms. For example, TRIM15 restriction of MLV was shown to require only its B-Box2, which can immunoprecipitate the MLV Gag polyprotein (Uchil et al. 2008).

TRIM19/PML is a component of nuclear bodies and is thought to be a restriction factor against various RNA and DNA viruses. PML has been suggested to exit the nucleus upon HIV-1 infection and interact with the HIV-1 preintegration complex (PIC) (Turelli et al. 2001), but this work has not yet been replicated. TRIM32 was identified by yeast two-hybrid as a binding partner of HIV-1 Tat (Fridell et al. 1995) and has since been shown to target STING/MITA, the signal adaptor, for K63-linked ubiquitination, enhancing type I IFN production (Zhang et al. 2012). TRIM25 stimulates another signaling protein, the triphosphorylated viral RNA receptor retinoic acid-inducible gene 1 (RIG-I), also polyubiquitinating it with K63-linked chains (Gack et al. 2007). A growing number of TRIM proteins are thought to influence signaling pathways through their E3 ligase activity (Kawai and Akira 2011; McNab et al. 2011), and a large number are shown to interact with ubiquitination machinery (Napolitano et al. 2011). Moreover, stimulation of innate signaling pathways by TRIM over-expression has been demonstrated for a large number of TRIM members (Versteeg et al. 2013), which in some cases might contribute to restriction activities, as suggested for MLV restriction by TRIM1 and TRIM62 (Uchil et al. 2013). TRIM1 has been shown to be both a weak (Yap et al. 2004) and strong (Uchil et al. 2008) CA-specific MLV restriction factor in similar assays, although TRIM1 depletion in human cells had no effect on the infectivity of a wide panel of retroviruses (Zhang et al. 2006), arguing against an important antiviral role. Significantly more TRIMs affect MLV replication over HIV-1 in these screens, which has been ascribed to *cis*-acting sequence differences between the two viral LTRs (Uchil et al. 2013). Specifically, NFκB sites in the HIV-1 LTR might confer sensitivity to the signaling induced by various TRIMs (Uchil et al. 2013). Whether this elaborate mechanism turns out to explain other retroviral tropism remains to be seen. However, the generation of an antiviral state by many TRIM family proteins might mean that a far larger number of these have antiretroviral activity than we are currently aware.

6 Concluding Remarks

The recurrent evolutionary adoption of TRIM proteins for defense against diverse retroviruses is a testament to the versatility and utility of this protein superfamily. Yet even within the small number of TRIM proteins that have been well characterized to target retroviral replication, there is significant variation in antiviral mechanism. For example, there appears to be a differential requirement for enzymatic activity. TRIM5α requires RING-dependent E3 ligase activity for complete restriction, while TRIM22 does not. However, a role for ubiquitination in restriction by either of these proteins has not been formally demonstrated. TRIM28 requires SUMO ligase activity, but the E3 is its C-terminal PHD rather than N-terminal RING domain (Zeng et al. 2008). The TRIM28 RBCC is necessary for interaction with a KRAB-ZFP, but the role of the RING in this relationship is not yet defined. TRIM24 binds ERV LTRs and cellular genes bearing all-*trans* RARα response elements, interfering with RAR activity, and in this manner indirectly inhibits VL30 ERV transcription (Herquel et al. 2013). Whether TRIM24 catalytic activity contributes to RAR antagonism is not yet known. Clearly, there is much scope for further research into the role of TRIM enzymatic activity in restriction.

TRIM5 and TRIM22 demonstrate signs of positive selection that, in the case of TRIM5, is a result of direct contact with an evolving viral protein (Sawyer et al. 2007). The TRIM22 PRYSPRY has not yet been implicated in restriction *per se*, although it does seem to influence the protein's cellular localization (Gao et al. 2009).

TRIM28 instead interacts exclusively with host factors and is a constitutive mediator of host transcriptional regulation, suggesting purifying selection has dominated its evolution. Indeed, it is the KRAB-ZFPs that evolve through positive selection in their DNA-binding domains (Rowe et al. 2010), indicative of an interaction with viral targets. Potentially, the regulatory KRAB-ZFP network evolved primarily as an antiviral defense before its adoption by the host for constitutive gene control.

The involvement of TRIMs in signaling cascades also appears to be a feature of the family that remains understudied. TRIM5 and perhaps TRIM22 are able to regulate specific signaling cascades and thereby contribute to the antiviral state in an indirect manner. An increasing number of TRIMs have been shown to possess similar activities (Kawai and Akira 2011). Whether TRIM22, like TRIM5, triggers innate signaling upon restriction has not yet been shown, but will require further understanding of the TRIM22 restriction mechanism. TRIM24 directly binds and represses the *Stat1* promoter, which contains RAR response elements, and thus the expression of many STAT1-regulated ISGs is upregulated upon TRIM24 depletion (Tisserand et al. 2011), although depression of VL30 ERVs might also contribute to this. Whether TRIM28 suppresses the stimulation of innate immunity due to prevention of cytoplasmic retroviral cDNA accumulation remains to be seen. Or perhaps, TRIM28 might trigger and potentiate an antiviral state upon recognition of recently endogenized retroviruses. That TRIM22, TRIM24, and TRIM28 are nuclear proteins reinforces the notion that any effect they have on innate signaling will be quite distinct from that of cytoplasmic TRIM5α.

TRIM5 and TRIM22 appear to be largely autonomous restriction factors, able to confer a restriction phenotype upon permissive cells (Kajaste-Rudnitski et al. 2011; Stremlau et al. 2004). Presumably, this is because their cofactors for restriction, for example the ubiquitination machinery for TRIM5, are ubiquitous in eukaryotic cells. Exceptions have been suggested, however—TRIM5 might be a less effective restriction factor in 293T cells owing to suboptimal expression of SUMO-1 (Arriagada et al. 2011). Conversely, TRIM28 requires expression of the correct KRAB-ZFP in order to target a retrovirus (Wolf and Goff 2009), suggesting the specificity of its repressive activity can be redirected by altering expression of various KRAB-ZFPs. Custom-designed KRAB-ZFPs might represent a future avenue of therapy for the repression of exogenous retroviruses like HIV-1.

Acknowledgments We are very grateful to Helen Rowe, Leo James, and Stéphane Hué for critically reading the manuscript. Work in the Towers lab is funded by the Wellcome Trust, the Medical Research Council, and the UCL/UCLH National Institute of Health Research Biomedical Research Centre.

References

Abrink M, Ortiz JA, Mark C, Sanchez C, Looman C, Hellman L, Chambon P, Losson R (2001) Conserved interaction between distinct Kruppel-associated box domains and the transcriptional intermediary factor 1 beta. Proc Natl Acad Sci U S A 98(4):1422–1426

Allouch A, Cereseto A (2009) Identification of cellular factors binding to acetylated HIV-1 integrase. Amino Acids 41:1137–1145

Allouch A, Di Primio C, Alpi E, Lusic M, Arosio D, Giacca M, Cereseto A (2011) The TRIM family protein KAP1 inhibits HIV-1 integration. Cell Host Microbe 9(6):484–495

Arriagada G, Muntean LN, Goff SP (2011) SUMO-interacting motifs of human TRIM5α are important for antiviral activity. PLoS Pathog 7(4):e1002019

Asaoka K, Ikeda K, Hishinuma T, Horie-Inoue K, Takeda S, Inoue S (2005) A retrovirus restriction factor TRIM5α is transcriptionally regulated by interferons. Biochem Biophys Res Commun 338(4):1950–1956

Barklis E, Mulligan RC, Jaenisch R (1986) Chromosomal position or virus mutation permits retrovirus expression in embryonal carcinoma cells. Cell 47(3):391–399

Barr SD, Smiley JR, Bushman FD (2008) The interferon response inhibits HIV particle production by induction of TRIM22. PLoS Pathog 4(2):e1000007

Barreiro LB, Quintana-Murci L (2010) From evolutionary genetics to human immunology: how selection shapes host defence genes. Nat Rev Genet 11(1):17–30

Bartova E, Pachernik J, Harnicarova A, Kovarik A, Kovarikova M, Hofmanova J, Skalnikova M, Kozubek M, Kozubek S (2005) Nuclear levels and patterns of histone H3 modification and HP1 proteins after inhibition of histone deacetylases. J Cell Sci 118(Pt 21):5035–5046

Bellefroid EJ, Poncelet DA, Lecocq PJ, Revelant O, Martial JA (1991) The evolutionarily conserved Kruppel-associated box domain defines a subfamily of eukaryotic multifingered proteins. Proc Natl Acad Sci USA 88(9):3608–3612

Biris N, Yang Y, Taylor AB, Tomashevski A, Guo M, Hart PJ, Diaz-Griffero F, Ivanov DN (2012) Structure of the rhesus monkey TRIM5alpha PRYSPRY domain, the HIV capsid recognition module. Proc Natl Acad Sci USA 109(33):13278–13283

Biswas P, Mengozzi M, Mantelli B, Delfanti F, Brambilla A, Vicenzi E, Poli G (1998) 1,25-Dihydroxyvitamin D3 upregulates functional CXCR4 human immunodeficiency virus type 1 coreceptors in U937 minus clones: NF-kappaB-independent enhancement of viral replication. J Virol 72(10):8380–8383

Black LR, Aiken C (2010) TRIM5alpha disrupts the structure of assembled HIV-1 capsid complexes in vitro. J Virol 84(13):6564–6569

Boudinot P, van der Aa LM, Jouneau L, Du Pasquier L, Pontarotti P, Briolat V, Benmansour A, Levraud JP (2011) Origin and evolution of TRIM proteins: new insights from the complete TRIM repertoire of zebrafish and pufferfish. PLoS One 6(7):e22022

Boutell C, Orr A, Everett RD (2003) PML residue lysine 160 is required for the degradation of PML induced by herpes simplex virus type 1 regulatory protein ICP0. J Virol 77(16):8686–8694

Brennan G, Kozyrev Y, Hu SL (2008) TRIMCyp expression in old world primates Macaca nemestrina and Macaca fascicularis. Proc Natl Acad Sci USA 105(9):3569–3574

Caines ME, Bichel K, Price AJ, McEwan WA, Towers GJ, Willett BJ, Freund SM, James LC (2012) Diverse HIV viruses are targeted by a conformationally dynamic antiviral. Nat Struct Mol Biol 19(4):411–416

Cammas F, Mark M, Dolle P, Dierich A, Chambon P, Losson R (2000) Mice lacking the transcriptional corepressor TIF1beta are defective in early postimplantation development. Development 127(13):2955–2963

Campbell EM, Dodding MP, Yap MW, Wu X, Gallois-Montbrun S, Malim MH, Stoye JP, Hope TJ (2007) TRIM5 alpha cytoplasmic bodies are highly dynamic structures. Mol Biol Cell 18(6):2102–2111

Campbell EM, Perez O, Anderson JL, Hope TJ (2008) Visualization of a proteasome-independent intermediate during restriction of HIV-1 by rhesus TRIM5alpha. J Cell Biol 180(3):549–561

Carlson KA, Leisman G, Limoges J, Pohlman GD, Horiba M, Buescher J, Gendelman HE, Ikezu T (2004) Molecular characterization of a putative antiretroviral transcriptional factor, OTK18. J Immunol 172(1):381–391

Carthagena L, Bergamaschi A, Luna JM, David A, Uchil PD, Margottin-Goguet F, Mothes W, Hazan U, Transy C, Pancino G, Nisole S (2009) Human TRIM gene expression in response to interferons. PLoS One 4(3):e4894

Cereseto A, Manganaro L, Gutierrez MI, Terreni M, Fittipaldi A, Lusic M, Marcello A, Giacca M (2005) Acetylation of HIV-1 integrase by p300 regulates viral integration. EMBO J 24(17):3070–3081

Chatterji U, Bobardt MD, Gaskill P, Sheeter D, Fox H, Gallay PA (2006) Trim5alpha accelerates degradation of cytosolic capsid associated with productive HIV-1 entry. J Biol Chem 281(48):37025–37033

Chu Y, Yang X (2011) SUMO E3 ligase activity of TRIM proteins. Oncogene 30(9):1108–1116

Cowan S, Hatziioannou T, Cunningham T, Muesing MA, Gottlinger HG, Bieniasz PD (2002) Cellular inhibitors with Fv1-like activity restrict human and simian immunodeficiency virus tropism. Proc Natl Acad Sci USA 99(18):11914–11919

Danielson CM, Cianci GC, Hope TJ (2012) Recruitment and dynamics of proteasome association with rhTRIM5α cytoplasmic complexes during HIV-1 infection. Traffic 13(9):1206–1217

Daugherty MD, Malik HS (2012) Rules of engagement: molecular insights from host-virus arms races. Annu Rev Genet 46:677–700

Di Pietro A, Kajaste-Rudnitski A, Oteiza A, Nicora L, Towers GJ, Mechti N, Vicenzi E (2013) TRIM22 inhibits influenza A virus infection by targeting the viral nucleoprotein for degradation. J Virol 87(8):4523–4533

Diaz-Griffero F (2011) Caging the beast: TRIM5alpha binding to the HIV-1 core. Viruses 3(5):423–428

Diaz-Griffero F, Kar A, Lee M, Stremlau M, Poeschla E, Sodroski J (2007) Comparative requirements for the restriction of retrovirus infection by TRIM5α and TRIMCyp. Virology 369(2):400–410

Diaz-Griffero F, Li X, Javanbakht H, Song B, Welikala S, Stremlau M, Sodroski J (2006a) Rapid turnover and polyubiquitylation of the retroviral restriction factor TRIM5. Virology 349(2):300–315

Diaz-Griffero F, Vandegraaff N, Li Y, McGee-Estrada K, Stremlau M, Welikala S, Si Z, Engelman A, Sodroski J (2006b) Requirements for capsid-binding and an effector function in TRIMCyp-mediated restriction of HIV-1. Virology 351(2):404–419

Diehl WE, Stansell E, Kaiser SM, Emerman M, Hunter E (2008) Identification of postentry restrictions to Mason-Pfizer monkey virus infection in New World monkey cells. J Virol 82(22):11140–11151

Dixit E, Boulant S, Zhang Y, Lee AS, Odendall C, Shum B, Hacohen N, Chen ZJ, Whelan SP, Fransen M, Nibert ML, Superti-Furga G, Kagan JC (2010) Peroxisomes are signaling platforms for antiviral innate immunity. Cell 141(4):668–681

Dodding MP, Bock M, Yap MW, Stoye JP (2005) Capsid processing requirements for abrogation of Fv1 and Ref1 restriction. J Virol 79(16):10571–10577

du Chene I, Basyuk E, Lin YL, Triboulet R, Knezevich A, Chable-Bessia C, Mettling C, Baillat V, Reynes J, Corbeau P, Bertrand E, Marcello A, Emiliani S, Kiernan R, Benkirane M (2007) Suv39H1 and HP1gamma are responsible for chromatin-mediated HIV-1 transcriptional silencing and post-integration latency. EMBO J 26(2):424–435

Eldin P, Papon L, Oteiza A, Brocchi E, Lawson TG, Mechti N (2009) TRIM22 E3 ubiquitin ligase activity is required to mediate antiviral activity against encephalomyocarditis virus. J Gen Virol 90(Pt 3):536–545

Emerson RO, Thomas JH (2009) Adaptive evolution in zinc finger transcription factors. PLoS Genet 5(1):e1000325

Everett RD, Rechter S, Papior P, Tavalai N, Stamminger T, Orr A (2006) PML contributes to a cellular mechanism of repression of herpes simplex virus type 1 infection that is inactivated by ICP0. J Virol 80(16):7995–8005

Feng S, Jacobsen SE, Reik W (2010) Epigenetic reprogramming in plant and animal development. Science 330(6004):622–627

Feuer G, Taketo M, Hanecak RC, Fan H (1989) Two blocks in Moloney murine leukemia virus expression in undifferentiated F9 embryonal carcinoma cells as determined by transient expression assays. J Virol 63(5):2317–2324

Forshey BM, Shi J, Aiken C (2005) Structural requirements for recognition of the human immunodeficiency virus type 1 core during host restriction in owl monkey cells. J Virol 79(2):869–875

Franke EK, Yuan HE, Luban J (1994) Specific incorporation of cyclophilin A into HIV-1 virions. Nature 372(6504):359–362

Franzoso G, Biswas P, Poli G, Carlson LM, Brown KD, Tomita-Yamaguchi M, Fauci AS, Siebenlist UK (1994) A family of serine proteases expressed exclusively in myelomonocytic cells specifically processes the nuclear factor-kappa B subunit p65 in vitro and may impair human immunodeficiency virus replication in these cells. J Exp Med 180(4):1445–1456

Fridell RA, Harding LS, Bogerd HP, Cullen BR (1995) Identification of a novel human zinc finger protein that specifically interacts with the activation domain of lentiviral Tat proteins. Virology 209(2):347–357

Friedman JR, Fredericks WJ, Jensen DE, Speicher DW, Huang XP, Neilson EG, Rauscher FJ 3rd (1996) KAP-1, a novel corepressor for the highly conserved KRAB repression domain. Genes Dev 10(16):2067–2078

Gack MU, Shin YC, Joo CH, Urano T, Liang C, Sun L, Takeuchi O, Akira S, Chen Z, Inoue S, Jung JU (2007) TRIM25 RING-finger E3 ubiquitin ligase is essential for RIG-I-mediated antiviral activity. Nature 446(7138):916–920

Ganser BK, Cheng A, Sundquist WI, Yeager M (2003) Three-dimensional structure of the M-MuLV CA protein on a lipid monolayer: a general model for retroviral capsid assembly. EMBO J 22(12):2886–2892

Ganser BK, Li S, Klishko VY, Finch JT, Sundquist WI (1999) Assembly and analysis of conical models for the HIV-1 core. Science 283(5398):80–83

Ganser-Pornillos BK, Chandrasekaran V, Pornillos O, Sodroski JG, Sundquist WI, Yeager M (2011) Hexagonal assembly of a restricting TRIM5alpha protein. Proc Natl Acad Sci USA 108(2):534–539

Ganser-Pornillos BK, Cheng A, Yeager M (2007) Structure of full-length HIV-1 CA: a model for the mature capsid lattice. Cell 131(1):70–79

Ganser-Pornillos BK, von Schwedler UK, Stray KM, Aiken C, Sundquist WI (2004) Assembly properties of the human immunodeficiency virus type 1 CA protein. J Virol 78(5):2545–2552

Gao B, Duan Z, Xu W, Xiong S (2009) Tripartite motif-containing 22 inhibits the activity of hepatitis B virus core promoter, which is dependent on nuclear-located RING domain. Hepatology 50(2):424–433

Goldstone DC, Yap MW, Robertson LE, Haire LF, Taylor WR, Katzourakis A, Stoye JP, Taylor IA (2010) Structural and functional analysis of prehistoric lentiviruses uncovers an ancient molecular interface. Cell Host Microbe 8(3):248–259

Groner AC, Meylan S, Ciuffi A, Zangger N, Ambrosini G, Denervaud N, Bucher P, Trono D (2010) KRAB-zinc finger proteins and KAP1 can mediate long-range transcriptional repression through heterochromatin spreading. PLoS Genet 6(3):e1000869

Groner AC, Tschopp P, Challet L, Dietrich JE, Verp S, Offner S, Barde I, Rodriguez I, Hiiragi T, Trono D (2012) The Kruppel-associated box repressor domain can induce reversible heterochromatization of a mouse locus in vivo. J Biol Chem 287(30):25361–25369

Han K, Lou DI, Sawyer SL (2011) Identification of a genomic reservoir for new TRIM genes in primate genomes. PLoS Genet 7(12):e1002388

Hatziioannou T, Cowan S, Von Schwedler UK, Sundquist WI, Bieniasz PD (2004) Species-specific tropism determinants in the human immunodeficiency virus type 1 capsid. J Virol 78(11):6005–6012

Herchenroder O, Hahne JC, Meyer WK, Thiesen HJ, Schneider J (1999) Repression of the human immunodeficiency virus type 1 promoter by the human KRAB domain results in inhibition of virus production. Biochim Biophys Acta 1445(2):216–223

Herquel B, Ouararhni K, Khetchoumian K, Ignat M, Teletin M, Mark M, Bechade G, Van Dorsselaer A, Sanglier-Cianferani S, Hamiche A, Cammas F, Davidson I, Losson R (2011) Transcription cofactors TRIM24, TRIM28, and TRIM33 associate to form regulatory complexes that suppress murine hepatocellular carcinoma. Proc Natl Acad Sci USA 108(20):8212–8217

Herquel B, Ouararhni K, Martianov I, Le Gras S, Ye T, Keime C, Lerouge T, Jost B, Cammas F, Losson R, Davidson I (2013) Trim24-repressed VL30 retrotransposons regulate gene expression by producing noncoding RNA. Nat Struct Mol Biol

Herr AM, Dressel R, Walter L (2009) Different subcellular localisations of TRIM22 suggest species-specific function. Immunogenetics 61(4):271–280

Hofmann W, Schubert D, LaBonte J, Munson L, Gibson S, Scammell J, Ferrigno P, Sodroski J (1999) Species-specific, postentry barriers to primate immunodeficiency virus infection. J Virol 73(12):10020–10028

Huntley S, Baggott DM, Hamilton AT, Tran-Gyamfi M, Yang S, Kim J, Gordon L, Branscomb E, Stubbs L (2006) A comprehensive catalog of human KRAB-associated zinc finger genes: insights into the evolutionary history of a large family of transcriptional repressors. Genome Res 16(5):669–677

Hwang CY, Holl J, Rajan D, Lee Y, Kim S, Um M, Kwon KS, Song B (2010) Hsp70 interacts with the retroviral restriction factor TRIM5alpha and assists the folding of TRIM5alpha. J Biol Chem 285(10):7827–7837

Ivanov AV, Peng H, Yurchenko V, Yap KL, Negorev DG, Schultz DC, Psulkowski E, Fredericks WJ, White DE, Maul GG, Sadofsky MJ, Zhou MM, Rauscher FJ 3rd (2007) PHD domain-mediated E3 ligase activity directs intramolecular sumoylation of an adjacent bromodomain required for gene silencing. Mol Cell 28(5):823–837

Iyengar S, Ivanov AV, Jin VX, Rauscher FJ 3rd, Farnham PJ (2011) Functional analysis of KAP1 genomic recruitment. Mol Cell Biol 31(9):1833–1847

James LC, Keeble AH, Khan Z, Rhodes DA, Trowsdale J (2007) Structural basis for PRYSPRY-mediated tripartite motif (TRIM) protein function. Proc Natl Acad Sci USA 104(15):6200–6205

Javanbakht H, Diaz-Griffero F, Stremlau M, Si Z, Sodroski J (2005) The contribution of RING and B-box 2 domains to retroviral restriction mediated by monkey TRIM5α. J Biol Chem 280(29):26933–26940

Kajaste-Rudnitski A, Marelli SS, Pultrone C, Pertel T, Uchil PD, Mechti N, Mothes W, Poli G, Luban J, Vicenzi E (2011) TRIM22 inhibits HIV-1 transcription independently of its E3 ubiquitin ligase activity, Tat, and NF-kappaB-responsive long terminal repeat elements. J Virol 85(10):5183–5196

Kar AK, Diaz-Griffero F, Li Y, Li X, Sodroski J (2008) Biochemical and biophysical characterization of a chimeric TRIM21–TRIM5α protein. J Virol 82(23):11669–11681

Katzourakis A, Tristem M, Pybus OG, Gifford RJ (2007) Discovery and analysis of the first endogenous lentivirus. Proc Natl Acad Sci USA 104(15):6261–6265

Kawai T, Akira S (2011) Regulation of innate immune signalling pathways by the tripartite motif (TRIM) family proteins. EMBO Mol Med 3(9):513–527

Keckesova Z, Ylinen LM, Towers GJ (2004) The human and African green monkey TRIM5α genes encode Ref1 and Lv1 retroviral restriction factor activities. Proc Natl Acad Sci USA 101(29):10780–10785

Keckesova Z, Ylinen LM, Towers GJ, Gifford RJ, Katzourakis A (2009) Identification of a RELIK orthologue in the European hare (Lepus europaeus) reveals a minimum age of 12 million years for the lagomorph lentiviruses. Virology 384(1):7–11

Kempler G, Freitag B, Berwin B, Nanassy O, Barklis E (1993) Characterization of the moloney murine leukemia virus stem cell-specific repressor binding site. Virology 193(2):690–699

Khetchoumian K, Teletin M, Tisserand J, Mark M, Herquel B, Ignat M, Zucman-Rossi J, Cammas F, Lerouge T, Thibault C, Metzger D, Chambon P, Losson R (2007) Loss of Trim24 (Tif1α) gene function confers oncogenic activity to retinoic acid receptor alpha. Nat Genet 39(12):1500–1506

Kim J, Tipper C, Sodroski J (2011) Role of TRIM5α RING domain E3 ubiquitin ligase activity in capsid disassembly, reverse transcription blockade, and restriction of simian immunodeficiency virus. J Virol 85(16):8116–8132

Kim SS, Chen YM, O'Leary E, Witzgall R, Vidal M, Bonventre JV (1996) A novel member of the RING finger family, KRIP-1, associates with the KRAB-A transcriptional repressor domain of zinc finger proteins. Proc Natl Acad Sci USA 93(26):15299–15304

Kim SY, Wang TK, Singh RD, Wheatley CL, Marks DL, Pagano RE (2009) Proteomic identification of proteins translocated to membrane microdomains upon treatment of fibroblasts with the glycosphingolipid, C8-beta-D-lactosylceramide. Proteomics 9(18):4321–4328

Kirmaier A, Wu F, Newman RM, Hall LR, Morgan JS, O'Connor S, Marx PA, Meythaler M, Goldstein S, Buckler-White A, Kaur A, Hirsch VM, Johnson WE (2010) TRIM5 suppresses cross-species transmission of a primate immunodeficiency virus and selects for emergence of resistant variants in the new species. PLoS Biol 8(8):e1000462

Kozak CA, Chakraborti A (1996) Single amino acid changes in the murine leukemia virus capsid protein gene define the target of Fv1 resistance. Virology 225(2):300–305

Kwon SH, Workman JL (2011) The changing faces of HP1: From heterochromatin formation and gene silencing to euchromatic gene expression: HP1 acts as a positive regulator of transcription. BioEssays 33(4):280–289

Langelier CR, Sandrin V, Eckert DM, Christensen DE, Chandrasekaran V, Alam SL, Aiken C, Olsen JC, Kar AK, Sodroski JG, Sundquist WI (2008) Biochemical characterization of a recombinant TRIM5alpha protein that restricts human immunodeficiency virus type 1 replication. J Virol 82(23):11682–11694

Le Douarin B, Nielsen AL, Garnier JM, Ichinose H, Jeanmougin F, Losson R, Chambon P (1996) A possible involvement of TIF1 alpha and TIF1 beta in the epigenetic control of transcription by nuclear receptors. EMBO J 15(23):6701–6715

Li J, Han K, Xing J, Kim HS, Rogers J, Ryder OA, Disotell T, Yue B, Batzer MA (2009) Phylogeny of the macaques (Cercopithecidae: Macaca) based on Alu elements. Gene 448(2):242–249

Li S, Hill CP, Sundquist WI, Finch JT (2000) Image reconstructions of helical assemblies of the HIV-1 CA protein. Nature 407(6802):409–413

Li X, Ito M, Zhou F, Youngson N, Zuo X, Leder P, Ferguson-Smith AC (2008) A maternal-zygotic effect gene, Zfp57, maintains both maternal and paternal imprints. Dev Cell 15(4):547–557

Li X, Yeung DF, Fiegen AM, Sodroski J (2011) Determinants of the higher order association of the restriction factor TRIM5α and other tripartite motif (TRIM) proteins. J Biol Chem 286(32):27959–27970

Liao CH, Kuang YQ, Liu HL, Zheng YT, Su B (2007) A novel fusion gene, TRIM5-cyclophilin A in the pig-tailed macaque determines its susceptibility to HIV-1 infection. AIDS 21(Suppl 8):S19–S26

Lienlaf M, Hayashi F, Di Nunzio F, Tochio N, Kigawa T, Yokoyama S, Diaz-Griffero F (2011) Contribution of E3-ubiquitin ligase activity to HIV-1 restriction by TRIM5α(rh): structure of the RING domain of TRIM5alpha. J Virol 85(17):8725–8737

Lin TY, Emerman M (2006) Cyclophilin A interacts with diverse lentiviral capsids. Retrovirology 3:70

Lin TY, Emerman M (2008) Determinants of cyclophilin A-dependent TRIM5αrestriction against HIV-1. Virology 379(2):335–341

Loh TP, Sievert LL, Scott RW (1988) Negative regulation of retrovirus expression in embryonal carcinoma cells mediated by an intragenic domain. J Virol 62(11):4086–4095

Loh TP, Sievert LL, Scott RW (1990) Evidence for a stem cell-specific repressor of moloney murine leukemia virus expression in embryonal carcinoma cells. Mol Cell Biol 10(8):4045–4057

Lukic Z, Hausmann S, Sebastian S, Rucci J, Sastri J, Robia SL, Luban J, Campbell EM (2011) TRIM5alpha associates with proteasomal subunits in cells while in complex with HIV-1 virions. Retrovirology 8:93

Maillard PV, Reynard S, Serhan F, Turelli P, Trono D (2007) Interfering residues narrow the spectrum of MLV restriction by human TRIM5α. PLoS Pathog 3(12):e200

Maillard PV, Zoete V, Michielin O, Trono D (2011) Homology-based identification of capsid determinants that protect HIV1 from human TRIM5α restriction. J Biol Chem 286(10):8128–8140

Malfavon-Borja R, Wu LI, Emerman M, Malik HS (2013) Birth, decay, and reconstruction of an ancient TRIMCyp gene fusion in primate genomes. Proc Natl Acad Sci USA 110:E583–592

Mallery DL, McEwan WA, Bidgood SR, Towers GJ, Johnson CM, James LC (2010) Antibodies mediate intracellular immunity through tripartite motif-containing 21 (TRIM21). Proc Natl Acad Sci USA 107(46):19985–19990

Marin I (2012) Origin and diversification of TRIM ubiquitin ligases. PLoS One 7(11):e50030

Mascle XH, Germain-Desprez D, Huynh P, Estephan P, Aubry M (2007) Sumoylation of the transcriptional intermediary factor 1beta (TIF1beta), the Co-repressor of the KRAB Multifinger proteins, is required for its transcriptional activity and is modulated by the KRAB domain. J Biol Chem 282(14):10190–10202

Matsui T, Leung D, Miyashita H, Maksakova IA, Miyachi H, Kimura H, Tachibana M, Lorincz MC, Shinkai Y (2010) Proviral silencing in embryonic stem cells requires the histone methyltransferase ESET. Nature 464(7290):927–931

McNab FW, Rajsbaum R, Stoye JP, O'Garra A (2011) Tripartite-motif proteins and innate immune regulation. Curr Opin Immunol 23(1):46–56

Meroni G, Diez-Roux G (2005) TRIM/RBCC, a novel class of 'single protein RING finger' E3 ubiquitin ligases. BioEssays 27(11):1147–1157

Messerschmidt DM, de Vries W, Ito M, Solter D, Ferguson-Smith A, Knowles BB (2012) Trim28 is required for epigenetic stability during mouse oocyte to embryo transition. Science 335(6075):1499–1502

Mikkelsen TS, Ku M, Jaffe DB, Issac B, Lieberman E, Giannoukos G, Alvarez P, Brockman W, Kim TK, Koche RP, Lee W, Mendenhall E, O'Donovan A, Presser A, Russ C, Xie X, Meissner A, Wernig M, Jaenisch R, Nusbaum C, Lander ES, Bernstein BE (2007) Genome-wide maps of chromatin state in pluripotent and lineage-committed cells. Nature 448(7153):553–560

Moosmann P, Georgiev O, Le Douarin B, Bourquin JP, Schaffner W (1996) Transcriptional repression by RING finger protein TIF1 beta that interacts with the KRAB repressor domain of KOX1. Nucleic Acids Res 24(24):4859–4867

Moriuchi H, Moriuchi M, Arthos J, Hoxie J, Fauci AS (1997) Promonocytic U937 subclones expressing CD4 and CXCR4 are resistant to infection with and cell-to-cell fusion by T-cell-tropic human immunodeficiency virus type 1. J Virol 71(12):9664–9671

Mortuza GB, Dodding MP, Goldstone DC, Haire LF, Stoye JP, Taylor IA (2008) Structure of B-MLV capsid amino-terminal domain reveals key features of viral tropism, gag assembly and core formation. J Mol Biol 376(5):1493–1508

Napolitano LM, Jaffray EG, Hay RT, Meroni G (2011) Functional interactions between ubiquitin E2 enzymes and TRIM proteins. Biochem J 434(2):309–319

Newman RM, Hall L, Connole M, Chen GL, Sato S, Yuste E, Diehl W, Hunter E, Kaur A, Miller GM, Johnson WE (2006) Balancing selection and the evolution of functional polymorphism in old world monkey TRIM5alpha. Proc Natl Acad Sci USA 103(50):19134–19139

Newman RM, Hall L, Kirmaier A, Pozzi LA, Pery E, Farzan M, O'Neil SP, Johnson W (2008) Evolution of a TRIM5-CypA splice isoform in old world monkeys. PLoS Pathog 4(2):e1000003

Nielsen AL, Ortiz JA, You J, Oulad-Abdelghani M, Khechumian R, Gansmuller A, Chambon P, Losson R (1999) Interaction with members of the heterochromatin protein 1 (HP1) family and histone deacetylation are differentially involved in transcriptional silencing by members of the TIF1 family. EMBO J 18(22):6385–6395

Nishitsuji H, Abe M, Sawada R, Takaku H (2012) ZBRK1 represses HIV-1 LTR-mediated transcription. FEBS Lett 586(20):3562–3568

Nisole S, Lynch C, Stoye JP, Yap MW (2004) A Trim5-cyclophilin A fusion protein found in owl monkey kidney cells can restrict HIV-1. Proc Natl Acad Sci U S A 101(36):13324–13328

Nisole S, Stoye JP, Saib A (2005) TRIM family proteins: retroviral restriction and antiviral defence. Nat Rev Microbiol 3(10):799–808

O'Geen H, Squazzo SL, Iyengar S, Blahnik K, Rinn JL, Chang HY, Green R, Farnham PJ (2007) Genome-wide analysis of KAP1 binding suggests autoregulation of KRAB-ZNFs. PLoS Genet 3(6):e89

Ocwieja KE, Brady TL, Ronen K, Huegel A, Roth SL, Schaller T, James LC, Towers GJ, Young JA, Chanda SK, Konig R, Malani N, Berry CC, Bushman FD (2011) HIV integration targeting: a pathway involving Transportin-3 and the nuclear pore protein RanBP2. PLoS Pathog 7(3):e1001313

Ohkura S, Goldstone DC, Yap MW, Holden-Dye K, Taylor IA, Stoye JP (2011) Novel escape mutants suggest an extensive TRIM5α binding site spanning the entire outer surface of the murine leukemia virus capsid protein. PLoS Pathog 7(3):e1002011

Ohmine S, Sakuma R, Sakuma T, Thatava T, Solis GP, Ikeda Y (2010) Cytoplasmic body component TRIM5α requires lipid-enriched microdomains for efficient HIV-1 restriction. J Biol Chem 285(45):34508–34517

Ohmine S, Sakuma R, Sakuma T, Thatava T, Takeuchi H, Ikeda Y (2011) The antiviral spectra of TRIM5alpha orthologues and human TRIM family proteins against lentiviral production. PLoS One 6(1):e16121

Owens CM, Song B, Perron MJ, Yang PC, Stremlau M, Sodroski J (2004) Binding and susceptibility to postentry restriction factors in monkey cells are specified by distinct regions of the human immunodeficiency virus type 1 capsid. J Virol 78(10):5423–5437

Ozato K, Shin DM, Chang TH, Morse HC 3rd (2008) TRIM family proteins and their emerging roles in innate immunity. Nat Rev Immunol 8(11):849–860

Patarca R, Freeman GJ, Schwartz J, Singh RP, Kong QT, Murphy E, Anderson Y, Sheng FY, Singh P, Johnson KA et al (1988) rpt-1, an intracellular protein from helper/inducer T cells that regulates gene expression of interleukin 2 receptor and human immunodeficiency virus type 1. Proc Natl Acad Sci USA 85(8):2733–2737

Peng H, Begg GE, Schultz DC, Friedman JR, Jensen DE, Speicher DW, Rauscher FJ 3rd (2000) Reconstitution of the KRAB-KAP-1 repressor complex: a model system for defining the molecular anatomy of RING-B box-coiled-coil domain-mediated protein–protein interactions. J Mol Biol 295(5):1139–1162

Pengue G, Caputo A, Rossi C, Barbanti-Brodano G, Lania L (1995) Transcriptional silencing of human immunodeficiency virus type 1 long terminal repeat-driven gene expression by the Kruppel-associated box repressor domain targeted to the transactivating response element. J Virol 69(10):6577–6580

Perez-Caballero D, Hatziioannou T, Yang A, Cowan S, Bieniasz PD (2005a) Human tripartite motif 5α domains responsible for retrovirus restriction activity and specificity. J Virol 79(14):8969–8978

Perez-Caballero D, Hatziioannou T, Zhang F, Cowan S, Bieniasz PD (2005b) Restriction of human immunodeficiency virus type 1 by TRIM-CypA occurs with rapid kinetics and independently of cytoplasmic bodies, ubiquitin, and proteasome activity. J Virol 79(24):15567–15572

Perron MJ, Stremlau M, Lee M, Javanbakht H, Song B, Sodroski J (2007) The human TRIM5α restriction factor mediates accelerated uncoating of the N-tropic murine leukemia virus capsid. J Virol 81(5):2138–2148

Pertel T, Hausmann S, Morger D, Zuger S, Guerra J, Lascano J, Reinhard C, Santoni FA, Uchil PD, Chatel L, Bisiaux A, Albert ML, Strambio-De-Castillia C, Mothes W, Pizzato M, Grutter MG, Luban J (2011) TRIM5 is an innate immune sensor for the retrovirus capsid lattice. Nature 472(7343):361–365

Petersen R, Kempler G, Barklis E (1991) A stem cell-specific silencer in the primer-binding site of a retrovirus. Mol Cell Biol 11(3):1214–1221

Poole E, Groves I, MacDonald A, Pang Y, Alcami A, Sinclair J (2009) Identification of TRIM23 as a cofactor involved in the regulation of NF-kappaB by human cytomegalovirus. J Virol 83(8):3581–3590

Pornillos O, Ganser-Pornillos BK, Yeager M (2011) Atomic-level modelling of the HIV capsid. Nature 469(7330):424–427

Price AJ, Marzetta F, Lammers M, Ylinen LM, Schaller T, Wilson SJ, Towers GJ, James LC (2009) Active site remodeling switches HIV specificity of antiretroviral TRIMCyp. Nat Struct Mol Biol 16(10):1036–1042

Quenneville S, Verde G, Corsinotti A, Kapopoulou A, Jakobsson J, Offner S, Baglivo I, Pedone PV, Grimaldi G, Riccio A, Trono D (2011) In embryonic stem cells, ZFP57/KAP1 recognize a methylated hexanucleotide to affect chromatin and DNA methylation of imprinting control regions. Mol Cell 44(3):361–372

Rahm N, Yap M, Snoeck J, Zoete V, Munoz M, Radespiel U, Zimmermann E, Michielin O, Stoye JP, Ciuffi A, Telenti A (2011) Unique spectrum of activity of prosimian TRIM5α against exogenous and endogenous retroviruses. J Virol 85(9):4173–4183

Rajsbaum R, Stoye JP, O'Garra A (2008) Type I interferon-dependent and -independent expression of tripartite motif proteins in immune cells. Eur J Immunol 38(3):619–630

Reddy BA, Kloc M, Etkin L (1991) The cloning and characterization of a maternally expressed novel zinc finger nuclear phosphoprotein (xnf7) in Xenopus laevis. Dev Biol 148(1):107–116

Reichelt M, Wang L, Sommer M, Perrino J, Nour AM, Sen N, Baiker A, Zerboni L, Arvin AM (2011) Entrapment of viral capsids in nuclear PML cages is an intrinsic antiviral host defense against varicella-zoster virus. PLoS Pathog 7(2):e1001266

Reymond A, Meroni G, Fantozzi A, Merla G, Cairo S, Luzi L, Riganelli D, Zanaria E, Messali S, Cainarca S, Guffanti A, Minucci S, Pelicci PG, Ballabio A (2001) The tripartite motif family identifies cell compartments. EMBO J 20(9):2140–2151

Reynolds L, Ullman C, Moore M, Isalan M, West MJ, Clapham P, Klug A, Choo Y (2003) Repression of the HIV-1 5′ LTR promoter and inhibition of HIV-1 replication by using engineered zinc-finger transcription factors. Proc Natl Acad Sci USA 100(4):1615–1620

Roa A, Hayashi F, Yang Y, Lienlaf M, Zhou J, Shi J, Watanabe S, Kigawa T, Yokoyama S, Aiken C, Diaz-Griffero F (2012) RING domain mutations uncouple TRIM5alpha restriction of HIV-1 from inhibition of reverse transcription and acceleration of uncoating. J Virol 86(3):1717–1727

Rold CJ, Aiken C (2008) Proteasomal degradation of TRIM5α during retrovirus restriction. PLoS Pathog 4(5):e1000074

Rowe HM, Friedli M, Offner S, Verp S, Mesnard D, Marquis J, Aktas T, Trono D (2013a) De novo DNA methylation of endogenous retroviruses is shaped by KRAB-ZFPs/KAP1 and ESET. Development 140(3):519–529

Rowe HM, Jakobsson J, Mesnard D, Rougemont J, Reynard S, Aktas T, Maillard PV, Layard-Liesching H, Verp S, Marquis J, Spitz F, Constam DB, Trono D (2010) KAP1 controls endogenous retroviruses in embryonic stem cells. Nature 463(7278):237–240

Rowe HM, Kapopoulou A, Corsinotti A, Fasching L, Macfarlan TS, Tarabay Y, Viville S, Jakobsson J, Pfaff SL, Trono D (2013b) TRIM28 repression of retrotransposon-based enhancers is necessary to preserve transcriptional dynamics in embryonic stem cells. Genome Res 23:452–461

Ryan RF, Schultz DC, Ayyanathan K, Singh PB, Friedman JR, Fredericks WJ, Rauscher FJ 3rd (1999) KAP-1 corepressor protein interacts and colocalizes with heterochromatic and euchromatic HP1 proteins: a potential role for Kruppel-associated box-zinc finger proteins in heterochromatin-mediated gene silencing. Mol Cell Biol 19(6):4366–4378

Sardiello M, Cairo S, Fontanella B, Ballabio A, Meroni G (2008) Genomic analysis of the TRIM family reveals two groups of genes with distinct evolutionary properties. BMC Evol Biol 8:225

Sawyer SL, Emerman M, Malik HS (2007) Discordant evolution of the adjacent antiretroviral genes TRIM22 and TRIM5 in mammals. PLoS Pathog 3(12):e197

Sayah DM, Sokolskaja E, Berthoux L, Luban J (2004) Cyclophilin A retrotransposition into TRIM5 explains owl monkey resistance to HIV-1. Nature 430(6999):569–573

Schaller T, Ocwieja KE, Rasaiyaah J, Price AJ, Brady TL, Roth SL, Hue S, Fletcher AJ, Lee K, KewalRamani VN, Noursadeghi M, Jenner RG, James LC, Bushman FD, Towers GJ (2011) HIV-1 capsid-cyclophilin interactions determine nuclear import pathway, integration targeting and replication efficiency. PLoS Pathog 7(12):e1002439

Schaller T, Ylinen LM, Webb BL, Singh S, Towers GJ (2007) Fusion of cyclophilin A to Fv1 enables cyclosporine-sensitive restriction of human and feline immunodeficiency viruses. J Virol 81(18):10055–10063

Schultz DC, Ayyanathan K, Negorev D, Maul GG, Rauscher FJ 3rd (2002) SETDB1: a novel KAP-1-associated histone H3, lysine 9-specific methyltransferase that contributes to

HP1-mediated silencing of euchromatic genes by KRAB zinc-finger proteins. Genes Dev 16(8):919–932
Schultz DC, Friedman JR, Rauscher FJ 3rd (2001) Targeting histone deacetylase complexes via KRAB-zinc finger proteins: the PHD and bromodomains of KAP-1 form a cooperative unit that recruits a novel isoform of the Mi-2α subunit of NuRD. Genes Dev 15(4):428–443
Sebastian S, Luban J (2005) TRIM5α selectively binds a restriction-sensitive retroviral capsid. Retrovirology 2:40
Segal DJ, Goncalves J, Eberhardy S, Swan CH, Torbett BE, Li X, Barbas CF 3rd (2004) Attenuation of HIV-1 replication in primary human cells with a designed zinc finger transcription factor. J Biol Chem 279(15):14509–14519
Shank PR, Varmus HE (1978) Virus-specific DNA in the cytoplasm of avian sarcoma virus-infected cells is a precursor to covalently closed circular viral DNA in the nucleus. J Virol 25(1):104
Shi J, Aiken C (2006) Saturation of TRIM5α-mediated restriction of HIV-1 infection depends on the stability of the incoming viral capsid. Virology 350(2):493–500
Shuai K, Liu B (2005) Regulation of gene-activation pathways by PIAS proteins in the immune system. Nat Rev Immunol 5(8):593–605
Singh R, Gaiha G, Werner L, McKim K, Mlisana K, Luban J, Walker BD, Karim SS, Brass AL, Ndung'u T (2011) Association of TRIM22 with the type 1 interferon response and viral control during primary HIV-1 infection. J Virol 85(1):208–216
Sivaramakrishnan G, Sun Y, Rajmohan R, Lin VC (2009) B30.2/SPRY domain in tripartite motif-containing 22 is essential for the formation of distinct nuclear bodies. FEBS Lett 583(12):2093–2099
Sodeik B (2002) Unchain my heart, baby let me go–the entry and intracellular transport of HIV. J Cell Biol 159(3):393–395
Sorin M, Cano J, Das S, Mathew S, Wu X, Davies KP, Shi X, Cheng SW, Ott D, Kalpana GV (2009) Recruitment of a SAP18-HDAC1 complex into HIV-1 virions and its requirement for viral replication. PLoS Pathog 5(6):e1000463
Sripathy SP, Stevens J, Schultz DC (2006) The KAP1 corepressor functions to coordinate the assembly of de novo HP1-demarcated microenvironments of heterochromatin required for KRAB zinc finger protein-mediated transcriptional repression. Mol Cell Biol 26(22):8623–8638
Stetson DB, Ko JS, Heidmann T, Medzhitov R (2008) Trex1 prevents cell-intrinsic initiation of autoimmunity. Cell 134(4):587–598
Stewart CL, Stuhlmann H, Jahner D, Jaenisch R (1982) De novo methylation, expression, and infectivity of retroviral genomes introduced into embryonal carcinoma cells. Proc Natl Acad Sci USA 79(13):4098–4102
Stremlau M, Owens CM, Perron MJ, Kiessling M, Autissier P, Sodroski J (2004) The cytoplasmic body component TRIM5alpha restricts HIV-1 infection in old world monkeys. Nature 427(6977):848–853
Stremlau M, Perron M, Lee M, Li Y, Song B, Javanbakht H, Diaz-Griffero F, Anderson DJ, Sundquist WI, Sodroski J (2006) Specific recognition and accelerated uncoating of retroviral capsids by the TRIM5α restriction factor. Proc Natl Acad Sci USA 103(14):5514–5519
Szulc J, Wiznerowicz M, Sauvain MO, Trono D, Aebischer P (2006) A versatile tool for conditional gene expression and knockdown. Nat Methods 3(2):109–116
Tareen SU, Emerman M (2011) Human Trim5α has additional activities that are uncoupled from retroviral capsid recognition. Virology 409(1):113–120
Tareen SU, Sawyer SL, Malik HS, Emerman M (2009) An expanded clade of rodent Trim5 genes. Virology 385(2):473–483
Taylor RT, Lubick KJ, Robertson SJ, Broughton JP, Bloom ME, Bresnahan WA, Best SM (2011) TRIM79alpha, an interferon-stimulated gene product, restricts tick-borne encephalitis virus replication by degrading the viral RNA polymerase. Cell Host Microbe 10(3):185–196
Teich NM, Weiss RA, Martin GR, Lowy DR (1977) Virus infection of murine teratocarcinoma stem cell lines. Cell 12(4):973–982

Terreni M, Valentini P, Liverani V, Gutierrez MI, Di Primio C, Di Fenza A, Tozzini V, Allouch A, Albanese A, Giacca M, Cereseto A (2010) GCN5-dependent acetylation of HIV-1 integrase enhances viral integration. Retrovirology 7:18
Thali M, Bukovsky A, Kondo E, Rosenwirth B, Walsh CT, Sodroski J, Gottlinger HG (1994) Functional association of cyclophilin A with HIV-1 virions. Nature 372(6504):363–365
Thiesen HJ, Bellefroid E, Revelant O, Martial JA (1991) Conserved KRAB protein domain identified upstream from the zinc finger region of Kox 8. Nucleic Acids Res 19(14):3996
Tisserand J, Khetchoumian K, Thibault C, Dembele D, Chambon P, Losson R (2011) Tripartite motif 24 (Trim24/Tif1α) tumor suppressor protein is a novel negative regulator of interferon (IFN)/signal transducers and activators of transcription (STAT) signaling pathway acting through retinoic acid receptor alpha (Raralpha) inhibition. J Biol Chem 286(38):33369–33379
Tissot C, Mechti N (1995) Molecular cloning of a new interferon-induced factor that represses human immunodeficiency virus type 1 long terminal repeat expression. J Biol Chem 270(25):14891–14898
Topper M, Luo Y, Zhadina M, Mohammed K, Smith L, Muesing MA (2007) Posttranslational acetylation of the human immunodeficiency virus type 1 integrase carboxyl-terminal domain is dispensable for viral replication. J Virol 81(6):3012–3017
Towers G, Bock M, Martin S, Takeuchi Y, Stoye JP, Danos O (2000) A conserved mechanism of retrovirus restriction in mammals. Proc Natl Acad Sci USA 97(22):12295–12299
Towers GJ, Hatziioannou T, Cowan S, Goff SP, Luban J, Bieniasz PD (2003) Cyclophilin A modulates the sensitivity of HIV-1 to host restriction factors. Nat Med 9(9):1138–1143
Triantafilou M, Lepper PM, Olden R, Dias Ide S, Triantafilou K (2011) Location, location, location: is membrane partitioning everything when it comes to innate immune activation? Mediators Inflamm 2011:186093
Tsai WW, Wang Z, Yiu TT, Akdemir KC, Xia W, Winter S, Tsai CY, Shi X, Schwarzer D, Plunkett W, Aronow B, Gozani O, Fischle W, Hung MC, Patel DJ, Barton MC (2010) TRIM24 links a non-canonical histone signature to breast cancer. Nature 468(7326):927–932
Tsuchida T, Zou J, Saitoh T, Kumar H, Abe T, Matsuura Y, Kawai T, Akira S (2010) The ubiquitin ligase TRIM56 regulates innate immune responses to intracellular double-stranded DNA. Immunity 33(5):765–776
Turelli P, Doucas V, Craig E, Mangeat B, Klages N, Evans R, Kalpana G, Trono D (2001) Cytoplasmic recruitment of INI1 and PML on incoming HIV preintegration complexes: interference with early steps of viral replication. Mol Cell 7(6):1245–1254
Uchil PD, Hinz A, Siegel S, Coenen-Stass A, Pertel T, Luban J, Mothes W (2013) TRIM protein-mediated regulation of inflammatory and innate immune signaling and its association with antiretroviral activity. J Virol 87(1):257–272
Uchil PD, Quinlan BD, Chan WT, Luna JM, Mothes W (2008) TRIM E3 ligases interfere with early and late stages of the retroviral life cycle. PLoS Pathog 4(2):e16
Urrutia R (2003) KRAB-containing zinc-finger repressor proteins. Genome Biol 4(10):231
van der Aa LM, Levraud JP, Yahmi M, Lauret E, Briolat V, Herbomel P, Benmansour A, Boudinot P (2009) A large new subset of TRIM genes highly diversified by duplication and positive selection in teleost fish. BMC Biol 7:7
Van Lint C, Emiliani S, Ott M, Verdin E (1996) Transcriptional activation and chromatin remodeling of the HIV-1 promoter in response to histone acetylation. EMBO J 15(5):1112–1120
Versteeg GA, Rajsbaum R, Sanchez-Aparicio MT, Maestre AM, Valdiviezo J, Shi M, Inn KS, Fernandez-Sesma A, Jung J, Garcia-Sastre A (2013) The E3-ligase TRIM family of proteins regulates signaling pathways triggered by innate immune pattern-recognition receptors. Immunity 38(2):384–398
Virgen CA, Kratovac Z, Bieniasz PD, Hatziioannou T (2008) Independent genesis of chimeric TRIM5-cyclophilin proteins in two primate species. Proc Natl Acad Sci USA 105(9):3563–3568
Vogel MJ, Guelen L, de Wit E, Peric-Hupkes D, Loden M, Talhout W, Feenstra M, Abbas B, Classen AK, van Steensel B (2006) Human heterochromatin proteins form large domains containing KRAB-ZNF genes. Genome Res 16(12):1493–1504

Welker R, Hohenberg H, Tessmer U, Huckhagel C, Krausslich HG (2000) Biochemical and structural analysis of isolated mature cores of human immunodeficiency virus type 1. J Virol 74(3):1168–1177
Wilson SJ, Webb BL, Maplanka C, Newman RM, Verschoor EJ, Heeney JL, Towers GJ (2008a) Rhesus macaque TRIM5 alleles have divergent antiretroviral specificities. J Virol 82(14):7243–7247
Wilson SJ, Webb BL, Ylinen LM, Verschoor E, Heeney JL, Towers GJ (2008b) Independent evolution of an antiviral TRIMCyp in rhesus macaques. Proc Natl Acad Sci USA 105(9):3557–3562
Wiznerowicz M, Jakobsson J, Szulc J, Liao S, Quazzola A, Beermann F, Aebischer P, Trono D (2007) The Kruppel-associated box repressor domain can trigger de novo promoter methylation during mouse early embryogenesis. J Biol Chem 282(47):34535–34541
Wolf D, Cammas F, Losson R, Goff SP (2008a) Primer binding site-dependent restriction of murine leukemia virus requires HP1 binding by TRIM28. J Virol 82(9):4675–4679
Wolf D, Goff SP (2007) TRIM28 mediates primer binding site-targeted silencing of murine leukemia virus in embryonic cells. Cell 131(1):46–57
Wolf D, Goff SP (2009) Embryonic stem cells use ZFP809 to silence retroviral DNAs. Nature 458(7242):1201–1204
Wolf D, Hug K, Goff SP (2008b) TRIM28 mediates primer binding site-targeted silencing of Lys1,2 tRNA-utilizing retroviruses in embryonic cells. Proc Natl Acad Sci USA 105(34):12521–12526
Woo JS, Imm JH, Min CK, Kim KJ, Cha SS, Oh BH (2006a) Structural and functional insights into the B30.2/SPRY domain. EMBO J 25(6):1353–1363
Woo JS, Suh HY, Park SY, Oh BH (2006b) Structural basis for protein recognition by B30.2/SPRY domains. Mol Cell 24(6):967–976
Wu X, Anderson JL, Campbell EM, Joseph AM, Hope TJ (2006) Proteasome inhibitors uncouple rhesus TRIM5α restriction of HIV-1 reverse transcription and infection. Proc Natl Acad Sci USA 103(19):7465–7470
Xia ZP, Sun L, Chen X, Pineda G, Jiang X, Adhikari A, Zeng W, Chen ZJ (2009) Direct activation of protein kinases by unanchored polyubiquitin chains. Nature 461(7260):114–119
Xu L, Yang L, Moitra PK, Hashimoto K, Rallabhandi P, Kaul S, Meroni G, Jensen JP, Weissman AM, D'Arpa P (2003) BTBD1 and BTBD2 colocalize to cytoplasmic bodies with the RBCC/tripartite motif protein, TRIM5delta. Exp Cell Res 288(1):84–93
Yamauchi K, Wada K, Tanji K, Tanaka M, Kamitani T (2008) Ubiquitination of E3 ubiquitin ligase TRIM5α and its potential role. FEBS J 275(7):1540–1555
Yamauchi M, Freitag B, Khan C, Berwin B, Barklis E (1995) Stem cell factor binding to retrovirus primer binding site silencers. J Virol 69(2):1142–1149
Yap MW, Lindemann D, Stanke N, Reh J, Westphal D, Hanenberg H, Ohkura S, Stoye JP (2008) Restriction of foamy viruses by primate Trim5α. J Virol 82(11):5429–5439
Yap MW, Mortuza GB, Taylor IA, Stoye JP (2007) The design of artificial retroviral restriction factors. Virology 365(2):302–314
Yap MW, Nisole S, Lynch C, Stoye JP (2004) Trim5α protein restricts both HIV-1 and murine leukemia virus. Proc Natl Acad Sci USA 101(29):10786–10791
Ylinen LM, Price AJ, Rasaiyaah J, Hue S, Rose NJ, Marzetta F, James LC, Towers GJ (2010) Conformational adaptation of Asian macaque TRIMCyp directs lineage specific antiviral activity. PLoS Pathog 6(8):e1001062
Yu S, Gao B, Duan Z, Xu W, Xiong S (2011) Identification of tripartite motif-containing 22 (TRIM22) as a novel NF-kappaB activator. Biochem Biophys Res Commun 410(2):247–251
Yueh A, Leung J, Bhattacharyya S, Perrone LA, de los Santos K, Pu SY, Goff SP (2006) Interaction of moloney murine leukemia virus capsid with Ubc9 and PIASy mediates SUMO-1 addition required early in infection. J Virol 80(1):342–352
Zeng L, Yap KL, Ivanov AV, Wang X, Mujtaba S, Plotnikova O, Rauscher FJ 3rd, Zhou MM (2008) Structural insights into human KAP1 PHD finger-bromodomain and its role in gene silencing. Nat Struct Mol Biol 15(6):626–633
Zhang F, Hatziioannou T, Perez-Caballero D, Derse D, Bieniasz PD (2006) Antiretroviral potential of human tripartite motif-5 and related proteins. Virology 353(2):396–409

Zhang J, Hu MM, Wang YY, Shu HB (2012) TRIM32 protein modulates type I interferon induction and cellular antiviral response by targeting MITA/STING protein for K63-linked ubiquitination. J Biol Chem 287(34):28646–28655

Zhang Y, Maksakova IA, Gagnier L, van de Lagemaat LN, Mager DL (2008) Genome-wide assessments reveal extremely high levels of polymorphism of two active families of mouse endogenous retroviral elements. PLoS Genet 4(2):e1000007

Zhao G, Ke D, Vu T, Ahn J, Shah VB, Yang R, Aiken C, Charlton LM, Gronenborn AM, Zhang P (2011) Rhesus TRIM5α disrupts the HIV-1 capsid at the inter-hexamer interfaces. PLoS Pathog 7(3):e1002009

Zuo X, Sheng J, Lau HT, McDonald CM, Andrade M, Cullen DE, Bell FT, Iacovino M, Kyba M, Xu G, Li X (2012) Zinc finger protein ZFP57 requires its co-factor to recruit DNA methyltransferases and maintains DNA methylation imprint in embryonic stem cells via its transcriptional repression domain. J Biol Chem 287(3):2107–2118

The Antiviral Activities of Tetherin

Stuart J. D. Neil

Abstract Tetherin (BST2/CD317) has emerged as a key host cell defense molecule, inhibiting the release and spread of diverse enveloped virions from infected cells. In this chapter, I review the molecular and cellular basis for tetherin's antiviral activities and the function of virally encoded countermeasures that disrupt its function. I further describe recent advances in our understanding of tetherin's associated role in viral pattern recognition and the evidence for its role in limiting viral pathogenesis in vivo.

Contents

S. J. D. Neil (✉)
Department of Infectious Disease, King's College London School of Medicine Guy's Hospital, London SE1 9RT, UK
e-mail: stuart.neil@kcl.ac.uk

B. R. Cullen (ed.), *Intrinsic Immunity*, Current Topics in Microbiology and Immunology 371, DOI: 10.1007/978-3-642-37765-5_3,

1 Introduction

Over the last decade, there has been an explosion in the identification of mammalian proteins that specifically block stages of viral replication cycles in infected cells (see intro chapter). While these can be constitutively expressed, they are often regulated by type 1 interferons, and their expression acts as an intrinsic barrier to the replication and spread of viruses both in vitro and in vivo (Neil and Bieniasz 2009). A common theme among these proteins is that they target molecular structures of the virus, its replicative intermediates, or the cellular environment required for viral replication, in a manner that cannot simply be avoided by point mutations in a given viral protein. Furthermore, such molecular targets are often common to various diverse viral families, rendering these antiviral proteins broadly specific effectors of host defense. To subvert these host defenses, many mammalian viruses have in turn evolved to encode specific countermeasures that allow their evasion. One such immutable viral structure is the lipid bilayer derived from the host cell that forms the membrane of enveloped viruses. The interferon-induced membrane protein tetherin (also known as CD317 or bone marrow stromal cell antigen 2—BST2) was identified as an antiviral factor with broad activity against enveloped viruses that targets this virion component.

Tetherin first came to the attention of virologists as the target of virally encoded accessory proteins. It was identified in a proteomic screen as a plasma membrane (PM) protein downregulated from the surface of cells expressing the membrane-associated RING-CH (MARCH) ubiquitin ligase of human herpesvirus 8 (HHV8) K5 (Bartee et al. 2006), although its relevance at the time was unclear. Shortly afterward, it was found to be a target of the human immunodeficiency virus type 1 (HIV-1) accessory gene Vpu (Neil et al. 2008; Van Damme et al. 2008).

Vpu-defective mutants of HIV-1 were well known to be impaired in their ability to release newly assembled virions from infected CD4+ T cells and macrophages (Bour and Strebel 2003). Viral particles remained associated with the cell surface and accumulated in endosomes (Klimkait et al. 1990; Gottlinger et al. 1993). However, a requirement for Vpu in HIV-1 release was cell type and species specific (Varthakavi et al. 2003; Neil et al. 2006). Heterokaryons between cells that were permissive and restrictive to Vpu-defective viral release revealed that the Vpu requirement was dominant, suggesting an active restriction (Varthakavi et al. 2003). Moreover, in restrictive cell types, Vpu was capable of enhancing the release of other retroviruses and filovirus-like particles (Gottlinger et al. 1993; Neil et al. 2006; Neil et al. 2007). This phenotype could also be induced by type 1 interferons (Neil et al. 2007), suggesting that such a restriction might be a general antiviral defense targeting diverse enveloped viruses. At the biochemical level, Vpu-defective HIV-1 particles assemble at the PM and formed mature particles that are subsequently endocytosed to accumulate in endosomes (Neil et al. 2006). These virions could be released by protease treatment, indicating that a protein-based tether (*a tetherin*) prevented cell-free viral release, and such virion tethering could be induced by interferon alpha in both permissive human and monkey cells

(Neil et al. 2007). These observations allowed the identification of BST2 as *tetherin* by comparative gene arrays of permissive and restrictive cell types (Neil et al. 2008). Ectopic expression of tetherin in permissive cells induced accumulations of virions adhered to the PM and to each other and was necessary and sufficient to confer a requirement for Vpu in HIV-1 assembly and release (Neil et al. 2008; Van Damme et al. 2008). Subsequently, tetherin has been found to have antiviral activity against a diverse array of enveloped RNA and DNA viruses, and in several cases, countermeasures encoded by these viruses have been identified (Le Tortorec et al. 2011). Accumulating evidence suggests that the adaptation of primate immunodeficiency virus countermeasures to species-specific differences in tetherin was a key event in the cross-species transmission of HIV-1 and HIV-2 (Sharp and Hahn 2011) and that in murine systems, tetherin can modulate viral pathogenesis and/or immune responses to viral infection (Liberatore and Bieniasz 2011; Barrett et al. 2012; Swiecki et al. 2012).

2 Tetherin

2.1 Structure and Subcellular Localization

Tetherin is a single-pass type II membrane protein with a short cytoplasmic tail, a transmembrane domain, and an extracellular domain, the majority of which forms an extended alpha-helical rod of approximately 17 nm in length (Fig. 1a, b) (Hinz et al. 2010; Schubert et al. 2010; Yang et al. 2010a; Swiecki et al. 2011). The immature protein contains a predicted second membrane-spanning helix (Andrew et al. 2011a) that is processed to a glycophosphatidylinositol (GPI) anchor so that the C-terminal of the mature protein is covalently anchored to the outer leaflet of the cellular membrane (Kupzig et al. 2003). Mature tetherin exists as a disulfide-linked dimer linked by three conserved cysteine residues in the N-terminal half of the ectodomain (Andrew et al. 2009; Perez-Caballero et al. 2009). Tetherin is also glycosylated on two conserved extracellular asparagine residues, leading to the mature protein running as a variably sized (30–40 kDa) endoglycosidase H-resistant smear in western blots (Perez-Caballero et al. 2009; Andrew et al. 2009).

The crystal structure of the tetherin ectodomain has been determined for both the human and murine proteins (Fig. 1c) (Hinz et al. 2010; Schubert et al. 2010; Yang et al. 2010a; Swiecki et al. 2011). In both species, two tetherin ectodomains form a parallel dimeric coiled-coil formed by 6–7 heptad repeats in C-terminal portion. This structure exists in crystals both in oxidized and in reduced forms, implying that the disulfide linkages stabilize the dimer rather than them being required for its establishment. The ectodomain dimer is hinged at two positions (A88 and G109 in the human protein), giving the N-terminal a degree of rotational flexibility (Schubert et al. 2010; Yang et al. 2010a). In its reduced form, the ectodomain dimer of human tetherin forms a head-to-head tetramer with a second

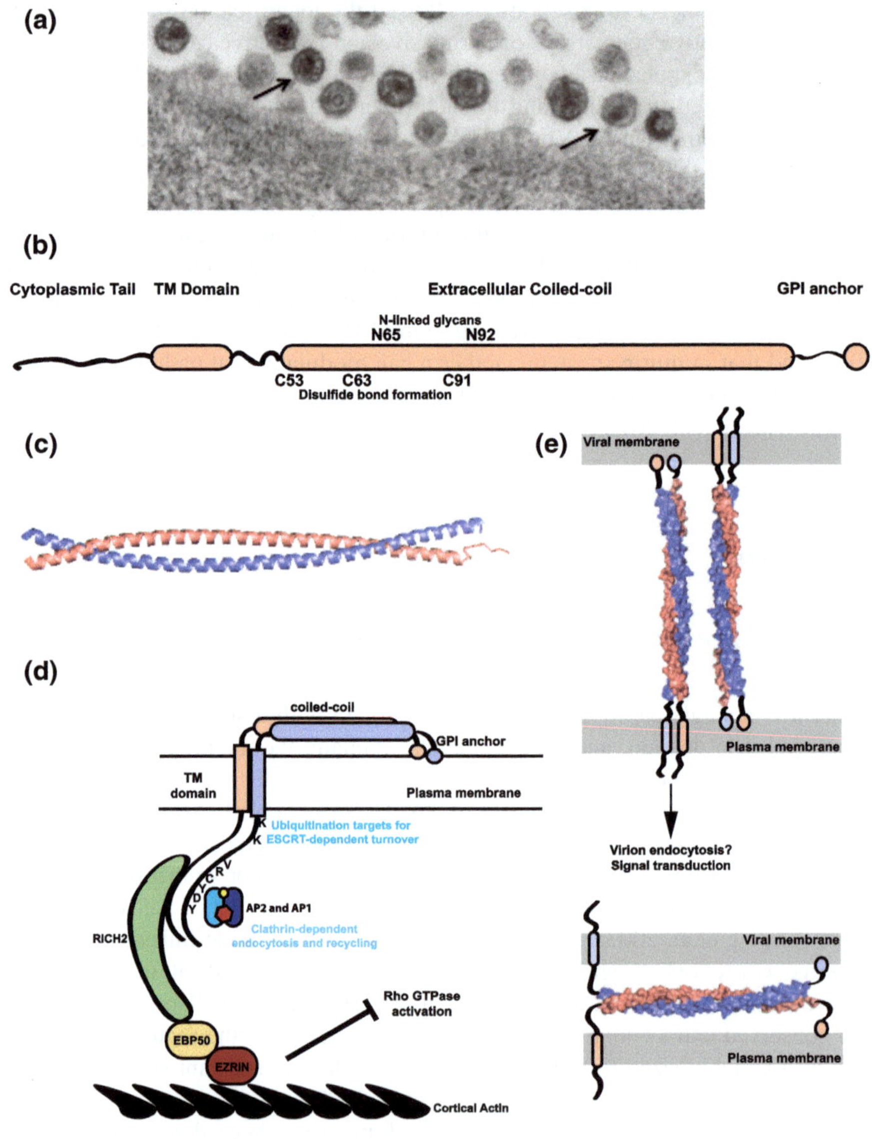

dimer (Schubert et al. 2010). This form is unlikely to accommodate disulfide linkages, and mutations in the contact residues have no effect on tetherin's biological function. However, analysis of the packing of dimers in the crystal of the murine protein suggests overlapping interactions reminiscent of BAR domains in intracellular trafficking proteins that mediate membrane curvature (Swiecki et al. 2013).

Tetherin localizes to the PM, endocytic compartments, and the trans-Golgi network (TGN) (Fig. 1d) (Kupzig et al. 2003). The cytoplasmic tail contains a conserved dual tyrosine motif that has been implicated in binding clathrin adaptors

◀**Fig. 1** Structure and antiviral function of tetherin. **a** Surface accumulations of Vpu-defective HIV-1 in tetherin-expressing HT1080 cells. **b** Schematic of human tetherin indicating the major domains, the cysteine residues involved in dimerization, and the glycosylated asparagines. **c** X-ray crystal structure of a dimer of the ectodomain of human tetherin adapted from (Yang et al. 2010a, b) PDB accession 3MQB. **d** Schematic representation of a human tetherin dimer at the plasma membrane. Tetherin is endocytosed and recycled via interactions between a tyrosine-based endocytic motif (YDYCRV) and clathrin adaptors AP2 and AP1, respectively. The tetherin is also associated with the cortical actin cytoskeleton by an interaction between its cytoplasmic tail and RICH2. This complex also includes EBP50 and Ezrin. Disruption of tetherin RICH2 interactions activates the Rho GTPase Rac1. **e** Potential configurations of tetherin dimers directly inhibiting virus release. Parallel cross-links in which both TM domains partition into the virion can be biochemically demonstrated. The reverse orientation is also suggested by both tetherin's ability to promote virion internalization and signaling which would require the cytoplasmic tail to remain in the cell. The third orientation (*lower panel*) is less favorable and may be structurally precluded due to dimerization of the tetherin TM domains

AP1 and AP2 and mediating the endocytic recycling of tetherin via the TGN (Rollason et al. 2007). The addition of the GPI anchor is essential for transport of newly synthesized tetherin to the surface from the endoplasmic reticulum (ER). Inhibition of GPI addition in cells that lack the *pigl* gene leads to tetherin retention in the ER, and surface expression can be restored by truncation at the GPI-processing signal (Perez-Caballero et al. 2009). By virtue of its GPI anchor, tetherin can be isolated in detergent-resistant cholesterol-rich microdomains (lipid rafts), although the relevance of raft domains in tetherin function is unclear at present (Kupzig et al. 2003). A recent study has suggested that tetherin may form "picket fence" demarcations between raft and non-raft domains at the PM, regulating the distribution of membrane proteins (Billcliff et al. 2013).

In polarized epithelial cells, tetherin localizes almost exclusively to the apical surface, where it interacts with ARHGAP44 (RICH2), a Rho GTPase-activating protein that links it to the subapical actin network (Rollason et al. 2009). Disruption of this interaction leads to Rac activation and dissolution of the polarized actin filaments without affecting permeability of the monolayer. What role tetherin plays in the integrity of polarized cell layers in vivo is not clear, but the lack of obvious developmental defects in tetherin knockout mice indicates that it is not essential for tissue integrity (Liberatore and Bieniasz 2011).

2.2 Mechanism of Tetherin's Antiviral Activity

The unusual structural topology of tetherin is the key to its antiviral function, and current structural, biochemical, and cell-biological observations implicate tetherin as forming physical cross-links between virion and cellular membranes. The dual-membrane anchors (one protein, the other lipid) are essential for tetherin-mediated virion retention (Perez-Caballero et al. 2009). Likewise, the ability of the tetherin to form stable disulfide-linked dimers (Perez-Caballero et al. 2009), and the integrity of the coiled-coil domain is also required for antiviral activity (Hammonds et al. 2012). That fact that these structural features, rather than the primary amino acid sequence, are major determinants of tetherin's antiviral activity was highlighted by the construction of an artificial tetherin (Perez-Caballero et al. 2009). This dimeric molecule had the same topological arrangement but was constructed from the N-terminus of the transferrin receptor, the coiled-coil of DMPK, and the GPI anchor of uPAR, when ectopically expressed restricted HIV-1 release. These data also strongly suggested that tetherin is capable of retaining virions without the need for a cofactor.

Tetherin-restricted virions can be released from cell surfaces by protease treatment (Neil et al. 2008), and immunolabeling of infected cells demonstrates that tetherin can be detected at the interface between the retained virion and the PM or between the virus particles themselves (Fitzpatrick et al. 2010; Hammonds et al. 2010). The two-membrane anchors of tetherin and its rod-like ectodomain suggest that tetherin dimers can simultaneously span virion and cellular membranes (Perez-Caballero et al. 2009; Lehmann et al. 2011). While several arrangements can be predicted (Fig. 1e), the presence of dimeric N-terminal stumps of tetherin in protease-released virion membranes demonstrates that parallel cross-links do exist. The flexibility of the ectodomain mediated by the hinge points may allow the curvature of the membrane of the budding virion to be accommodated as tetherin dimers partition into the membrane (Weissenhorn et al. 2012). Consistent with this, mutation of these hinges reduces tetherin's restrictive potency (Yang et al. 2010a). The N-terminal TM domain, rather than the GPI anchor, is required for tetherin incorporation into virion membranes (Perez-Caballero et al. 2009), although this does not necessarily mean that the cross-links are only composed of dimers with their N-termini incorporated in virion. The presence of virions tethered to each other as well as recent data identifying tetherin isoforms that have differential signaling capacities suggests that either orientation can be present (Cocka and Bates 2012; Galao et al. 2012; Tokarev et al. 2013).

A recent super-resolution microscopy study has quantified the number of tetherin molecules that cluster with a budding HIV-1 virion, revealing that the majority of assembling particles are associated with 4–7 tetherin dimers (Lehmann et al. 2011). Interestingly, budding events are adjacent to, rather than overlapping with, markers of lipid rafts. What further signals control the recruitment of tetherin into nascent virions remains an open question, although a recent study suggests that the membrane curvature induced by viral assembly may itself be a driver (Grover et al. 2013).

Some EM images of tetherin-restricted virions sometimes show electron-dense membranous stalks between viral particles and the PM/other virions (Neil et al. 2007; Hammonds et al. 2010). What these structures are is unclear, but tetherin immunoreactivity can be detected along their length (Hammonds et al. 2010). An obvious explanation is that they are extensions of the PM, with a virion tethered by tetherin dimers at the tip. However, similarities in the packing of murine tetherin ectodomains in the crystal structure, with the interactions between BAR domain-containing proteins that mediate membrane tubulation, have led to speculations that such arrangements of tetherin may stabilize these structures (Swiecki et al. 2013). Their relevance to virion retention is unclear at present.

In tetherin-expressing cells, virions can often be observed accumulating in endosomal compartments due to their endocytosis from the surface (Neil et al. 2008). This process is not required for the restriction of virion release per se, but likely results in virion degradation that may play a role in tetherin's antiviral function in vivo (see Sect. 6). Whether this virion uptake is dependent on clathrin-mediated endocytosis has been debated, but the process can be completely blocked by dominant-negative mutants of the GTPase Rab5, which controls early endosome formation (Neil et al. 2006; Harila et al. 2007). Tetherin molecules that lack the Tyr-based recycling motif show a reduction (but not an absence) in endosomal virion accumulation, suggesting that tetherin itself may play a direct role in virus particle internalization (Galao et al. 2012). Rabring7 (BCA2), a Ras superfamily GTPase with putative E3 ubiquitin ligase activity that has been identified as interacting with tetherin's cytoplasmic tail, has been implicated in virion degradation in tetherin-expressing cells (Miyakawa et al. 2009). The association of tetherin with the cortical actin network via RICH2 (Rollason et al. 2009) has yet to be shown to be important in its antiviral activity, but conceivably, it could also be involved in virion uptake. Interestingly a rare single-nucleotide polymorphism (SNP) in RICH2 that is predicted to affect a splice acceptor site has been found in one study to be associated with HIV-1 disease progression (Le Clerc et al. 2011), suggesting that a role for RICH2 in tetherin's antiviral activity merits further study.

2.3 Genetic Diversity in Mammalian Tetherins

Tetherin is encoded on chromosome 19 in humans (Ishikawa et al. 1995) (chromosome 8 in the mouse), and so far, only mammalian orthologues have been described. While no splice variants of tetherin have yet been identified, recent evidence suggests that the protein exists in two isoforms (Cocka and Bates 2012). All mammalian tetherins so far sequenced have a methionine at the position 13, downstream of the conserved tyrosine-based endocytic motif. Both this AUG codon and the +1 translational initiating codon have weak Kozak consensuses. In the translation of the natural human mRNA, this appears to lead to leaky ribosomal scanning and expression of both a long and a short isoforms. While both are able to restrict virion release, the shorter isoform lacks amino acids required for both sensitivity to HIV-1 Vpu and signal transduction, the implications for which will be discussed later. Furthermore, the 5′ initiation codon is polymorphic in at least two species: both the domestic cat sequence (Celestino et al. 2012) and the tetherin gene in NZW mice (Barrett et al. 2012) lack the +1 AUG, leading to expression of the shorter isoform only. Whether further such polymorphisms occur in other species is unknown. Additionally, the tetherin gene was duplicated during the evolution of ruminants prior to the divergence of sheep, goats, and cows (Arnaud et al. 2010). The two sheep tetherins (oBST2A and oBST2B) have antiviral activity against retroviruses, but differ in their potency and potentially display cell-type specificity.

A major difference between human tetherin and those of primates including chimpanzees is the deletion of a 5 amino acid patch (G/DIWKK) in the cytoplasmic tail (Sauter et al. 2011b). These amino acids are also missing in the recently sequenced Neanderthal and Denisovan genes, implying that it occurred at least 0.8–1 Mya. This deletion is of major importance. Not only does it encompass a determinant of primate tetherin sensitivity to at least one viral countermeasure [SIV Nef (Jia et al. 2009; Zhang et al. 2009)], but it also may confer an augmentative effect to tetherin-mediated signal transduction (Galao et al. 2012). In humans, there is one rare coding SNP, but there are several polymorphisms in putative promoter/enhancer and 3′ untranslated regions of tetherin (Laplana et al. 2013).

In common with other antiviral genes, tetherin displays evidence of positive evolutionary selection suggestive of host–pathogen conflict during mammalian evolution (Patel et al. 2011). There is some debate about the significance of some of the amino acid positions identified as being positively selected depending on the amount of mammalian tetherin sequences used and the algorithms/statistical methodology used in the calculation. In the cytoplasmic tail of primate tetherins, C9 and the W of the G/DIWKK motif have the highest positive selection signatures (Lim et al. 2010). Some studies have also found sites in the TM and extracellular domains (Gupta et al. 2009a, b; McNatt et al. 2009). These include residues relevant for counteraction by HIV-1 Vpu and SIVtan Env.

2.4 Tissue expression and Transcriptional Regulation of Tetherin

Prior to its characterization as an antiviral factor, tetherin had been identified as a cell surface marker of bone marrow stromal cells and mature B cells (Ishikawa et al. 1995; Ohtomo et al. 1999) and had also been defined as a tumor antigen (HM1.24) that is highly expressed on multiple myeloma cells. Furthermore, murine tetherin corresponds to the plasmacytoid dendritic cell (pDC) marker PDCA1 and is additionally found to various levels on myeloid cells and activated T cells (Blasius et al. 2006). A more recent immunohistopathological study of human tissues revealed wide constitutive expression on vascular endothelium, hepatocytes, renal cells, and epithelia, particularly on secretory tissues in lung and the gastrointestinal tract (Erikson et al. 2011). In tetherin negative cell lines, its expression can be induced by IFN-I, and both IFN-I and IFN-II as well as other proinflammatory stimuli increase expression levels on myeloid and lymphoid cells to varying degrees (Blasius et al. 2006; Neil et al. 2008; Van Damme et al. 2008; Bego et al. 2012; Homann et al. 2011). In ruminants, expression of both tetherin orthologues is induced by IFN-tau, and as a consequence, tetherin is highly expressed in stromal cells of the sheep reproductive tract surrounding the implanted conceptus (Arnaud et al. 2010).

Initial characterizations of the tetherin promoter have yielded insight into its tissue expression and response to inflammatory stimuli. In addition to putative NFAT-, STAT-, and NFkB-binding sites in the 1.5-Kb region upstream of the transcriptional start site, analysis of the promoter region has revealed an overlapping sequence containing an interferon-stimulated response element (ISRE), gamma-interferon activation site (GAS), and interferon regulatory factor (IRF)-binding site that are conserved across species (Bego et al. 2012). While this sequence is required for IFN-I-mediated upregulation of tetherin, expression can also be induced directly via IRFs 1, 3, and 7 independent of autocrine IFN-I induction, implying that tetherin is an early target gene of host pattern recognition receptor signaling. Consistent with this, tetherin is upregulated upon rhabdovirus infection of murine embryonic fibroblasts in a manner dependent predominantly on IRF7. Furthermore, tetherin expression can be directly induced in human lymphoid and myeloid cells via Toll-like receptors TLR3 and TLR8, respectively (Homann et al. 2011; Bego et al. 2012). Thus, in addition to interferon inducibility, tetherin expression is regulated by exposure of cells to viral PAMPs and its expression, particularly on mucosal epithelium, suggests that it may have a frontline role in host defense.

3 Viral Countermeasures of Tetherin

3.1 Overview

By virtue of targeting a virion component derived from the host cell, evasion of tetherin among enveloped viruses has necessitated the development of virally encoded countermeasures (summarized in Fig. 2a). Tetherin has been shown to restrict the cell-free release of all members of the *Retroviridae* tested (Jouvenet et al. 2009) as well as paramyxoviruses (Nipah and Hendra) (Radoshitzky et al. 2010; Kong et al. 2012), filoviruses (Ebola and Marburg viruses) (Kuhl et al. 2011), rhabdoviruses (vesicular stomatitis virus—VSV) (Weidner et al. 2010), arenaviruses (Lassa) (Sakuma et al. 2009; Radoshitzky et al. 2010), flaviviruses (dengue virus) (Pan et al. 2012), and at least one herpesvirus (human herpesvirus 8—HHV8) (Mansouri et al. 2009; Pardieu et al. 2010). While in theory tetherin has the potential to restrict the release of any enveloped virus, the notable controversy as to whether influenza A virus is sensitive to tetherin or not suggests that some enveloped viruses may be intrinsically insensitive (Watanabe et al. 2011; Yondola et al. 2011; Bruce et al. 2012; Mangeat et al. 2012; Winkler et al. 2012).

A common theme among virally encoded tetherin countermeasures is that they block the incorporation of tetherin into nascent virions and remove it from the sites of viral budding and release (Tokarev and Guatelli 2011). This is often, but not always, coupled to a decrease in tetherin levels at the PM through the subversion of tetherin trafficking to or from the PM. Tetherin's endosomal degradation is a final outcome for at least two antagonists. However, this end point is not necessarily required if prior differential tetherin trafficking has put it "beyond use" in a compartment where it cannot reaccess assembling virions. Species-specific variation in tetherin also underlies the sensitivity of primate orthologues to lentiviral countermeasures Vpu, Nef, and Env (Fig. 2b).

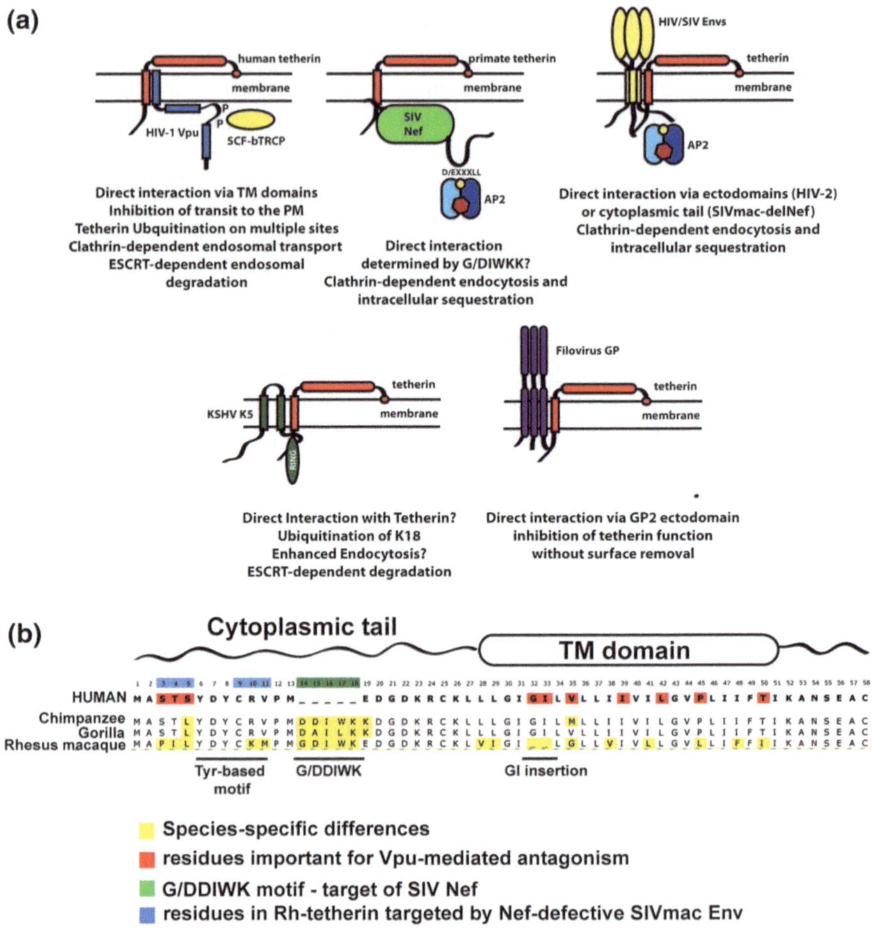

Fig. 2 Virally encoded tetherin antagonists. **a** Schematic representation of the five major virally encoded tetherin antagonists, HIV-1 Vpu, SIV Nef, HIV-2/SIV Envs, KSHV K5, and filovirus GP. **b** An alignment of the human, chimpanzee, gorilla, and rhesus macaque tetherins highlighting the determinants of sensitivity to primate lentiviral countermeasures

3.2 HIV-1 and SIV Vpu

The accessory protein viral protein U (Vpu) is a small type 1 membrane protein encoded by HIV-1, SIVcpz, and their evolutionary precursor SIVs found in several species of African guenons, notably SIVgsn, SIVmon, and SIVmon (Sharp and Hahn 2011). While the majority of studies on Vpu and tetherin have focused on the HIV-1 and SIVcpz proteins, the precursor Vpus all have antitetherin activity at least against their host species orthologues (Sauter et al. 2009).

Structurally, Vpu consists of an N-terminal transmembrane domain and a cytoplasmic tail that comprises two alpha helices separated by a conserved casein kinase II phosphorylation site (DSxxES) (Bour and Strebel 2003). This acts as a recruitment signal for an SCF E3 ubiquitin ligase complex containing the adaptors βTRCP1 or 2 that directly interact with the phosphoserine residues via their F-box domains (Margottin et al. 1998). Vpu can form TM domain–driven oligomers, probably pentamers, in infected cells that have been proposed to form a cation channel (Schubert et al. 1996; Park et al. 2003). While the oligomerization status of Vpu may play a role in regulating its function (Magadan and Bonifacino 2012), mutations that abolish the putative ion channel activity do not affect tetherin counteraction (Vigan and Neil 2010; Bolduan et al. 2011). HIV-1 clade B Vpu localizes primarily to the TGN, although some can also be detected in endosomal compartments at the ER (Varthakavi et al. 2006). Some clade C HIV-1 Vpus are more prominently localized at the PM, and this may be due to variability in the presence and the position of putative acidic dileucine (D/ExxxL(L/I/V/M)) trafficking in the cytoplasmic tail (Ruiz et al. 2008).

3.2.1 Tetherin/Vpu Interactions

A highly conserved function of HIV-1 and SIV Vpu proteins is the ability to induce the degradation of newly synthesized CD4 in the ER (Bour and Strebel 2003). This degradation is via a variant of ER-associated degradation (ERAD), whereby $SCF^{\beta TRCP}$ recruitment by Vpu mediates ubiquitination of the CD4 cytoplasmic tail on both lysine and serine/threonine residues (Magadan et al. 2010). This triggers CD4 extraction from the ER membrane, deglycosylation, and degradation by cytosolic proteasomes. Although Vpu utilizes the same ubiquitin ligase complex in its antagonism of tetherin, the mechanism by which tetherin is counteracted is distinct (Douglas et al. 2009; Mangeat et al. 2009; Mitchell et al. 2009). In particular, it had long been known that Vpu's ability to promote virus release from CD4+ T cells took place outside the ER (Schubert and Strebel 1994).

HIV-1 Vpu and tetherin interact in virus-infected cells via their respective TM domains, and this interaction is essential for the subsequent antagonism of tetherin function (Iwabu et al. 2009; Mangeat et al. 2009; Dube et al. 2010; Vigan and Neil 2010). Species-specific amino acid differences in the TM domain of tetherin, which explain why only human and chimpanzee tetherins interact with HIV-1 Vpu, lie along one face of the transmembrane helix, suggestive of a binding interface (Dube et al. 2010; Kobayashi et al. 2011). In addition, a threonine at the extracellular membrane interface in human and chimpanzee tetherins is essential for tetherin/HIV-1 Vpu interactions (Gupta et al. 2009a; McNatt et al. 2009; Kobayashi et al. 2011). For HIV-1 group M Vpus, a similar picture has emerged. A highly conserved "alanine face" and tryptophan residue in the TM helix of group M, $A_{11}xxx(A/V)_{15}xxxA_{19}xxxW_{23}$ (numbered as per the consensus HIV-1 clade B Vpu), is critical for tetherin antagonism (Vigan and Neil 2010; Skasko et al. 2012). Replacement of the alanines, particularly A_{15}, with bulky hydrophobic residues specifically impairs Vpu's physical interaction with human tetherin and its subsequent antagonism without affecting CD4 targeting. The tryptophan, which is invariant in HIV-1 and SIVcpz Vpu proteins, plays a role in the interaction with tetherin as well as CD4. NMR structures of both Vpu and tetherin TM domains have been solved, and molecular modeling of interaction between the two yields an energetically favorable binding between these two helical faces (Fig. 3a) (Skasko et al. 2012). While the stoichiometry of Vpu/tetherin interactions in vivo is unclear, such a monomeric interaction is consistent with the virological phenotypes discussed above. In the predicted multimeric form of Vpu, the alanine face would be occluded by adjacent monomers (Park et al. 2003) and none of the Vpu/tetherin interaction mutants described dominantly interfere with wild-type Vpu function (Vigan and Neil 2010). Recent studies on the Vpu/CD4 interaction have suggested that the invariant tryptophan may regulate the transition of Vpu between monomeric and multimeric forms in the ER (Magadan and Bonifacino 2012). In the monomeric form, Vpu can engage CD4 and trigger its degradation; mutations of W23 appear to stabilize Vpu multimers and inhibit this process. Whether this is true for other Vpu targets such as tetherin remains to be determined.

3.2.2 Vpu-Mediated Subversion of Tetherin Trafficking

Unlike CD4 degradation, Vpu's effects on tetherin can be inhibited by its artificial retention in the ER (Schubert and Strebel 1994; Andrew et al. 2011b; Skasko et al. 2011; Vigan and Neil 2011). This primarily reflects the differences in outcome that Vpu exerts on these proteins. Vpu reduces the steady-state levels of tetherin at the cell surface (Van Damme et al. 2008). Concomitant with that, tetherin is degraded by an ubiquitin-dependent process. While initially proposed to be an ERAD-like proteasome-dependent process (Goffinet et al. 2009; Mangeat et al. 2009), it is now clear that in virally infected cells, Vpu-mediated degradation of tetherin takes place in endolysosomal compartments (Douglas et al. 2009; Mitchell et al. 2009; Andrew et al. 2011b).

Current data suggest that Vpu only has a minor effect on the rate of tetherin endocytosis from the surface (Dube et al. 2011; Lau et al. 2011; Schmidt et al. 2011), nor does it remove tetherin from lipid rafts (Fritz et al. 2012; Lopez et al. 2012). Although inhibition of endocytosis by dynamin-2 or AP2 knockdown does rescue some tetherin levels at the surface in Vpu-expressing cells (Iwabu et al. 2009; Mitchell et al. 2009; Lau et al. 2011), this process also interferes with natural tetherin turnover and recycling, potentially confounding these observations by stabilizing pool of surface tetherin which is not exposed to Vpu. By contrast, Vpu expression potently blocks the transport both of newly synthesized tetherin to the PM and that which is recycling to and from the surface via the TGN (Dube et al. 2011; Schmidt et al. 2011), confirming the importance of recycling compartment for Vpu function that had been proposed before tetherin's identification as its cellular target (Varthakavi et al. 2006). At steady state in virally infected cells, tetherin and Vpu can be seen colocalizing to both TGN-associated compartments and some early and late endosomes, consistent with a Vpu-mediated retargeting of

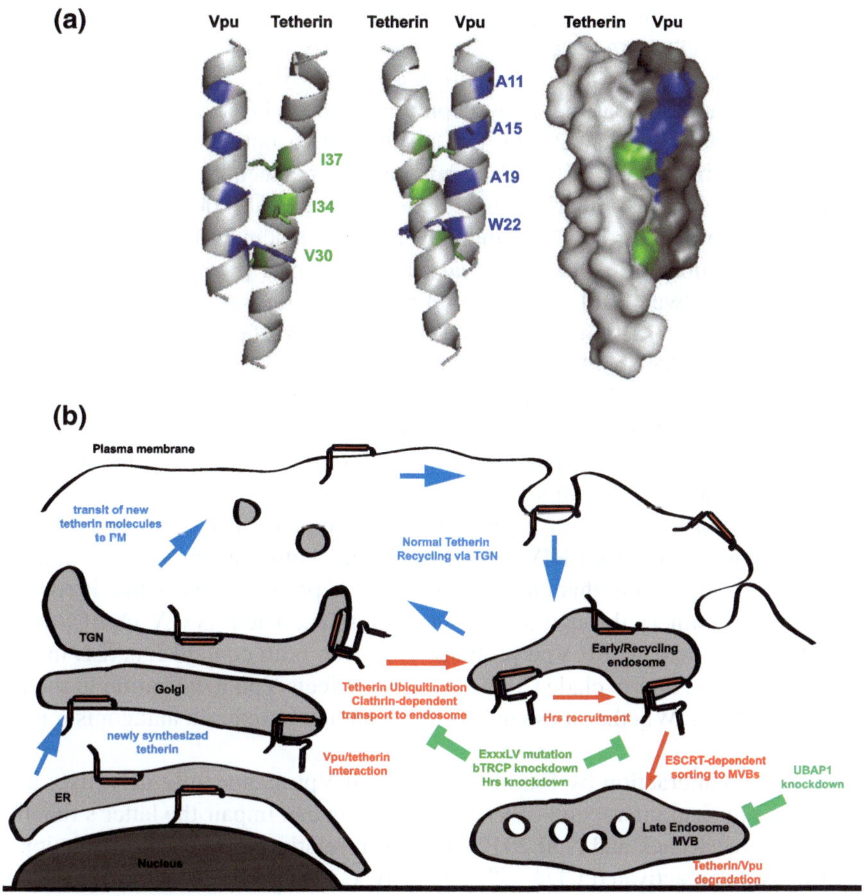

Fig. 3 Counteraction of tetherin by HIV-1 Vpu. **a** A model for human tetherin and HIV-1 Vpu interaction based on the solid-state NMR structures (PDB accessions 2LK9 and 1PI7). Highlighted in *green* are the tetherin TM residues at the interface that when mutated effect Vpu/tetherin interaction. In *blue* are the residues in the Vpu TM domain that form the "alanine face," AxxxAxxxAxxxW. **b** Schematic model of Vpu-mediated tetherin antagonism. Vpu engages tetherin in Golgi compartments, thereby targeting either newly synthesized tetherin or that which is recycling to the plasma membrane via the TGN. Vpu/tetherin complexes are targeted to endosomal compartments by a clathrin-dependent transport mechanism. At the same, Vpu recruitment of $SCF^{\beta TRCP2}$ leads to ubiquitination of tetherin on multiple cytoplasmic tail residues. In sorting endosome compartments, recruitment of Hrs promotes sorting of Vpu/tetherin complexes for ESCRT-dependent degradation in multivesicular bodies of late endosomes. Blue arrows indicate normal tetherin trafficking, *red arrows* indicate Vpu-induced tetherin transport, and *green bars* indicate Vpu mutations or siRNA treatments that perturb this process

tetherin for endosomal destruction from the TGN (Dube et al. 2009; Dube et al. 2010). This inhibition of tetherin transit to the PM and its subsequent routing for endosomal degradation is dependent on determinants in both the Vpu and tetherin cytoplasmic tails (Dube et al. 2009; Tokarev et al. 2011; Kueck and Neil 2012).

Vpu-mediated tetherin counteraction is dependent on clathrin, implying that differential trafficking of Vpu/tetherin complexes is a fundamental requirement (Dube et al. 2009; Lau et al. 2011; Kueck and Neil 2012). Moreover, dysregulation of endosomal compartments by depletion of the GTPase Rab7a also inhibits Vpu function (Caillet et al. 2011). In Vpu itself, two putative trafficking motifs have been described, one of which forms part of the second alpha helix, which when truncated inhibits tetherin counteraction (Dube et al. 2009, Kueck and Neil 2012). This motif, ExxxLV, resembles an acidic dileucine sorting signal which binds to the σ-subunits of the canonical clathrin adaptor proteins AP1, AP2, and AP3 (Kelly and Owen 2011). Mutation of the key residues does not inhibit physical tetherin/Vpu interaction, but blocks efficient cell surface downregulation and degradation, with Vpu and tetherin localizing both to the PM and to the early endosomes (Kueck and Neil 2012). That Vpu ExxxLV mutants are compromised in post-binding trafficking of tetherin for degradation is evidenced their ability to become incorporated into nascent virions in complex with tetherin—that is, in the absence of an ExxxLV motif, Vpu is dragged to the cell surface by tetherin. While the ExxxLV motif can be functionally complemented by a heterologous acidic dileucine motif from HIV-1 Nef, no single clathrin adaptor is essential for Vpu function, suggesting that either the ExxxLV is promiscuous or that it engages the clathrin machinery through a non-canonical route. The ExxxLV motif is missing in HIV-1 clade C and F Vpu proteins. However, both contain ExxxLL in their first alpha helices, and in clade C, this sequence effects Vpu localization to the PM (Ruiz et al. 2008). Whether these motifs also function in tetherin antagonism is not yet known.

Tetherin counteraction by ExxxLV mutants of Vpu is not fully defective, indicating that physical interaction of Vpu with tetherin can impair the latter's function to some degree(Kueck and Neil 2012). However, relatively modest upregulation of tetherin levels on activated CD4+ T cells by type 1 Interferon is sufficient to abolish this residual activity. Furthermore, tetherin lacking its endocytic tyrosines is fully resistant to an ExxxLV mutant, suggesting that there may be functional cooperation between Vpu and tetherin trafficking motifs.

3.2.3 Ubiquitination of Tetherin Induced by Vpu

Tetherin is poly-ubiquitinated in the presence of Vpu (Tokarev et al. 2011; Gustin et al. 2012). This modification is essential for tetherin degradation and requires the recruitment of the E3 ligase $SCF^{\beta TRCP2}$ by the conserved phosphoserine motif in the Vpu cytoplasmic tail. Again, there is some controversy over how important $SCF^{\beta TRCP2}$ recruitment is for tetherin antagonism as opposed to its final degradation. Mutation of the phosphorylated serines or depletion of βTRCP2 can have variable effects in some studies (Douglas et al. 2009; Mangeat et al. 2009; Mitchell et al. 2009; Schmidt et al. 2011), leading to suggestion that serine phosphorylation of Vpu has effects on tetherin antagonism that cannot explained by Ub-ligase recruitment (Tervo et al. 2011). However, like the ExxxLV motif, the importance of the phosphoserines in tetherin counteraction becomes more apparent at higher tetherin expression levels (Goffinet et al. 2009). Furthermore, serine phosphorylation of Vpu is essential for HIV-1 replication in macrophages and tonsillar explants (Schindler et al. 2010).

In human tetherin, two juxtamembrane lysine residues (K18 and K21) have been shown to be ubiquitinated and proposed to be important for tetherin's final degradation (Goffinet et al. 2010; Pardieu et al. 2010). However, mutation of these lysine residues has minimal impact on the ability of Vpu to counteract tetherin-mediated restriction of viral release. Other studies have detected further ubiquitination on a serine/threonine motif (STS) in tetherin cytoplasmic tail (Tokarev et al. 2011), as well as potential ubiquitination on the N-terminal methionine residues (Gustin et al. 2012). Of particular interest, the STS motif has been shown to be required for efficient reduction in tetherin surface expression and counteraction of tetherin's activity (Tokarev et al. 2011). Moreover, this data is consistent both with the reduced sensitivity of the short isoform of human tetherin to Vpu (Cocka and Bates 2012), and also a comparable result of partial sensitivity of monkey tetherin chimeras bearing the human protein's TM domain that permit Vpu interaction (McNatt et al. 2009). Since ubiquitination of tetherin may also be a determinant of endosomal trafficking as well as final degradation, the differential effects of ubiquitin addition to tetherin cytoplasmic tail residues in the presence of Vpu may underlie the discrepancies in data regarding the functional importance of βTRCP recruitment.

3.2.4 Degradation of Tetherin in Response to Vpu

The final degradation of tetherin in response to Vpu in HIV-1-infected cells takes place in late endosomes and requires the endosomal sorting complex required for transport (ESCRT) pathway, which mediates the sorting of ubiquitinated membrane proteins into intralumenal vesicles that bud into endosomal compartments (Janvier et al. 2012). Vpu enhances the recruitment of Hrs (an ESCRT-0 subunit) that binds to ubiquitinated targets (Janvier et al. 2011). Hrs is thought to sort proteins so targeted for degradation in early/recycling endosomal compartments toward the late endosome, and in one study, Hrs depletion inhibited Vpu activity in HeLa cells. Interestingly, Hrs itself binds clathrin and may provide a link between Vpu and the clathrin machinery important for trafficking as well as recruiting ESCRT-I. TSG101, an essential ESCRT-I complex component, is required for tetherin degradation, both in the presence of Vpu and also in its steady-state levels, implying that Vpu enhances tetherin's natural turnover (Janvier et al. 2011). While the core ESCRT-I complex of TSG101/VPS28/VPS37 is also essential for HIV-1 assembly (Martin-Serrano and Neil 2011), the recently identified UBAP1 is a subunit of ESCRT-I involved only in endosomal degradation (Stefani et al. 2011; Agromayor et al. 2012). Depletion of UBAP1 completely abolishes Vpu-induced tetherin degradation, but has no effect on Vpu-mediated tetherin antagonism (Agromayor et al. 2012). Thus, by the time UBAP1-containing ESCRT-I complexes are sorting tetherin into invaginations of the late endosomal membrane, it cannot be retrieved to restrict viral release. Vpu expression levels are also enhanced by UBAP1 depletion to levels similar to those of ExxxLV mutants (Kueck and Neil 2012). Since this has no effect on Vpu/tetherin interactions, it seems likely that Vpu chaperones tetherin all the way to the late endosomal membrane and is degraded with it.

Together the overall picture of Vpu's function suggests that it interacts with tetherin at the latest in TGN, thus accessing both newly synthesized and recycling tetherin pools (Fig. 3b). A clathrin-dependent transport shunts Vpu/tetherin complexes directly to endosomal compartments. This is coupled to complex ubiquitination of the tetherin cytoplasmic tail and recruitment of ESCRT-0, ultimately leading to a final degradation that is itself not essential. In which temporal order these trafficking and ubiquitination events occur is not clear, nor what additional cellular factors are involved in the process. In the absence of this trafficking subversion, however, Vpu/tetherin complexes can traffic to the PM. At lower levels of tetherin expression, this physical association of Vpu is sufficient to abrogate its function and permit viral release. However, physiological upregulation of tetherin levels by interferon requires that Vpu/tetherin complexes be targeted into a degradative pathway from which they cannot escape for efficient antagonism. In this regard, it is interesting to note that IFN also upregulates SCYL2, a clathrin-associated cofactor for protein phosphatase 2A, which has recently been shown to dephosphorylate Vpu and impair tetherin counteraction (Miyakawa et al. 2012).

3.2.5 Impaired or Absent Tetherin Counteraction in SIVcpz and HIV-1 Non-group M Vpu Proteins: A Determinant of Pandemic Spread?

Diverse SIVcpz strains have been zoonotically transferred to humans at least 4 times (Sharp and Hahn 2011). This has resulted in the HIV-1 major group (M) primarily responsible for the AIDS pandemic, the outlier group (O) that constitutes a minority of infections and is largely restricted to Cameroon, and the non-M/O group (N) and P group for which only a handful of sequences exist. Unlike HIV-1 group M Vpu proteins, HIV-1 group O Vpu proteins have no tetherin-counteracting activity although they do maintain an ability to induce CD4 degradation (Sauter et al. 2009; Yang et al. 2011). The TM domains of O-Vpu proteins lack the AxxxAxxxAxxxW face important for tetherin interaction (Vigan and Neil 2010). However, when binding to tetherin is conferred upon O-Vpu proteins, they are still poor tetherin antagonists (Vigan and Neil 2011). Most O-Vpus are retained in the ER, primarily due to an acidic residue in the hinge region between the TM domain and the first alpha helix of the cytoplasmic tail that is basic in M-Vpu (Petit et al. 2011; Vigan and Neil 2011). It may be that this affects the conformation of the cytoplasmic tail of O-Vpus such that first alpha helix does not lie parallel with the membrane or interact with cellular cofactors. Mutation of the same position in M-Vpu proteins to a hydrophobic residue also affects its localization to the TGN with a concomitant effect on tetherin antagonism (Dube et al. 2009). Group

N Vpu-mediated antagonism of human tetherin is variable (Sauter et al. 2009). In this case, the TM domain has the alanine face, likely permitting tetherin interaction (Vigan and Neil 2010). In addition, most N-Vpu proteins lack both an ExxxLV motif and on a DSGNES (Sauter et al. 2009; Sauter et al. 2012). Thus, they cannot interact with βTRCP1/2 and degrade CD4 or tetherin. HIV-1 group P, for which only one case has been documented, also encodes a Vpu unable to target tetherin (Sauter et al. 2011a; Yang et al. 2011).

For reasons discussed below (Sect. 3.3), Vpu proteins found in the SIVcpz lineage are unable to counteract chimpanzee tetherin, again primarily because the TM domain does not physically interact (Lim et al. 2010). HIV-1 groups M, N, O, and P have arisen from distinct SIVcpz zoonoses, and the Vpu protein found in SIVcpz isolates from different chimpanzee subspecies are highly variable at the amino acid level (Sauter et al. 2010). Groups M, N, and O are derived from SIVcpz strains found in *Pan troglodytes troglodytes* (SIVcpz*Ptt*), and recent evidence suggests that the Vpu proteins from SIVcpz*Ptt* strains most closely related to HIV-1 group M are likely to have required the least amount of amino acid changes to adapt their TM domains to human tetherin. HIV-1 group O is more distantly related to the rest of the SIVcpz*Ptt* family and may have come to humans from gorillas, which are also infected with an SIVcpz*Ptt*-derived virus. Within individual clades of HIV-1 group M, there may be some variability in the efficiency of Vpu function, although whether this is a reflection of the small number thus far tested is not clear. However, it has been speculated that adaptation in group M Vpu function to target tetherin upon zoonosis was a major factor in its ability to spread pandemically in humans. Although group O HIV-1 has been isolated from individuals with AIDS, its person-to-person spread is less efficient, and whether there are differences in pathogenicity and disease progression is unclear. For group N, there are even less data, with only a few isolates described. However, a recent study has suggested that adaptation to tetherin in group N Vpu is associated with pathogenesis (Sauter et al. 2012). A French national returning from a visit to Togo (a country with no prior record of non-group M HIV-1) and suffering from symptoms of acute HIV-1 infection was found to be infected by a new group N isolate. Unlike those known previously, this N-Vpu was highly efficient at counteracting and degrading tetherin, and these adaptations mapped to residues flanking the alanine face, repair of the DSGNES, and the presence of an ExxxLV motif in the second alpha helix.

3.3 SIV Nef

In SIV lineages that lack a Vpu gene, the Nef accessory protein counteracts the tetherin orthologue of the natural primate host species (Jia et al. 2009; Zhang et al. 2009; Lim et al. 2010). As with Vpu, tetherin is downregulated from the surface of Nef-expressing cells, but apparently, this does not lead to endosomal degradation at steady state. Although direct interaction between Nef and tetherin has yet to be formally demonstrated, the sensitivity of monkey tetherins maps to the G/DIWKK motif in the cytoplasmic tail (Jia et al. 2009; Zhang et al. 2009). In particular, the W residue that is predicted to be under positive selection in non-human primates is essential for counteraction by Nef, suggesting that Nef-like proteins may have exerted selective pressure on tetherin during primate evolution (Lim et al. 2010).

The Nef proteins of SIVcpz counteract their chimpanzee tetherin, consistent with the lack of this function associated with SIVcpz Vpu (Yang et al. 2010b; Sauter and Kirchhoff 2011). The explanation for this is likely to be the evolutionary heritage of SIVcpz, which is a recombinant virus derived from SIVs that infect chimpanzee prey species (Sharp and Hahn 2011). In this recombination event, the prototypic SIVcpz acquired the 5′ portion of its genome including *vpu* from a guenon SIV and the 3′ part encoding a tetherin-antagonizing *nef* from a virus similar to that which infects red-capped mangabeys (SIVrcm) and which lacks *vpu*. Therefore, SIVcpz originally had two potential tetherin antagonists, and the function was preserved in SIVcpz *nef* but lost in *vpu* (Sauter et al. 2010). Since the G/DIWKK motif was deleted in hominids, human tetherin is completely resistant to all SIV Nef proteins tested (Sauter et al. 2011b). Interestingly, HIV-1 Nef reacquired this ancestral function in one case of an experimentally infected chimpanzee that developed disease (Gotz et al. 2012). Although chimpanzee tetherin is sensitive to HIV-1 Vpu, the virus used to initially infect this animal was an isolate (SF2), subsequently shown to have a defective *vpu* gene. This observation strongly suggests that tetherin exerts a powerful selective pressure on SIV replication in vivo, a notion reinforced by pathogenic revertants of attenuated Nef-defective SIVmac infections (Serra-Moreno et al. 2011).

Nef association with the inner leaflet of the PM by myristoylation is essential for all its functions, including tetherin counteraction (Zhang et al. 2011). SIV and HIV Nefs contain a highly conserved acidic dileucine motif ((D/E)xxxL(L/M)) that binds to AP2 and mediates clathrin-dependent endocytosis of CD4 (Lindwasser et al. 2008). This is further stabilized by two acidic residues that are 8 amino acids downstream. These features are essential for tetherin counteraction, removal from viral budding zones, and surface downregulation, as is AP2 function itself (Zhang et al. 2011). Three further amino acids in a SIVcpz Nef C-terminal domain (Q152, N162, and C163) confer chimpanzee tetherin counteraction to HIV-1 Nef. Two of these (N162 and C163) fall between the acidic and hydrophobic residues in the AP2-binding site. Similarly, the key adaptations in HIV-1 SF2 Nef that led to reacquisition of tetherin antagonism map to this motif (Gotz et al. 2012). In this case, these mutations do not affect other AP2-dependent Nef functions. Whether they stabilize Nef/tetherin/AP2 complexes or play a role in direct tetherin/Nef interactions remains to be determined.

3.4 HIV-2 and Other Lentiviral Envelope Proteins

The lack of a G/DIWKK motif in human tetherin probably underlies the lack of adaptation of SIVcpz Nef to human tetherin and the revival of this function in HIV-1 Vpu. However, the other etiological agent of human AIDS, HIV-2, is a zoonotic infection of SIV from sooty mangabeys (SIVsm) that does not encode a Vpu (Sharp and Hahn 2011). It was known that laboratory isolates of HIV-2 had a Vpu-like activity associated with their envelope glycoproteins (Env—a trimer of surface gp105/transmembrane gp41 heterodimers), and in particular, single amino acid changes in both gp105 and the extracellular domain of gp41 in the HIV-2 ROD14 isolate lacked this attribute (Bour and Strebel 1996; Bour et al. 2003). HIV-2 ROD10 and HIV-2 ST Env interact with tetherin and mediate its removal from the cell surface, whereas that of HIV-2 ROD14 or the Env of SIVmac does not (Jia et al. 2009; Le Tortorec and Neil 2009; Hauser et al. 2010). The determinants of interaction are likely to be extracellular as exchange of the gp41 TM domain and cytoplasmic tail of HIV-2 Env with that of SIVmac does not abolish function(Le Tortorec and Neil 2009). However, the proteolytic processing of Env from its precursor gp140 to gp105/gp41 is essential (Bour et al. 2003; Le Tortorec and Neil 2009). Counteraction of tetherin and its cell surface removal is again dependent on a clathrin-dependent endocytic process and requires the membrane-proximal binding site for AP-2μ, GYXXΦ, common to HIV and SIV envelopes (Abada et al. 2005; Noble et al. 2006; Le Tortorec and Neil 2009). Again, as with SIV Nef, there is no evidence of tetherin degradation in HIV-2-infected cells, and tetherin appears to accumulate in TGN-associated compartments, implying intracellular sequestration as the major mechanism (Le Tortorec and Neil 2009). HIV-2 Env displays a broader species specificity for primate tetherins than Vpu, although in vitro its antagonism is less potent than either Vpu or Nef (Hauser et al. 2010). At present, the data regarding HIV-2 Env function are limited to a few laboratory isolates of HIV-2 clade A, and how widespread this attribute is in vivo is not known. However, since tetherin counteraction is associated with an essential viral gene that confers infectivity to the virion, it is of interest to speculate about whether such adaptations have come with a fitness cost (Evans et al. 2010). HIV-2 appears to be less pathogenic that HIV-1 and has spread less efficiently in humans (Sharp and Hahn 2011). Whether some of this difference reflects variable adaptations to host restriction factors is an area of active research.

Tetherin counteraction has also been ascribed to several other lentiviral envelopes. Similar to HIV-2, tantalus SIV (SIVtan) Env counteracts several primate tetherins (Gupta et al. 2009b). Again, an AP2-binding site is required in the cytoplasmic tail, and interaction via extracellular domain is implied by a mammalian species-specific change (A100D) that confers sensitivity of primate tetherins to SIVtan. One caveat to this data is that the infectious molecular clone of SIVtan is derived from a virus passaged through human T cells, and so the variable function of either Env or Nef in several studies may not reflect tetherin targeting in the natural host. Similarly, feline tetherin has been suggested to be a target for FIV Env in one study (Celestino et al. 2012), but not another (Dietrich et al. 2011). But perhaps the most interesting example of Env/tetherin interactions has been described in Nef-deleted SIVmac mutants that revert to pathogenicity in infected Rhesus macaques (Serra-Moreno et al. 2011). In these viruses, one of the determinants of pathogenicity is a group of amino acid changes in the cytoplasmic tail of gp41 that confer the ability to counteract rhesus macaque tetherin.

Mechanistically, this again leads to surface tetherin reduction. However, unlike HIV-2 Env, interaction with tetherin is driven by these cytoplasmic tail changes and is determined by species-specific differences between the rhesus and human tetherins in the residues that flank the dual tyrosines—PIL**YDY**(R/C)KM versus STS**YDY**CRV. Furthermore, there are hints of differential sensitivity of rhesus tetherins conferred by a polymorphism at position 9 (C/R). This residue has been identified as being under positive selection, perhaps suggesting that similar mechanisms of viral antagonism have targeted this region in the past. These studies, however, underscore the selective pressure that tetherin can exert in vivo to drive viral adaptation.

3.5 HHV8 K5

Human herpesvirus 8 (HHV8 or KSHV), a γ2-herpesvirus and the causative agent of Kaposi's sarcoma, encodes two immunomodulatory membrane-bound RING-CH (MARCH) E3 ubiquitin ligases K3 and K5 (also known as vMIR1 and vMIR2) (Boname and Lehner 2011). Like their cellular homologue MARCH-VIII, K3 and K5 both contain two TM domains linked by a short extracellular loop. The N-terminal cytoplasmic tail consists of the RING-CH domain essential for the proteins' function; the C-terminal cytoplasmic tail contains motifs likely involved in subcellular trafficking and endosomal targeting. Both K3 and K5 mediate the ubiquitination and endosomal degradation of target PM proteins. Unlike K3, which primarily promotes endosomal degradation of class I MHC molecules, K5 has an extensive substrate specificity targeting MHC and MHC-related molecules, NK receptor ligands, adhesion molecules, and cytokine receptors (Boname and Lehner 2011). K5 was the first viral protein shown to target tetherin, which was identified in a mass-spectrometry-based proteomic screen as being degraded in HeLa cells upon K5 expression (Bartee et al. 2006). Subsequently, it was shown that tetherin overexpression, or K5 depletion in latently infected cells constitutively expressing tetherin, led to an inhibition of cell-free virion production upon induction of lytic HHV8 replication (Mansouri et al. 2009; Pardieu et al. 2010). Moreover, K5 expression rescued Vpu-defective HIV-1 release from tetherin-expressing cells (Hauser et al. 2010; Pardieu et al. 2010). Herpesviruses have a complex assembly process thought to involve primary envelopment at the nuclear membrane followed by virion reentry to the cytoplasm (Mettenleiter et al. 2006). A secondary envelopment in membrane derived from TGN-associated compartments then leads to virion egress via the secretory pathway. Although visualization of tetherin-restricted herpes virions has yet to be determined, it seems likely that it is this latter stage that tetherin inhibits. HHV8 replicates primarily in mature B cells and lymphatic endothelium (Ganem 2006), both of which are known to express tetherin (Erikson et al. 2011). It is not yet known whether other herpesvirus families encode tetherin antagonists or whether tetherin's antiviral activity against them is widespread.

The substrate specificity of K3 and K5 is determined in part by their TM domains, but whether they directly interact with their targets is unclear (Ishido et al. 2000). K5 mediates the ubiquitination of membrane-proximal lysine residues in the cytoplasmic tails of its target proteins. In human tetherin, the determinant of K5-mediated counteraction, surface downregulation, and degradation are lysine 18, whereas lysine 21 appears dispensable (Mansouri et al. 2009; Pardieu et al. 2010). In the presence of K5, tetherin is ubiquitinated on K18 and degraded in lysosomes by an ESCRT-dependent mechanism (Mansouri et al. 2009; Pardieu et al. 2010) that involves ESCRT-I complexes composed of the core TSG101/VPS28/VPS37 and UBAP1(Agromayor et al. 2012). By analogy to K3- and K5-mediated MHC degradation, it is likely that K5 mediates tetherin ubiquitination during its transport to the PM, which then induces enhanced uptake and endosomal targeting. Both K3 and K5 have been shown to utilize E2 ligases of the UbcH5 family to mediate the addition of Ub to the target lysine, followed by Ubc13 to further promote poly-Ub chain formation (Duncan et al. 2006). In addition to K63 linkages, evidence suggests that the formation of branched complex poly-Ub chains also containing K11 linkages is important for K5 activity (Boname et al. 2010). Like Vpu, K5 targeting of tetherin appears to have some species specificity. Rhesus and murine tetherins are insensitive to K5, but human tetherin bearing macaque-specific TM mutations that abolish Vpu interaction can still be downregulated from cell surface, implying that the determinants of species specificity are different (Pardieu et al. 2010).

Virally encoded MARCH ligases are not limited to HHV8. Other mammalian γ2-herpesviruses encode K3 or K5 homologues, although whether any target host species tetherin is unknown. Furthermore, examples of putative MARCH ligases are found in the genomes of mammalian poxviruses (myxoma virus and several yatapoxviruses), although not in vaccinia or variola virus (Lehner et al. 2005). Vaccinia virus release itself is insensitive to tetherin, but whether it encodes an antagonist or whether poxvirus assembly is intrinsically resistant to tetherin is unclear (Sliva et al. 2012). Finally, the K5 cellular homologue, MARCH-VIII, can induce tetherin degradation, suggesting there may be an ESCRT-dependent cellular regulation of tetherin expression mediated by MARCH ligases (Bartee et al. 2006).

3.6 *Ebola Virus GP*

The observation that filovirus-like particle release was sensitive to tetherin led to the implication of the Ebola virus glycoprotein (EBOV-GP) as a tetherin antagonist (Kaletsky et al. 2009). EBOV and Marburg virus (MARV) GPs can rescue the release of Vpu-defective HIV-1, and this requires direct interaction of GP with tetherin via its GP2 subunit (Kaletsky et al. 2009; Kuhl et al. 2011). The mechanism of antagonism is unclear. Unlike most known tetherin antagonists, EBOV-GP can counteract human, primate, and murine tetherins (Lopez et al. 2010).

Furthermore, there is little or no cell surface removal of tetherin in the presence of EBOV-GP, and tetherin remains associated with lipid rafts (Lopez et al. 2010; Lopez et al. 2012). In the context of infectious Ebola virus replication, the role of tetherin is unclear, with publications indicating either inhibition or no effect on viral spread (Radoshitzky et al. 2010). However, GP may also have cell-type-specific role in filoviral particle morphogenesis that may affect these observations (Kuhl et al. 2011).

3.7 Other Viruses and Tetherin

So far, several viruses that are highly sensitive to tetherin expression in culture do not appear to have a direct tetherin antagonist. This suggests that either tetherin is not expressed in the target tissues in which they replicate or that tetherin expression or activity does not impose a significant restriction to viral spread. Furthermore, as tetherin is an IFN-/PRR-induced gene, viruses that globally target new host gene expression or modulate antiviral responses by inhibiting IFN or PRR signaling pathways may simply never be exposed to it (or other directly antiviral proteins). Alternatively, the assembly pathway of some enveloped viruses itself may be intrinsically insensitive to tetherin.

Influenza A virus (IAV) has been the subject of opposing data regarding its sensitivity to tetherin. Initial observations, however, revealed that while tetherin could restrict the release of IAV-like particles derived from transfected cells, infectious virus replication was unaffected (Watanabe et al. 2011). However, a study published at the same time showed that the neuraminidase (NA) protein of some IAV strains had a Vpu-like function (Yondola et al. 2011). Further studies have not yet resolved this issue. One study showed that restriction of IAV release by tetherin could be achieved in the absence of NA or a viral regulator of innate immune gene expression, NS1 (Mangeat et al. 2012). However, others have disputed these data (Bruce et al. 2012; Winkler et al. 2012). Microscopic examination of the spherical and filamentous assembly of IAV of various strains (including those that lack either the viral ion channel M2 or NS1) in polarized epithelial cells, suggested that not only were wild-type virions produced, but that tetherin appeared to be excluded from budding filaments (Bruce et al. 2012).

A final intriguing example of virus/tetherin interactions has been the recent implication that tetherin acts as an entry cofactor for the β-herpesvirus, human cytomegalovirus (HCMV). Tetherin depletion in monocytic cells reduced the replication of HCMV (Viswanathan et al. 2011). The authors speculate that tetherin molecules incorporated into HCMV particles promote interaction with others on the target cell to enhance virion adsorption/entry. Since herpesviral assembly is likely to be sensitive to inhibition by tetherin, the implication of these data is that HCMV not only modulates tetherin function, but also parasitizes it to promote virus replication. However, the mechanistic basis underlying these observations remains to be determined.

4 Tetherin and Cell-to-Cell Transmission of Retroviruses

The vast majority of studies on the antiviral activity of tetherin have focused on its ability to block cell-free virion release. However, in the case of retroviruses, the major route of viral transmission between cells in culture is through the induction of polarized cellular contacts at which viral assembly and transfer is concentrated by physical interaction between donor and target cells [reviewed in (Martin-Serrano and Neil 2011)]. These contacts, known as virological synapses (VS), are driven by viral envelope/receptor interactions and are further stabilized by cytoskeletal rearrangements and a variety of adhesion molecules. For HIV-1, the cryo-EM reconstructions of the VS in infected CD4+ T cells have revealed it to be a partially enclosed cleft consisting of interdigitating microvillous projections in which newly assembled virions are concentrated, providing greatly increased local multiplicity of infection. Since tetherin-restricted viruses are essentially mature and infectious, whether they can still be transferred by cell-to-cell contact is a key question.

Studies that have addressed the effects of Vpu on HIV-1 cell-to-cell transfer considerably predate the discovery of tetherin. Early studies of Vpu-defective HIV-1 replication in T-cell cultures suggested that while accumulation of virions in the supernatants was reduced, spread through the culture as measured by viral gene expression was unimpaired (Strebel et al. 1989; Terwilliger et al. 1989; Klimkait et al. 1990; Yao et al. 1992; Schubert et al. 1995). Furthermore, studies from the Emerman group demonstrated that selection of HIV-1 variants for enhanced reliance on cell-to-cell contact resulted in viruses with inactivating mutations in *vpu* (Gummuluru et al. 2000). More recent studies that have directly addressed the effects of tetherin on cell-to-cell transmission have yielded conflicting results (Casartelli et al. 2010; Jolly et al. 2010; Kuhl et al. 2010). Tetherin is readily observable at synaptic contacts (Casartelli et al. 2010; Jolly et al. 2010; Pais-Correia et al. 2010). While there is agreement that CD4+ T cells and T-cell lines infected with Vpu-defective HIV-1 establish more synaptic contacts than those infected with the wild-type virus, correlating with virion retention and hence more surface envelope protein to interact with target cell CD4, the consequences of tetherin-mediated virion retention have been controversial (Casartelli et al. 2010; Jolly et al. 2010). In broad agreement with historical data described above, one study showed that Vpu-defective HIV-1 spread more efficiently across VS contacts between infected Jurkat T cells and primary CD4+ T cells, and this could be reversed by RNAi-mediated depletion of tetherin (Jolly et al. 2010). In two other studies, evidence suggested that tetherin expression reduced cell-to-cell transfer, particularly in cells artificially expressing tetherin (Casartelli et al. 2010; Kuhl et al. 2010). By contrast, human T-lymphotropic virus type 1 (HTLV-1), which spreads between infected T cells almost exclusively by cell-to-cell contact, is insensitive to tetherin, despite it being recruited to the VS and blocking cell-free virion release in other cellular systems (Pais-Correia et al. 2010; Ilinskaya et al. 2013). The reason for the discrepancy for HIV-1 is not clear. One possible explanation highlighted in a recent study is that the dynamics of HIV-1 cell-to-cell transfer can be cell-type dependent (Zhong et al. 2013). Furthermore in adherent cells, replication of MLV, which spreads via cell-to-cell contacts called cytonemes (Sherer et al. 2007), is restricted by IFN-induced tetherin expression (Liberatore and Bieniasz 2011). In addition to the amount of tetherin at the surface, another parameter that may account for the discrepant observations is the speed at which tetherin-retained virions are removed from the PM by endocytosis. An important recent observation in macrophages showed that a pool of tetherin in infected cells that was not degraded by Vpu is localized to subcellular compartments where HIV-1 accumulated (Chu et al. 2012). The nature of this tetherin pool is unknown, particularly if it is made up of short-tetherin isoforms, which are less sensitive to Vpu (Cocka and Bates 2012). In these cultures, again tetherin does inhibit cell-to-cell transfer of HIV-1 to T-cell targets (Chu et al. 2012). It should also be noted that in all cases where tetherin was observed to affect HIV-1 spread between primary cells (either positively or negatively), the potency of the effect was weak.

5 Tetherin and Viral Pathogenesis In vivo

So far, studies in murine systems have lead to mixed observations of the role of tetherin in the antiviral immune response in vivo. Mice bearing a homozygous deletion of tetherin develop normally, with no apparent defects in their lymphoid or myeloid cell populations (Liberatore and Bieniasz 2011). However, replication of Moloney MLV in fibroblasts derived from these mice is much less sensitive to inhibition by type 1 IFN. Mo-MLV replication in vivo does not induce a systemic IFN response, but another MLV, LP-BM5, induces both a systemic immune activation and a splenomegaly/lymphadenopathy reminiscent of HIV/AIDS. In this model, LP-BM5 viral loads were enhanced and pathogenesis significantly exacerbated in tetherin -/- mice, indicating tetherin plays a direct antiviral role in vivo (Liberatore and Bieniasz 2011). In a second study, a homozygous polymorphism in the +1 ATG of tetherin in NZW mice that leads to only a short-tetherin isoform expressed (from the second methionine codon in the cytoplasmic tail and lacking the tyrosine-based endocytic motif) results in higher surface tetherin expression on lymphoid and myeloid cells (Barrett et al. 2012). This correlated with reduced viral loads and pathogenesis of Friend MLV, again suggesting that tetherin can play a role in controlling virus replication in vivo. Similarly, a short-term in vivo RNAi study targeting tetherin in mice infected with mouse mammary tumor virus (MMTV) led to an increase in viral loads (Jones et al. 2012). Paradoxically, in an independently generated tetherin knock-out mouse infected with influenza virus, viral titers in the lung were reduced correlating with an increase in CD8+ T-cell responses (Swiecki et al. 2012). Whether this is a reflection of the insensitivity of influenza to tetherin in some systems remains to be determined (Sect. 3.7), but highlights that the role of tetherin in viral pathogenesis is not straightforward.

In human disease, the best evidence that tetherin plays an important role in viral infection and spread is the selection on HIV-1 to adapt Vpu to targeting tetherin (Sauter et al. 2009). This may not simply be due to tetherin blocking efficient person-to-person transmission. In HIV-1 and HCV coinfected individuals treated with pegylated IFNα, there is evidence that HIV-1 viral load reduction correlates with tetherin and APOBEC3G expression levels in HIV target cells and a suggestion that sequence changes occur in Vpu that enhance tetherin antagonism (Pillai et al. 2012). Furthermore, in single-genome analyses of *vpu* sequences from HIV-1-infected individuals, we can show that despite high levels of what appears to be HLA-driven variation in viral quasi-species, mutants with impaired antitetherin function are rare and potentially selected against even many years after infection (Pickering et al. in prep). Therefore, maintenance of this activity may be important throughout HIV-1 infection. Finally, recent data have suggested an association between putative promoter polymorphisms in tetherin and the rate of HIV-1 disease progression (Laplana et al. 2013). Further investigations will reveal whether such polymorphisms also affect the pathogenesis of other enveloped viruses.

6 Tetherin as a Viral Sensor and Regulator of Antiviral Immune Responses

The superficial paradox between the weak effects of tetherin on retroviral cell-to-cell transfer and the strong genetic evidence in primate lentiviruses that its antagonism promotes viral transmission and replication suggests that the in vivo antiviral role of tetherin may not be limited to restriction of cell-free virion release or that the cell culture assays do not fully account for the downstream consequences of its activity in vivo (Fig. 4a). In particular, since tetherin-mediated restriction of virion release retains viral particles on cell surfaces, this raises the possibility that retention of virions by tetherin may augment further antiviral immune responses. For example, virion retention would increase viral antigen exposure on the surface. This could potentially enhance antibody opsonization, sensitizing infected cells to complement deposition and/or clearance by phagocytes and natural killer cells through Fc- and C'-receptor recognition, resulting in proinflammatory sequelae. Related to this, tetherin-retained viral particles may be sensitive to enhanced endosomal processing for antigen presentation. The accumulation of Vpu-defective HIV-1 particles in late endosomes of infected cells suggests that in long-lived primary targets such as macrophages, viral proteins degraded in a tetherin-dependent manner could be routed to MHC class II-presenting compartments, leading to augmentation of antiviral T-cell responses. Interestingly, direct targeting of ovalbumin to DCs by fusing it to an antitetherin antibody enhances the induction of antigen-specific CD8+ and CD4+ T-cell responses (Loschko et al. 2011; Moffat et al. 2013).

Another activity of tetherin has been the focus of three recent publications that provide potential for a further source of selective pressure imposed on viruses to

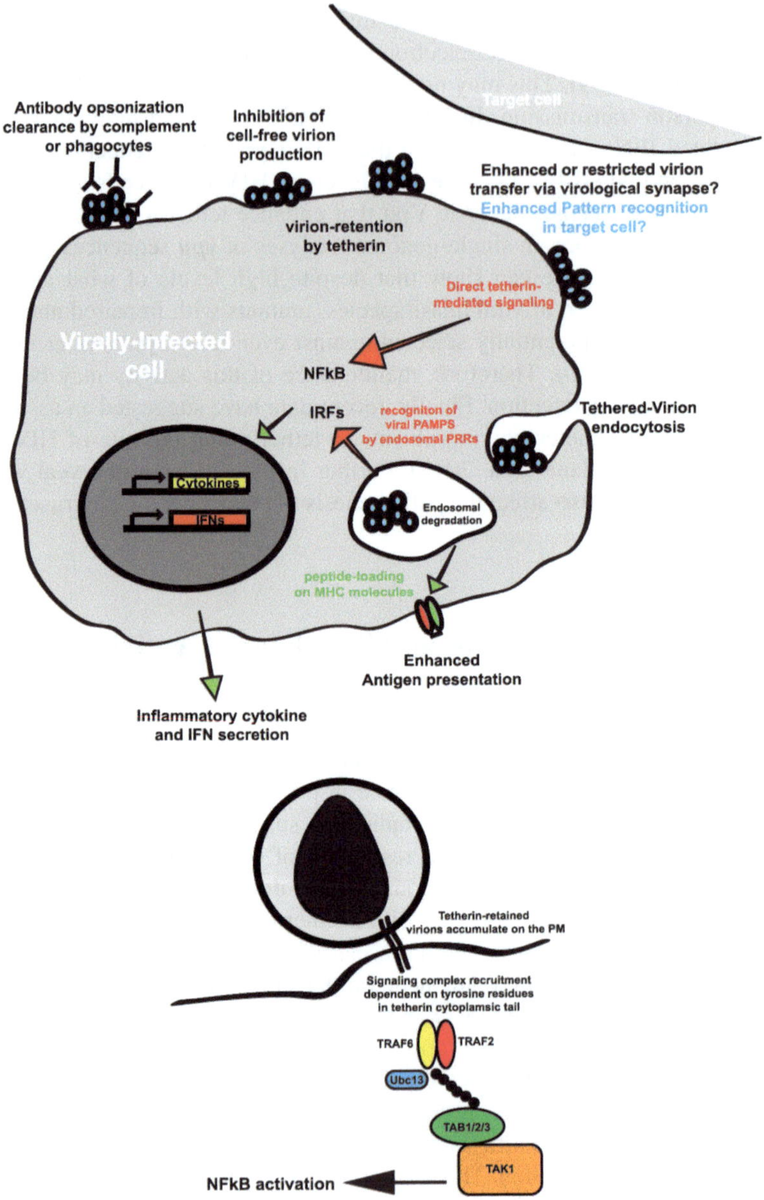

counteract its function (Cocka and Bates 2012; Galao et al. 2012; Tokarev et al. 2013). Prior to tetherin's characterization as an antiviral factor, it had been identified in a transfection screen for human proteins that induced NFκB-dependent reporter gene activation (Matsuda et al. 2003). This raised the possibility that tetherin had a proinflammatory signal transduction function that may be related

◀ **Fig. 4** **a** Potential immunological consequences of tetherin-mediated HIV-1 retention in vivo. Tetherin-mediated retention of Vpu-defective HIV-1 physically blocks cell-free virion release. These viral accumulations may promote opsonization by specific antiviral antibodies and lead to clearance of the virus infected cell by complement deposition and/or killing by phagocytes expressing Fc receptors. In addition, tetherin aggregation by viral particles induces NFκB activation and synthesis of proinflammatory cytokines. Internalization of tetherin-retained virions for destruction in late endosomes may further liberate particle-associated PAMPs for recognition by other endosomal pattern recognition receptors, promoting type 1 IFN expression. Endosomal processing of viral proteins in myeloid cells may also lead to enhanced presentation of viral antigens on class II MHC molecules to CD4+ T cells. Finally, enhanced pattern recognition may also occur *in trans*. Increased transfer of tetherin-retained viral particles across virological synapses may enhanced exposure of viral components to membrane and cytosolic PRRs in target cells. **b** Putative mechanism of signal transduction by tetherin. Clustering of tetherin at the PM by virions induces canonical NFκB activation dependent on the tyrosine-based motif in the tetherin cytoplasmic tail. This is mediated through the recruitment of complexes containing TRAF6 and TRAF2, and the associated E2 component Ubc13, which synthesize K63-linked ubiquitin chains. This in turn leads to recruitment of TABs 1–3 and activation of mitogen-activated protein kinase TAK1, which ultimately activates NFκB p50/p65 heterodimers

to its antiviral activity. Indeed, while overexpression of human tetherin by transient transfection can potently activate NFκB (Cocka and Bates 2012; Galao et al. 2012; Tokarev et al. 2013), cross-linking of surface tetherin by antibody has the same effect correlating with phosphorylation of IKKα/β and degradation of IκB (Galao et al. 2012). Most importantly, budding of tetherin-sensitive retroviral and filoviral particles from tetherin-expressing cells enhances NFκB activation (Galao et al. 2012; Tokarev et al. 2013). This demonstrated that direct virion retention can induce a signal transduction and suggested that tetherin-mediated virion retention by a virally infected cell might lead to that cell releasing higher levels of inflammatory mediators and thus promote its visibility to the immune system. In other words, tetherin may act as an innate sensor of enveloped virus assembly by recognizing the "pattern" or shape of a budding virion. Consistent with this, human CD4+ T cells infected with tetherin-sensitive HIV-1 produce higher levels of proinflammatory chemokines/cytokines CXCL10 and IL-6, as well as type 1 interferon, in a tetherin-dependent manner (Galao et al. 2012).

Mechanistically, tetherin-mediated signal transduction requires the structural characteristics in the ectodomain that are essential for virus restriction (Perez-Caballero et al. 2009; Galao et al. 2012). In addition, amino acids in the cytoplasmic tail of human tetherin that are not required for physical virion retention are essential for NFκB reporter gene activation. In particular, the tyrosine-based motif, YDYCRV, present only in the long-tetherin isoform, is a key requirement (Cocka and Bates 2012; Galao et al. 2012; Tokarev et al. 2013). While this has been described as an endocytic motif, virion endocytosis is not required for tetherin-mediated signaling. In fact, blockade of virion uptake potentiates the signaling (Galao et al. 2012). This evidence, alongside the ability of antibody cross-linking to induce the same signal, strongly suggests that virion (and hence tetherin) aggregation on the PM is responsible for triggering NFκB activation. Furthermore, this activity suggests that some tetherin dimers must remain with their N-termini in the cellular rather than the virion membrane.

Tetherin-mediated NFκB activation is dependent on the mitogen-activated protein kinase TAK1 (Galao et al. 2012). TAK1 and its associated factors TAB 1, 2, and 3 are central to the "classical" NFκB activation pathway from various inflammatory cytokine and pattern recognition receptors (Skaug et al. 2009). Its activation requires the action of one or more members of the TNF-receptor-associated factor (TRAF) E3 ubiquitin ligases, which synthesize poly-ubiquitin chains linked primarily through lysine 63 linkages. In one study, TRAF6 (Galao et al. 2012), and in another, TRAF2 and the TAB/TAK complex itself (Tokarev et al. 2013), associate in immunoprecipitates with the tetherin cytoplasmic tail dependent entirely on the cytoplasmic tail tyrosines. Furthermore, siRNA depletion of TRAF2, TRAF6, or the E2 ligase Ubc13 that is required for Lys-63 poly-ubiquitin synthesis impairs tetherin-mediated signal transduction (Galao et al. 2012). How this complex is recruited is yet to be determined. Human tetherin has a putative TRAF6-binding site, but while mutation of some of the key residues affects signaling, they do not block TRAF6 recruitment, suggesting that TRAF6 recruitment is indirect and is necessary but not sufficient to account for tetherin signal transduction. Interestingly, short isoforms of human tetherin that lack the YDYCRV motif dominantly interfere with tetherin signaling. Since dimerization of tetherin is required for signal transduction, these observations imply homodimers of the long-tetherin isoform only signal (Cocka and Bates 2012; Galao et al. 2012; Tokarev et al. 2013). This potentially indicates that tetherin's signaling capacity might be regulated at the level of differential isoform expression or sub-cellular compartmentalization. Peptides containing the YDYCRV motif have been identified by mass spectroscopy of the phosphoproteome of multiple myeloma cells (St-Germain et al. 2009). Thus, in addition to being an endocytic motif, the tyrosine residues may serve as targets for tyrosine kinases. Whether this underlies tetherin-mediated signaling and the recruitment of cellular components required to activate NFκB remains to be determined.

A further intriguing aspect of tetherin-mediated signal transduction is its apparent species specificity (Galao et al. 2012). While human tetherin is a highly potent inducer of NFκB, the chimpanzee orthologue is weaker and those of rhesus macaques, African green monkeys, and mice are unable to activate reporter genes when overexpressed. A major caveat here is that, thus far, all these studies have been carried out in human cells and could reflect species-specific differences in downstream effectors of signaling. However, examination of the determinants of signaling in human and chimpanzee tetherins suggests that this function may in part be a recent evolutionarily acquisition in hominoids. The RV residues in the YDYCRV motif, which some studies have suggested are under positive selection (Gupta et al. 2009a; McNatt et al. 2009), impair tetherin signaling when mutated to the KM found in macaques (Galao et al. 2012). Furthermore, the G/DIWKK motif that was deleted in tetherin after the ancestors of modern humans, Neanderthals, and Denisovans diverged from chimpanzees (Sauter et al. 2011b) is responsible for differences in the potency of signaling in human cells. Finally, the ability of chimpanzee/rhesus chimeric tetherins to signal requires the entire chimpanzee cytoplasmic tail up to and including the insertion of a glycine–isoleucine pair found in the TM domains of great apes (Galao et al. 2012). Since the above determinants overlie areas of either positive selection and/or of sensitivity to lentiviral countermeasures, these observations raise a provocative question: Did the ability to directly activate NFκB arise in ape tetherins as a result of pressures exerted by viruses? It is interesting to note that only known antagonists of *human* tetherin (Vpu and K5) induce its degradation. Since the threshold for efficient blockade of cell-free virus release and signal transduction induced by a degree of virion retention may be different, one might speculate that only permanent removal via degradation can ensure that direct tetherin signaling is inhibited. Might this account for the discrepancies over the importance of tetherin degradation in Vpu function as described earlier (Sect. 3.2.3)? Clearly, much further work to characterize the molecular mechanism of tetherin signaling and its species specificity is required to be able to answer these questions.

The direct activation of NFκB by restriction of virion release may not be the only way in which tetherin exerts an effect on viral pattern recognition. Although the induction of both proinflammatory cytokines and type 1 IFN in CD4+ T cells infected with Vpu-defective HIV-1 is tetherin-dependent, the IFN release appears to be indirect (Galao et al. 2012). In contrast to NFκB-dependent gene expression, tetherin cannot directly activate IRF3 or an IFNβ-promoter construct. The tetherin-dependent IFN responses in primary CD4+ T-cell infections can be blocked by inhibiting the TLR3 adaptor TRIF. TLR3, that recognizes double-stranded and highly structured RNAs (Takeuchi and Akira 2010) has been implicated in augmenting antiretroviral immunity in the mouse (Yu et al. 2012). Therefore, virus particle retention may promote the availability of virion components to other host PRRs, particularly following their endocytosis and endosomal degradation. Such a process might be separable from direct tetherin signaling and be mediated by either isoform. Importantly, such effects could occur *in cis* (enhancing responses in cells that are already infected) and also *in trans* by aggregated virions during cell-to-cell contact. Interestingly, in the case of HIV-1, a precedent for this exists. Cell-to-cell transfer of virions from infected T cells to primary target cells promotes pattern recognition of virion-associated nucleic acid in either a TLR7-mediated manner in pDCs or a non-TLR-mediated process in other target cells bearing HIV-1 receptors (Lepelley et al. 2011). This leads to type 1 IFN induction via IRF activation. Thus, under situations where tetherin enhances physical cell-to-cell transfer of Vpu-defective HIV-1 in culture systems, a concomitant augmentation of pattern recognition in the target cell may mean that it is detrimental to the virus in vivo and thus provide pressure to maintain an effective antagonist.

Finally, tetherin has been proposed to be a ligand and cofactor for ILT7, an inhibitory leukocyte receptor on pDCs that downregulates TLR7 and TLR9 responses to exogenous viral nucleic acids (Cao et al. 2009). Whether this interaction is modulated in the case of viral infection is unclear, but it has been proposed that tetherin acts in conjunction with ILT7 as a negative regulator of IFN responses by pDCs. However, in mixed leukocyte populations, blockade of tetherin/ILT7 interactions does not impair the ability of ILT7 on pDCs to dampen TLR9 responses (Tavano et al. 2013). Also, the pDCs of tetherin-deficient mice showed reduced IFN responses to TLR ligands, rather than the enhancement predicted if it were purely an ILT7 cofactor (Swiecki et al. 2012). Despite this, the potential of tetherin to exert further modulatory affects on antiviral responses requires further study.

7 Concluding Remarks

Since its identification as a target of HIV-1 Vpu, the study of tetherin has yielded a diverse array of data attesting to its potential role as an antiviral effector of the innate immune response. Its ability to interfere with the release of diverse enveloped viruses from the cell, and the discovery of other examples of virally encoded antagonists, suggests that it exerts a powerful selective pressure on viruses to adapt to avoid its activity, particularly in primate lentiviruses. However, discrepancies still exist regarding its potency at inhibiting cell-to-cell transfer of viruses whose cell-free release is efficiently blocked. Despite this, a nascent picture is emerging from tetherin-modulating antiviral immune responses either directly or indirectly through its ability to retain virions on cell surfaces. Thus, in the case of HIV-1, and other viruses that encode an antagonist, pharmacological targeting of tetherin counteraction may serve to augment and enhance antiviral immunity. The next few years of tetherin research will determine if this is a worthwhile strategy.

References

Abada P, Noble B, Cannon PM (2005) Functional domains within the human immunodeficiency virus type 2 envelope protein required to enhance virus production. J Virol 79:3627–3638

Agromayor M, Soler N, Caballe A, Kueck T, Freund SM, Allen MD et al (2012) The UBAP1 subunit of ESCRT-I interacts with ubiquitin via a SOUBA domain. Structure 20:414–428

Andrew AJ, Miyagi E, Kao S, Strebel K (2009) The formation of cysteine-linked dimers of BST-2/tetherin is important for inhibition of HIV-1 virus release but not for sensitivity to Vpu. Retrovirology 6:80

Andrew AJ, Kao S, Strebel K (2011a) C-terminal hydrophobic region in human bone marrow stromal cell antigen 2 (BST-2)/tetherin protein functions as second transmembrane motif. J Biol Chem 286:39967–39981

Andrew AJ, Miyagi E, Strebel K (2011b) Differential effects of human immunodeficiency virus type 1 Vpu on the stability of BST-2/tetherin. J Virol 85:2611–2619

Arnaud F, Black SG, Murphy L, Griffiths DJ, Neil SJ, Spencer TE et al (2010) Interplay between ovine bone marrow stromal cell antigen 2/tetherin and endogenous retroviruses. J Virol 84:4415–4425

Barrett BS, Smith DS, Li SX, Guo K, Hasenkrug KJ, Santiago ML (2012) A single nucleotide polymorphism in tetherin promotes retrovirus restriction in vivo. PLoS Pathog 8:e1002596

Bartee E, McCormack A, Fruh K (2006) Quantitative membrane proteomics reveals new cellular targets of viral immune modulators. PLoS Pathog 2:e107

Bego MG, Mercier J, Cohen EA (2012) Virus-activated interferon regulatory factor 7 upregulates expression of the interferon-regulated BST2 gene independently of interferon signaling. J Virol 86:3513–3527

Billcliff PG, Rollason R, Prior I, Owen DM, Gaus K, Banting G (2013) CD317/tetherin is an organiser of membrane microdomains. J Cell Sci [advancer online publication, 23378022]

Blasius AL, Giurisato E, Cella M, Schreiber RD, Shaw AS, Colonna M (2006) Bone marrow stromal cell antigen 2 is a specific marker of type I IFN-producing cells in the naive mouse, but a promiscuous cell surface antigen following IFN stimulation. J Immunol 177:3260–3265

Bolduan S, Votteler J, Lodermeyer V, Greiner T, Koppensteiner H, Schindler M et al (2011) Ion channel activity of HIV-1 Vpu is dispensable for counteraction of CD317. Virology 416:75–85

Boname JM, Lehner PJ (2011) What has the study of the K3 and K5 viral ubiquitin E3 ligases taught us about ubiquitin-mediated receptor regulation? Viruses 3:118–131

Boname JM, Thomas M, Stagg HR, Xu P, Peng J, Lehner PJ (2010) Efficient internalization of MHC I requires lysine-11 and lysine-63 mixed linkage polyubiquitin chains. Traffic 11:210–220

Bour S, Strebel K (1996) The human immunodeficiency virus (HIV) type 2 envelope protein is a functional complement to HIV type 1 Vpu that enhances particle release of heterologous retroviruses. J Virol 70:8285–8300

Bour S, Strebel K (2003) The HIV-1 Vpu protein: a multifunctional enhancer of viral particle release. Microbes Infect 5:1029–1039

Bour S, Akari H, Miyagi E, Strebel K (2003) Naturally occurring amino acid substitutions in the HIV-2 ROD envelope glycoprotein regulate its ability to augment viral particle release. Virology 309:85–98

Bruce EA, Abbink TE, Wise HM, Rollason R, Galao RP, Banting G et al (2012) Release of filamentous and spherical influenza A virus is not restricted by tetherin. J Gen Virol 93:963–969

Caillet M, Janvier K, Pelchen-Matthews A, Delcroix-Genete D, Camus G, Marsh M et al (2011) Rab7A is required for efficient production of infectious HIV-1. PLoS Pathog 7:e1002347

Cao W, Bover L, Cho M, Wen X, Hanabuchi S, Bao M et al (2009) Regulation of TLR7/9 responses in plasmacytoid dendritic cells by BST2 and ILT7 receptor interaction. J Exp Med 206:1603–1614

Casartelli N, Sourisseau M, Feldmann J, Guivel-Benhassine F, Mallet A, Marcelin AG et al (2010) Tetherin restricts productive HIV-1 cell-to-cell transmission. PLoS Pathog 6:e1000955

Celestino M, Calistri A, Del Vecchio C, Salata C, Chiuppesi F, Pistello M et al (2012) Feline tetherin is characterized by a short N-terminal region and is counteracted by the feline immunodeficiency virus envelope glycoprotein. J Virol 86:6688–6700

Chu H, Wang JJ, Qi M, Yoon JJ, Chen X, Wen X et al (2012) Tetherin/BST-2 is essential for the formation of the intracellular virus-containing compartment in HIV-infected macrophages. Cell Host Microbe 12:360–372

Cocka LJ, Bates P (2012) Identification of alternatively translated tetherin isoforms with differing antiviral and signaling activities. PLoS Pathog 8:e1002931

Dietrich I, McMonagle EL, Petit SJ, Vijayakrishnan S, Logan N, Chan CN et al (2011) Feline tetherin efficiently restricts release of feline immunodeficiency virus but not spreading of infection. J Virol 85:5840–5852

Douglas JL, Viswanathan K, McCarroll MN, Gustin JK, Fruh K, Moses AV (2009) Vpu directs the degradation of the human immunodeficiency virus restriction factor BST-2/tetherin via a {beta}TrCP-dependent mechanism. J Virol 83:7931–7947

Dube M, Roy BB, Guiot-Guillain P, Mercier J, Binette J, Leung G et al (2009) Suppression of tetherin-restricting activity upon human immunodeficiency virus type 1 particle release correlates with localization of Vpu in the trans-Golgi network. J Virol 83:4574–4590

Dube M, Roy BB, Guiot-Guillain P, Binette J, Mercier J, Chiasson A et al (2010) Antagonism of tetherin restriction of HIV-1 release by Vpu involves binding and sequestration of the restriction factor in a perinuclear compartment. PLoS Pathog 6:e1000856

Dube M, Paquay C, Roy BB, Bego MG, Mercier J, Cohen EA (2011) HIV-1 Vpu antagonizes BST-2 by interfering mainly with the trafficking of newly synthesized BST-2 to the cell surface. Traffic 12:1714–1729

Duncan LM, Piper S, Dodd RB, Saville MK, Sanderson CM, Luzio JP et al (2006) Lysine-63-linked ubiquitination is required for endolysosomal degradation of class I molecules. EMBO J 25:1635–1645

Erikson E, Adam T, Schmidt S, Lehmann-Koch J, Over B, Goffinet C et al (2011) In vivo expression profile of the antiviral restriction factor and tumor-targeting antigen CD317/BST-2/HM1.24/tetherin in humans. Proc Natl Acad Sci U S A 108:13688–13693

Evans DT, Serra-Moreno R, Singh RK, Guatelli JC (2010) BST-2/tetherin: a new component of the innate immune response to enveloped viruses. Trends Microbiol 18:388–396

Fitzpatrick K, Skasko M, Deerinck TJ, Crum J, Ellisman MH, Guatelli J (2010) Direct restriction of virus release and incorporation of the interferon-induced protein BST-2 into HIV-1 particles. PLoS Pathog 6:e1000701

Fritz JV, Tibroni N, Keppler OT, Fackler OT (2012) HIV-1 Vpu's lipid raft association is dispensable for counteraction of the particle release restriction imposed by CD317/tetherin. Virology 424:33–44

Galao RP, Le Tortorec A, Pickering S, Kueck T, Neil SJ (2012) Innate sensing of HIV-1 assembly by tetherin induces NFkappaB-dependent proinflammatory responses. Cell Host Microbe 12:633–644

Ganem D (2006) KSHV infection and the pathogenesis of Kaposi's sarcoma. Annu Rev Pathol 1:273–296

Goffinet C, Allespach I, Homann S, Tervo HM, Habermann A, Rupp D et al (2009) HIV-1 antagonism of CD317 is species specific and involves Vpu-mediated proteasomal degradation of the restriction factor. Cell Host Microbe 5:285–297

Goffinet C, Homann S, Ambiel I, Tibroni N, Rupp D, Keppler OT et al (2010) Antagonism of CD317 restriction of human immunodeficiency virus type 1 (HIV-1) particle release and depletion of CD317 are separable activities of HIV-1 Vpu. J Virol 84:4089–4094

Gottlinger HG, Dorfman T, Cohen EA, Haseltine WA (1993) Vpu protein of human immunodeficiency virus type 1 enhances the release of capsids produced by gag gene constructs of widely divergent retroviruses. Proc Natl Acad Sci U S A 90:7381–7385

Gotz N, Sauter D, Usmani SM, Fritz JV, Goffinet C, Heigele A et al (2012) Reacquisition of Nef-mediated tetherin antagonism in a single in vivo passage of HIV-1 through its original chimpanzee host. Cell Host Microbe 12:373–380

Grover JR, Llewellyn GN, Soheilian F, Nagashima K, Veatch SL, Ono A (2013) Roles played by capsid-dependent induction of membrane curvature and Gag-ESCRT interactions in tetherin recruitment to HIV-1 assembly sites. J Virol 87(8):4650–4664. doi:10.1128/JVI.03526-12. [Epub 2013 Feb 13]

Gummuluru S, Kinsey CM, Emerman M (2000) An in vitro rapid-turnover assay for human immunodeficiency virus type 1 replication selects for cell-to-cell spread of virus. J Virol 74:10882–10891

Gupta RK, Hue S, Schaller T, Verschoor E, Pillay D, Towers GJ (2009a) Mutation of a single residue renders human tetherin resistant to HIV-1 Vpu-mediated depletion. PLoS Pathog 5:e1000443

Gupta RK, Mlcochova P, Pelchen-Matthews A, Petit SJ, Mattiuzzo G, Pillay D et al (2009b) Simian immunodeficiency virus envelope glycoprotein counteracts tetherin/BST-2/CD317 by intracellular sequestration. Proc Natl Acad Sci U S A 106:20889–20894

Gustin JK, Douglas JL, Bai Y, Moses AV (2012) Ubiquitination of BST-2 protein by HIV-1 Vpu protein does not require lysine, serine, or threonine residues within the BST-2 cytoplasmic domain. J Biol Chem 287:14837–14850

Hammonds J, Wang JJ, Yi H, Spearman P (2010) Immunoelectron microscopic evidence for tetherin/BST2 as the physical bridge between HIV-1 virions and the plasma membrane. PLoS Pathog 6:e1000749

Hammonds J, Ding L, Chu H, Geller K, Robbins A, Wang JJ et al (2012) The tetherin/BST-2 coiled-coil ectodomain mediates plasma membrane microdomain localization and restriction of particle release. J Virol 86:2259–2272

Harila K, Salminen A, Prior I, Hinkula J, Suomalainen M (2007) The Vpu-regulated endocytosis of HIV-1 Gag is clathrin-independent. Virology 369:299–308

Hauser H, Lopez LA, Yang SJ, Oldenburg JE, Exline CM, Guatelli JC et al (2010) HIV-1 Vpu and HIV-2 Env counteract BST-2/tetherin by sequestration in a perinuclear compartment. Retrovirology 7:51

Hinz A, Miguet N, Natrajan G, Usami Y, Yamanaka H, Renesto P et al (2010) Structural basis of HIV-1 tethering to membranes by the BST-2/tetherin ectodomain. Cell Host Microbe 7:314–323

Homann S, Smith D, Little S, Richman D, Guatelli J (2011) Upregulation of BST-2/tetherin by HIV infection in vivo. J Virol 85:10659–10668

Ilinskaya A, Derse D, Hill S, Princler G, Heidecker G (2013) Cell-cell transmission allows human T-lymphotropic virus 1 to circumvent tetherin restriction. Virology 436:201–209

Ishido S, Wang C, Lee BS, Cohen GB, Jung JU (2000) Downregulation of major histocompatibility complex class I molecules by Kaposi's sarcoma-associated herpesvirus K3 and K5 proteins. J Virol 74:5300–5309

Ishikawa J, Kaisho T, Tomizawa H, Lee BO, Kobune Y, Inazawa J et al (1995) Molecular cloning and chromosomal mapping of a bone marrow stromal cell surface gene, BST2, that may be involved in pre-B-cell growth. Genomics 26:527–534

Iwabu Y, Fujita H, Kinomoto M, Kaneko K, Ishizaka Y, Tanaka Y et al (2009) HIV-1 accessory protein Vpu internalizes cell-surface BST-2/tetherin through transmembrane interactions leading to lysosomes. J Biol Chem 284:35060–35072

Janvier K, Pelchen-Matthews A, Renaud JB, Caillet M, Marsh M, Berlioz-Torrent C (2011) The ESCRT-0 component HRS is required for HIV-1 Vpu-mediated BST-2/tetherin down-regulation. PLoS Pathog 7:e1001265

Janvier K, Roy N, Berlioz-Torrent C (2012) Role of the endosomal ESCRT machinery in HIV-1 Vpu-induced down-regulation of BST2/tetherin. Curr HIV Res 10:315–320

Jia B, Serra-Moreno R, Neidermyer W, Rahmberg A, Mackey J, Fofana IB et al (2009) Species-specific activity of SIV Nef and HIV-1 Vpu in overcoming restriction by tetherin/BST2. PLoS Pathog 5:e1000429

Jolly C, Booth NJ, Neil SJ (2010) Cell-cell spread of human immunodeficiency virus type 1 overcomes tetherin/BST-2-mediated restriction in T cells. J Virol 84:12185–12199

Jones PH, Mehta HV, Maric M, Roller RJ, Okeoma CM (2012) Bone marrow stromal cell antigen 2 (BST-2) restricts mouse mammary tumor virus (MMTV) replication in vivo. Retrovirology 9:10

Jouvenet N, Neil SJ, Zhadina M, Zang T, Kratovac Z, Lee Y et al (2009) Broad-spectrum inhibition of retroviral and filoviral particle release by tetherin. J Virol 83:1837–1844

Kaletsky RL, Francica JR, Agrawal-Gamse C, Bates P (2009) Tetherin-mediated restriction of filovirus budding is antagonized by the Ebola glycoprotein. Proc Natl Acad Sci U S A 106:2886–2891

Kelly BT, Owen DJ (2011) Endocytic sorting of transmembrane protein cargo. Curr Opin Cell Biol 23:404–412

Klimkait T, Strebel K, Hoggan MD, Martin MA, Orenstein JM (1990) The human immunodeficiency virus type 1-specific protein vpu is required for efficient virus maturation and release. J Virol 64:621–629

Kobayashi T, Ode H, Yoshida T, Sato K, Gee P, Yamamoto SP et al (2011) Identification of amino acids in the human tetherin transmembrane domain responsible for HIV-1 Vpu interaction and susceptibility. J Virol 85:932–945

Kong WS, Irie T, Yoshida A, Kawabata R, Kadoi T, Sakaguchi T (2012) Inhibition of virus-like particle release of Sendai virus and Nipah virus, but not that of mumps virus, by tetherin/CD317/BST-2. Hiroshima J Med Sci 61:59–67

Kueck T, Neil SJ (2012) A cytoplasmic tail determinant in HIV-1 Vpu mediates targeting of tetherin for endosomal degradation and counteracts interferon-induced restriction. PLoS Pathog 8:e1002609

Kuhl BD, Sloan RD, Donahue DA, Bar-Magen T, Liang C, Wainberg MA (2010) Tetherin restricts direct cell-to-cell infection of HIV-1. Retrovirology 7:115

Kuhl A, Banning C, Marzi A, Votteler J, Steffen I, Bertram S et al (2011) The Ebola virus glycoprotein and HIV-1 Vpu employ different strategies to counteract the antiviral factor tetherin. J Infect Dis 204(Suppl 3):S850–S860

Kupzig S, Korolchuk V, Rollason R, Sugden A, Wilde A, Banting G (2003) Bst-2/HM1.24 is a raft-associated apical membrane protein with an unusual topology. Traffic 4:694–709

Laplana M, Caruz A, Pineda JA, Puig T, Fibla J (2013) Association of BST-2 gene variants with HIV disease progression underscores the role of BST-2 in HIV type 1 infection. J Infect Dis 207:411–419

Lau D, Kwan W, Guatelli J (2011) Role of the endocytic pathway in the counteraction of BST-2 by human lentiviral pathogens. J Virol 85:9834–9846

Le Clerc S, Coulonges C, Delaneau O, Van Manen D, Herbeck JT, Limou S et al (2011) Screening low-frequency SNPS from genome-wide association study reveals a new risk allele for progression to AIDS. J Acquir Immune Defic Syndr 56:279–284

Le Tortorec A, Neil SJ (2009) Antagonism to and intracellular sequestration of human tetherin by the human immunodeficiency virus type 2 envelope glycoprotein. J Virol 83:11966–11978

Le Tortorec A, Willey S, Neil SJ (2011) Antiviral inhibition of enveloped virus release by tetherin/BST-2: action and counteraction. Viruses 3:520–540

Lehmann M, Rocha S, Mangeat B, Blanchet F, Uji IH, Hofkens J et al (2011) Quantitative multicolor super-resolution microscopy reveals tetherin HIV-1 interaction. PLoS Pathog 7:e1002456

Lehner PJ, Hoer S, Dodd R, Duncan LM (2005) Downregulation of cell surface receptors by the K3 family of viral and cellular ubiquitin E3 ligases. Immunol Rev 207:112–125

Lepelley A, Louis S, Sourisseau M, Law HK, Pothlichet J, Schilte C et al (2011) Innate sensing of HIV-infected cells. PLoS Pathog 7:e1001284

Liberatore RA, Bieniasz PD (2011) Tetherin is a key effector of the antiretroviral activity of type I interferon in vitro and in vivo. Proc Natl Acad Sci U S A 108:18097–18101

Lim ES, Malik HS, Emerman M (2010) Ancient adaptive evolution of tetherin shaped the functions of Vpu and Nef in human immunodeficiency virus and primate lentiviruses. J Virol 84:7124–7134

Lindwasser OW, Smith WJ, Chaudhuri R, Yang P, Hurley JH, Bonifacino JS (2008) A diacidic motif in human immunodeficiency virus type 1 Nef is a novel determinant of binding to AP-2. J Virol 82:1166–1174

Lopez LA, Yang SJ, Hauser H, Exline CM, Haworth KG, Oldenburg J et al (2010) Ebola virus glycoprotein counteracts BST-2/tetherin restriction in a sequence-independent manner that does not require tetherin surface removal. J Virol 84:7243–7255

Lopez LA, Yang SJ, Exline CM, Rengarajan S, Haworth KG, Cannon PM (2012) Anti-tetherin activities of HIV-1 Vpu and Ebola virus glycoprotein do not involve removal of tetherin from lipid rafts. J Virol 86:5467–5480

Loschko J, Schlitzer A, Dudziak D, Drexler I, Sandholzer N, Bourquin C et al (2011) Antigen delivery to plasmacytoid dendritic cells via BST2 induces protective T cell-mediated immunity. J Immunol 186:6718–6725

Magadan JG, Bonifacino JS (2012) Transmembrane domain determinants of CD4 downregulation by HIV-1 Vpu. J Virol 86:757–772

Magadan JG, Perez-Victoria FJ, Sougrat R, Ye Y, Strebel K, Bonifacino JS (2010) Multilayered mechanism of CD4 downregulation by HIV-1 Vpu involving distinct ER retention and ERAD targeting steps. PLoS Pathog 6:e1000869

Mangeat B, Gers-Huber G, Lehmann M, Zufferey M, Luban J, Piguet V (2009) HIV-1 Vpu neutralizes the antiviral factor tetherin/BST-2 by binding it and directing its beta-TrCP2-dependent degradation. PLoS Pathog 5:e1000574

Mangeat B, Cavagliotti L, Lehmann M, Gers-Huber G, Kaur I, Thomas Y et al (2012) Influenza virus partially counteracts restriction imposed by tetherin/BST-2. J Biol Chem 287:22015–22029

Mansouri M, Viswanathan K, Douglas JL, Hines J, Gustin J, Moses AV et al (2009) Molecular mechanism of BST2/tetherin downregulation by K5/MIR2 of Kaposi's sarcoma-associated herpesvirus. J Virol 83:9672–9681

Margottin F, Bour SP, Durand H, Selig L, Benichou S, Richard V et al (1998) A novel human WD protein, h-beta TrCp, that interacts with HIV-1 Vpu connects CD4 to the ER degradation pathway through an F-box motif. Mol Cell 1:565–574

Martin-Serrano J, Neil SJ (2011) Host factors involved in retroviral budding and release. Nat Rev Microbiol 9:519–531

Matsuda A, Suzuki Y, Honda G, Muramatsu S, Matsuzaki O, Nagano Y et al (2003) Large-scale identification and characterization of human genes that activate NF-kappaB and MAPK signaling pathways. Oncogene 22:3307–3318

McNatt MW, Zang T, Hatziioannou T, Bartlett M, Fofana IB, Johnson WE et al (2009) Species-specific activity of HIV-1 Vpu and positive selection of tetherin transmembrane domain variants. PLoS Pathog 5:e1000300

Mettenleiter TC, Klupp BG, Granzow H (2006) Herpesvirus assembly: a tale of two membranes. Curr Opin Microbiol 9:423–429

Mitchell RS, Katsura C, Skasko MA, Fitzpatrick K, Lau D, Ruiz A et al (2009) Vpu antagonizes BST-2-mediated restriction of HIV-1 release via beta-TrCP and endo-lysosomal trafficking. PLoS Pathog 5:e1000450

Miyakawa K, Ryo A, Murakami T, Ohba K, Yamaoka S, Fukuda M et al (2009) BCA2/Rabring7 promotes tetherin-dependent HIV-1 restriction. PLoS Pathog 5:e1000700

Miyakawa K, Sawasaki T, Matsunaga S, Tokarev A, Quinn G, Kimura H et al (2012) Interferon-induced SCYL2 limits release of HIV-1 by triggering PP2A-mediated dephosphorylation of the viral protein Vpu. Sci Signal 5:ra73

Moffat JM, Segura E, Khoury G, Caminschi I, Cameron PU, Lewin SR et al. (2013) Targeting antigen to bone marrow stromal cell-2 expressed by conventional and plasmacytoid dendritic cells elicits efficient antigen presentation. Eur J Immunol 43(3):595–605. doi:10.1002/eji.201242799. [Epub 2013 Feb 6]

Neil S, Bieniasz P (2009) Human immunodeficiency virus, restriction factors, and interferon. J Interferon Cytokine Res 29:569–580

Neil SJ, Eastman SW, Jouvenet N, Bieniasz PD (2006) HIV-1 Vpu promotes release and prevents endocytosis of nascent retrovirus particles from the plasma membrane. PLoS Pathog 2:e39

Neil SJ, Sandrin V, Sundquist WI, Bieniasz PD (2007) An interferon-alpha-induced tethering mechanism inhibits HIV-1 and Ebola virus particle release but is counteracted by the HIV-1 Vpu protein. Cell Host Microbe 2:193–203

Neil SJ, Zang T, Bieniasz PD (2008) Tetherin inhibits retrovirus release and is antagonized by HIV-1 Vpu. Nature 451:425–430

Noble B, Abada P, Nunez-Iglesias J, Cannon PM (2006) Recruitment of the adaptor protein 2 complex by the human immunodeficiency virus type 2 envelope protein is necessary for high levels of virus release. J Virol 80:2924–2932

Ohtomo T, Sugamata Y, Ozaki Y, Ono K, Yoshimura Y, Kawai S et al (1999) Molecular cloning and characterization of a surface antigen preferentially overexpressed on multiple myeloma cells. Biochem Biophys Res Commun 258:583–591

Pais-Correia AM, Sachse M, Guadagnini S, Robbiati V, Lasserre R, Gessain A et al (2010) Biofilm-like extracellular viral assemblies mediate HTLV-1 cell-to-cell transmission at virological synapses. Nat Med 16:83–89

Pan XB, Han JC, Cong X, Wei L (2012) BST2/tetherin inhibits dengue virus release from human hepatoma cells. PLoS ONE 7:e51033

Pardieu C, Vigan R, Wilson SJ, Calvi A, Zang T, Bieniasz P et al (2010) The RING-CH ligase K5 antagonizes restriction of KSHV and HIV-1 particle release by mediating ubiquitin-dependent endosomal degradation of tetherin. PLoS Pathog 6:e1000843

Park SH, Mrse AA, Nevzorov AA, Mesleh MF, Oblatt-Montal M, Montal M et al (2003) Three-dimensional structure of the channel-forming trans-membrane domain of virus protein "u" (Vpu) from HIV-1. J Mol Biol 333:409–424

Patel MR, Emerman M, Malik HS (2011) Paleovirology—ghosts and gifts of viruses past. Curr Opin Virol 1:304–309

Perez-Caballero D, Zang T, Ebrahimi A, McNatt MW, Gregory DA, Johnson MC et al (2009) Tetherin inhibits HIV-1 release by directly tethering virions to cells. Cell 139:499–511

Petit SJ, Blondeau C, Towers GJ (2011) Analysis of the human immunodeficiency virus type 1 M group Vpu domains involved in antagonizing tetherin. J Gen Virol 92:2937–2948

Pillai SK, Abdel-Mohsen M, Guatelli J, Skasko M, Monto A, Fujimoto K et al (2012) Role of retroviral restriction factors in the interferon-alpha-mediated suppression of HIV-1 in vivo. Proc Natl Acad Sci U S A 109:3035–3040
Radoshitzky SR, Dong L, Chi X, Clester JC, Retterer C, Spurgers K et al (2010) Infectious Lassa virus, but not filoviruses, is restricted by BST-2/tetherin. J Virol 84:10569–10580
Rollason R, Korolchuk V, Hamilton C, Schu P, Banting G (2007) Clathrin-mediated endocytosis of a lipid-raft-associated protein is mediated through a dual tyrosine motif. J Cell Sci 120:3850–3858
Rollason R, Korolchuk V, Hamilton C, Jepson M, Banting G (2009) A CD317/tetherin-RICH2 complex plays a critical role in the organization of the subapical actin cytoskeleton in polarized epithelial cells. J Cell Biol 184:721–736
Ruiz A, Hill MS, Schmitt K, Guatelli J, Stephens EB (2008) Requirements of the membrane proximal tyrosine and dileucine-based sorting signals for efficient transport of the subtype C Vpu protein to the plasma membrane and in virus release. Virology 378:58–68
Sakuma T, Noda T, Urata S, Kawaoka Y, Yasuda J (2009) Inhibition of Lassa and Marburg virus production by tetherin. J Virol 83:2382–2385
Sauter D, Kirchhoff F (2011) Tetherin antagonism by primate lentiviral nef proteins. Curr HIV Res 9:514–523
Sauter D, Schindler M, Specht A, Landford WN, Munch J, Kim KA et al (2009) Tetherin-driven adaptation of Vpu and Nef function and the evolution of pandemic and nonpandemic HIV-1 strains. Cell Host Microbe 6:409–421
Sauter D, Specht A, Kirchhoff F (2010) Tetherin: holding on and letting go. Cell 141:392–398
Sauter D, Hue S, Petit SJ, Plantier JC, Towers GJ, Kirchhoff F et al (2011a) HIV-1 Group P is unable to antagonize human tetherin by Vpu, Env or Nef. Retrovirology 8:103
Sauter D, Vogl M, Kirchhoff F (2011b) Ancient origin of a deletion in human BST2/tetherin that confers protection against viral zoonoses. Hum Mutat 32:1243–1245
Sauter D, Unterweger D, Vogl M, Usmani SM, Heigele A, Kluge SF et al (2012) Human tetherin exerts strong selection pressure on the HIV-1 group N Vpu protein. PLoS Pathog 8:e1003093
Schindler M, Rajan D, Banning C, Wimmer P, Koppensteiner H, Iwanski A et al (2010) Vpu serine 52 dependent counteraction of tetherin is required for HIV-1 replication in macrophages, but not in ex vivo human lymphoid tissue. Retrovirology 7:1
Schmidt S, Fritz JV, Bitzegeio J, Fackler OT, Keppler OT (2011) HIV-1 Vpu blocks recycling and biosynthetic transport of the intrinsic immunity factor CD317/tetherin to overcome the virion release restriction. MBio 2:e00036-11
Schubert U, Strebel K (1994) Differential activities of the human immunodeficiency virus type 1-encoded Vpu protein are regulated by phosphorylation and occur in different cellular compartments. J Virol 68:2260–2271
Schubert U, Clouse KA, Strebel K (1995) Augmentation of virus secretion by the human immunodeficiency virus type 1 Vpu protein is cell type independent and occurs in cultured human primary macrophages and lymphocytes. J Virol 69:7699–7711
Schubert U, Ferrer-Montiel AV, Oblatt-Montal M, Henklein P, Strebel K, Montal M (1996) Identification of an ion channel activity of the Vpu transmembrane domain and its involvement in the regulation of virus release from HIV-1-infected cells. FEBS Lett 398:12–18
Schubert HL, Zhai Q, Sandrin V, Eckert DM, Garcia-Maya M, Saul L et al (2010) Structural and functional studies on the extracellular domain of BST2/tetherin in reduced and oxidized conformations. Proc Natl Acad Sci U S A 107:17951–17956
Serra-Moreno R, Jia B, Breed M, Alvarez X, Evans DT (2011) Compensatory changes in the cytoplasmic tail of gp41 confer resistance to tetherin/BST-2 in a pathogenic nef-deleted SIV. Cell Host Microbe 9:46–57
Sharp PM, Hahn BH (2011) Origins of HIV and the AIDS pandemic. Cold Spring Harb Perspect Med 1:a006841
Sherer NM, Lehmann MJ, Jimenez-Soto LF, Horensavitz C, Pypaert M, Mothes W (2007) Retroviruses can establish filopodial bridges for efficient cell-to-cell transmission. Nat Cell Biol 9:310–315

Skasko M, Tokarev A, Chen CC, Fischer WB, Pillai SK, Guatelli J (2011) BST-2 is rapidly down-regulated from the cell surface by the HIV-1 protein Vpu: evidence for a post-ER mechanism of Vpu-action. Virology 411:65–77

Skasko M, Wang Y, Tian Y, Tokarev A, Munguia J, Ruiz A et al (2012) HIV-1 Vpu protein antagonizes innate restriction factor BST-2 via lipid-embedded helix–helix interactions. J Biol Chem 287:58–67

Skaug B, Jiang X, Chen ZJ (2009) The role of ubiquitin in NF-kappaB regulatory pathways. Annu Rev Biochem 78:769–796

Sliva K, Resch T, Kraus B, Goffinet C, Keppler OT, Schnierle BS (2012) The cellular antiviral restriction factor tetherin does not inhibit poxviral replication. J Virol 86:1893–1896

Stefani F, Zhang L, Taylor S, Donovan J, Rollinson S, Doyotte A et al (2011) UBAP1 is a component of an endosome-specific ESCRT-I complex that is essential for MVB sorting. Curr Biol 21:1245–1250

St-Germain JR, Taylor P, Tong J, Jin LL, Nikolic A, Stewart II et al (2009) Multiple myeloma phosphotyrosine proteomic profile associated with FGFR3 expression, ligand activation, and drug inhibition. Proc Natl Acad Sci U S A 106:20127–20132

Strebel K, Klimkait T, Maldarelli F, Martin MA (1989) Molecular and biochemical analyses of human immunodeficiency virus type 1 vpu protein. J Virol 63:3784–3791

Swiecki M, Scheaffer SM, Allaire M, Fremont DH, Colonna M, Brett TJ (2011) Structural and biophysical analysis of BST-2/tetherin ectodomains reveals an evolutionary conserved design to inhibit virus release. J Biol Chem 286:2987–2997

Swiecki M, Wang Y, Gilfillan S, Lenschow DJ, Colonna M (2012) Cutting edge: paradoxical roles of BST2/tetherin in promoting type I IFN response and viral infection. J Immunol 188:2488–2492

Swiecki M, Omattage NS, Brett TJ (2013) BST-2/tetherin: Structural biology, viral antagonism, and immunobiology of a potent host antiviral factor. Mol Immunol 54:132–139

Takeuchi O, Akira S (2010) Pattern recognition receptors and inflammation. Cell 140:805–820

Tavano B, Galao RP, Graham DR, Neil SJ, Aquino VN, Fuchs D et al (2013) Ig-like Transcript 7, but Not Bone Marrow Stromal Cell Antigen 2 (Also Known as HM1.24, tetherin, or CD317), Modulates Plasmacytoid Dendritic Cell Function in Primary Human Blood Leukocytes. J Immunol 190(6):2622–2630

Tervo HM, Homann S, Ambiel I, Fritz JV, Fackler OT, Keppler OT (2011) beta-TrCP is dispensable for Vpu's ability to overcome the CD317/tetherin-imposed restriction to HIV-1 release. Retrovirology 8:9

Terwilliger EF, Cohen EA, Lu YC, Sodroski JG, Haseltine WA (1989) Functional role of human immunodeficiency virus type 1 vpu. Proc Natl Acad Sci U S A 86:5163–5167

Tokarev A, Guatelli J (2011) Misdirection of membrane trafficking by HIV-1 Vpu and Nef: Keys to viral virulence and persistence. Cell Logist 1:90–102

Tokarev AA, Munguia J, Guatelli JC (2011) Serine-threonine ubiquitination mediates downregulation of BST-2/tetherin and relief of restricted virion release by HIV-1 Vpu. J Virol 85:51–63

Tokarev A, Suarez M, Kwan W, Fitzpatrick K, Singh R, Guatelli J (2013) Stimulation of NF-kappaB activity by the HIV restriction factor BST2. J Virol 87:2046–2057

Van Damme N, Goff D, Katsura C, Jorgenson RL, Mitchell R, Johnson MC et al (2008) The interferon-induced protein BST-2 restricts HIV-1 release and is downregulated from the cell surface by the viral Vpu protein. Cell Host Microbe 3:245–252

Varthakavi V, Smith RM, Bour SP, Strebel K, Spearman P (2003) Viral protein U counteracts a human host cell restriction that inhibits HIV-1 particle production. Proc Natl Acad Sci U S A 100:15154–15159

Varthakavi V, Smith RM, Martin KL, Derdowski A, Lapierre LA, Goldenring JR et al (2006) The pericentriolar recycling endosome plays a key role in Vpu-mediated enhancement of HIV-1 particle release. Traffic 7:298–307

Vigan R, Neil SJ (2010) Determinants of tetherin antagonism in the transmembrane domain of the human immunodeficiency virus type 1 Vpu protein. J Virol 84:12958–12970

Vigan R, Neil SJ (2011) Separable determinants of subcellular localization and interaction account for the inability of group O HIV-1 Vpu to counteract tetherin. J Virol 85:9737–9748

Viswanathan K, Smith MS, Malouli D, Mansouri M, Nelson JA, Fruh K (2011) BST2/tetherin enhances entry of human cytomegalovirus. PLoS Pathog 7:e1002332

Watanabe R, Leser GP, Lamb RA (2011) Influenza virus is not restricted by tetherin whereas influenza VLP production is restricted by tetherin. Virology 417:50–56

Weidner JM, Jiang D, Pan XB, Chang J, Block TM, Guo JT (2010) Interferon-induced cell membrane proteins, IFITM3 and tetherin, inhibit vesicular stomatitis virus infection via distinct mechanisms. J Virol 84:12646–12657

Weissenhorn W, Miguet N, Aschman N, Renesto P, Usami Y, Gottlinger HG (2012) Structural basis of tetherin function. Curr HIV Res 10:298–306

Winkler M, Bertram S, Gnirss K, Nehlmeier I, Gawanbacht A, Kirchhoff F et al (2012) Influenza A virus does not encode a tetherin antagonist with Vpu-like activity and induces IFN-dependent tetherin expression in infected cells. PLoS ONE 7:e43337

Yang H, Wang J, Jia X, McNatt MW, Zang T, Pan B et al (2010a) Structural insight into the mechanisms of enveloped virus tethering by tetherin. Proc Natl Acad Sci U S A 107:18428–18432

Yang SJ, Lopez LA, Hauser H, Exline CM, Haworth KG, Cannon PM (2010b) Anti-tetherin activities in Vpu-expressing primate lentiviruses. Retrovirology 7:13

Yang SJ, Lopez LA, Exline CM, Haworth KG, Cannon PM (2011) Lack of adaptation to human tetherin in HIV-1 group O and P. Retrovirology 8:78

Yao XJ, Gottlinger H, Haseltine WA, Cohen EA (1992) Envelope glycoprotein and CD4 independence of vpu-facilitated human immunodeficiency virus type 1 capsid export. J Virol 66:5119–5126

Yondola MA, Fernandes F, Belicha-Villanueva A, Uccelini M, Gao Q, Carter C et al (2011) Budding capability of the influenza virus neuraminidase can be modulated by tetherin. J Virol 85:2480–2491

Yu P, Lubben W, Slomka H, Gebler J, Konert M, Cai C et al (2012) Nucleic acid-sensing Toll-like receptors are essential for the control of endogenous retrovirus viremia and ERV-induced tumors. Immunity 37:867–879

Zhang F, Wilson SJ, Landford WC, Virgen B, Gregory D, Johnson MC et al (2009) Nef proteins from simian immunodeficiency viruses are tetherin antagonists. Cell Host Microbe 6:54–67

Zhang F, Landford WN, Ng M, McNatt MW, Bieniasz PD, Hatziioannou T (2011) SIV Nef proteins recruit the AP-2 complex to antagonize tetherin and facilitate virion release. PLoS Pathog 7:e1002039

Zhong P, Agosto LM, Ilinskaya A, Dorjbal B, Truong R, Derse D et al (2013) Cell-to-cell transmission can overcome multiple donor and target cell barriers imposed on cell-free HIV. PLoS ONE 8:e53138

Restriction of Retroviral Infection of Macrophages

Mark Sharkey

Abstract Primate immunodeficiency viruses are highly specialized lentiviruses that have evolved to successfully infect and persist for the lifetime of the host. Despite encountering numerous potent antiviral factors, HIVs and SIVs are successful pathogens due to the acquisition of equally potent countermeasures in the form of accessory genes. The accessory gene Vpx encoded by HIV-2 and a subset of SIVs have a profound effect on the ability of lentiviruses to infect non-dividing cells, such as macrophages. Although most virus replication occurs in activated $CD4^+$ T cells, myeloid lineage cells are natural targets of infection and play a central role in virus transmission, dissemination, and persistence. However, myeloid lineage cells are poorly sensitive to lentiviral infection due partly to the high-level expression of a host protein that regulates nucleic acid metabolism named SAMHD1. Degradation of SAMHD1 is induced by Vpx to eliminate this intrinsic antiviral factor. Importantly, SAMHD1 has also been implicated as a negative regulator of the innate immune response, so the interplay between SAMHD1 and Vpx is likely to have significant consequences for virus replication, persistence, and immune control.

Contents

M. Sharkey (✉)
University of Miami Miller School of Medicine, 1420 Northwest 9th Avenue, Miami, FL 33136, USA
e-mail: msharkey@med.miami.edu

B. R. Cullen (ed.), *Intrinsic Immunity*, Current Topics in Microbiology and Immunology 371, DOI: 10.1007/978-3-642-37765-5_4,

1 Introduction

Throughout evolution, organisms have needed to adapt to their environments and develop defenses to pathogens in order to ensure survival and propagation. Although the skin and the epithelial lining of internal mucosal surfaces provide an effective physical barrier to exposure to most microorganisms, the immune system has developed in vertebrates to prevent or clear pathogenic infections through innate and adaptive immune responses. The conventional innate response combats invading pathogens through non-specific mechanisms such as inflammation, complement activation, and phagocytosis by leukocytes. This often results in activation of the adaptive response in which the immune system develops more potent, specific recognition of pathogens or pathogen-infected cells and immunological memory develops.

In addition to the classically defined immune effectors, intrinsic host factors have been acquired during evolution to inhibit viral replication. Also know as restriction factors, well-defined examples include TRIM5α, tetherin (BST2), and the APOBEC family of proteins. These are dominant genes that impart antiviral effects early after exposure to prevent or limit viral infections. The presence of intrinsic antiviral factors creates an unfavorable cellular environment for viruses, and they have developed mechanisms to overcome restrictions to replication and persistence. For example, Vpu downregulates CD4 from the plasma membrane to increase virion infectivity and enhances virus release by counteracting the effects of tetherin (Andrew and Strebel 2010; Dube et al. 2010). The Vif protein is essential in vivo to counteract the inhibition of reverse transcription and G to A hypermutation mediated by the APOBEC3G/3F proteins (Gillick et al. 2013). Vif recruits the cellular ubiquitin ligase machinery to target APOBEC3G/3F for ubiquitination and subsequent proteasomal degradation (Mehle et al. 2004; Yu et al. 2003). Vpr also utilizes a similar strategy to target an as yet undefined cellular target to create a more favorable cellular environment for virus replication (Belzile et al. 2007; Hrecka et al. 2007). Inducing degradation of a cellular protein, such as a restriction factor, is a common strategy employed by viruses to eliminate obstacles to efficient replication (Gustin et al. 2011). A recent addition to the growing list of viral restriction factors has been identified as the sterile alpha motif and HD domain-containing protein (SAMHD1). SAMHD1 expression in non-cycling myeloid cells potently inhibits infection by a divergent group of retroviruses including HIV-1 and MLV (Goujon et al. 2007; Kaushik et al. 2009). Vpx, an accessory gene encoded by some lentiviruses, is able to overcome this block to infection. This chapter will focus on the restriction to lentiviral replication in myeloid cells and the roles that Vpx and SAMHD1 have in promoting or inhibiting the process.

2 Restriction to Retroviral Infection in Myeloid Cells

The majority of HIV-1 replication occurs in activated $CD4^+$ T cells, and it is the progressive loss of these immune cells that results in immunodeficiency and associated disease. In contrast, non-proliferating, quiescent $CD4^+$ T cells exhibit a profound block to infection as evidenced by the labile nature of the products of reverse transcription (Pierson et al. 2002; Stevenson et al. 1990; Zack et al. 1990). Similarly, myeloid lineage cells of the immune system are susceptible to lentivirus infection, but the efficiency of infection is low. Peripheral blood monocytes are highly refractory to productive HIV-1 infection, but become more permissive upon differentiation into macrophages and dendritic cells (Neil et al. 2001; Rich et al. 1992; Sonza et al. 1996). In vitro, HIV-1 enters freshly isolated peripheral blood monocytes efficiently but fails to generate products of reverse transcription (Sonza et al. 1996; Triques and Stevenson 2004). Upon differentiation in culture, monocyte-derived dendritic cells and macrophages become more susceptible to infection (Bergamaschi and Pancino 2010), although virus replication is much slower and less efficient than in activated, cycling cells. Thus, cells such as macrophages are susceptible to HIV-1 infection, but restrict or limit the levels of replication through cellular processes. Observations that intracellular dNTP levels are low in monocytes suggest that viral cDNA synthesis in these cells is abortive (O'Brien et al. 1994; Triques and Stevenson 2004) and this might account for the low level of infection. A role for APOBEC3G in regulating the permissivity of myeloid cells to HIV-1 infection has also been proposed (Ellery et al. 2007; Peng et al. 2007; Pion et al. 2006). Although these previously described mechanisms may contribute to inhibition of virus replication in monocytes and macrophages, it has become evident through several studies that myeloid lineage cells harbor a dominant restriction to lentiviral infection (Bergamaschi et al. 2009; Goujon et al. 2007; Kaushik et al. 2009; Negri et al. 2012; Schule et al. 2009; Sharova et al. 2008; Srivastava et al. 2008; Triques and Stevenson 2004).

3 Vpr and Vpx

3.1 Vpr and Vpx are Virulence Factors

HIVs and SIVs are highly adapted primate lentiviruses that encode several accessory genes that have a general role in enhancing virus replication and persistence in the host cell (Fig. 1). While the Tat and Rev genes are essential regulators of viral expression, Vif, Vpu, Vpr, Vpx, and Nef are dispensable for virus growth in most cell culture systems in vitro. Originally thought to be marginally important

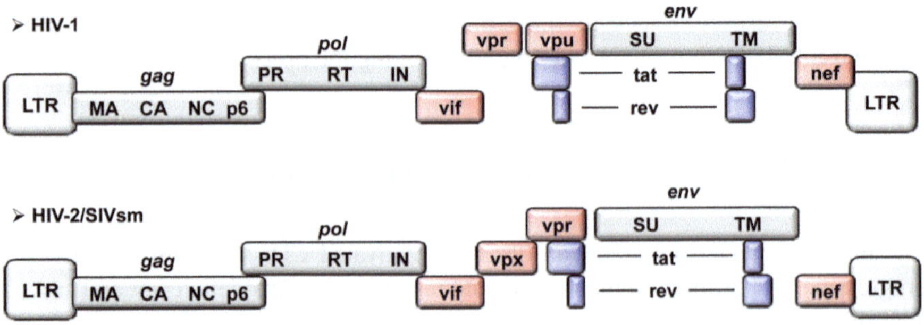

Fig. 1 Genomic organization of HIV-1 and HIV-2/SIVsm. In addition to the gag, pol, and env genes common to all retroviruses, primate lentiviruses encode regulatory and accessory genes that modify the host cell environment to benefit the virus. While the HIV-1 lineage encodes Vpr, the HIV-2/SIVsm lineage encodes both a Vpr gene and a Vpx gene. Reprinted from Ayinde et al. (2010)

for virus replication, recent advances are providing a greater understanding and appreciation for their importance in the setting of a natural infection.

Vpx and the genetically related accessory protein Vpr are virulence factors originally reported to exert a positive effect on virus replication in myeloid lineage cells. Although dispensable in vitro, Vpr and Vpx are important in vivo as demonstrated in non-human primate models (Gibbs et al. 1995; Hirsch et al. 1998). Rhesus macaques infected with Vpx-deficient SIV have reduced viral loads and delayed progression to AIDS. SIV defective for both genes is greatly attenuated, and infected animals have low viral burdens and lack symptoms of disease. The two accessory genes are not encoded by all HIVs and SIVs. HIV-1, HIV-2, and SIV isolated from sooty mangabeys (SIVsm) encode Vpr, but Vpx is specific to the HIV-2/SIVsm lineage (Cohen et al. 1990; Henderson et al. 1988; Kappes et al. 1988). SIV isolated from red-capped mangabeys and mandrills also encode both Vpr and Vpx, but all other primate lentiviruses have a single related gene that has been designated Vpr (Beer et al. 2001; Hu et al. 2003). Vpx encoded by the HIV-2/SIVsm lineage is thought to have arisen by recombination between SIV from sooty mangabey and SIV from African green monkeys, essentially resulting in the addition of Vpx to the ancestral SIVsm genome (Sharp et al. 1996; Tristem et al. 1998). The most recent common ancestor of modern day SIVs and HIV-1 was an SIV that encoded a Vpx gene (Zhang et al. 2012). The Vpx gene was likely lost during long-term evolution of SIV within chimpanzees, and the zoonotic transfer of virus from chimpanzees to humans gave rise to an HIV-1 that encodes Vpr, but not Vpx (Zhang et al. 2012).

3.2 *Vpr and Vpx are Packaged into Virions*

Both proteins are packaged into the virion core via an interaction with the structural p6 component of the Gag gene (Accola et al. 1999; Bachand et al. 1999; Kewalramani and Emerman 1996; Wu et al. 1994). As integral virion-encapsidated proteins, they are delivered intracellularly upon virion fusion with the target cell membrane, suggesting that they function early in the viral life cycle. Several properties have been described for Vpr, but a clear consensus on its role in virus replication and pathogenesis is lacking. In addition to its properties in promoting cell cycle arrest and transcriptional activation (Planelles and Benichou 2009), early studies suggested that Vpr facilitates macrophage infection by driving the nuclear import of the preintegration complex (Fassati 2006; Heinzinger et al. 1994). Retroviruses must integrate into the host cell chromosome to transition to a mode of productive infection, and lentiviruses are unique in their ability to traverse an intact nuclear membrane to allow infection of non-cycling cells. Several lines of evidence support the conclusion that Vpr, along with gagMA protein, are nucleophilic proteins that drive nuclear localization of the preintegration complex (Bukrinsky et al. 1993; Popov et al. 1998; von Schwedler et al. 1994). Virus in which Vpr was inactivated replicated poorly in primary macrophages, although a similar effect for the same virus was not observed in activated lymphocytes (Balliet et al. 1994). Although this implies that Vpr is required for optimal viral replication in terminally differentiated macrophages, the effect can vary significantly from donor to donor (Balliet et al. 1994). Subsequent studies have also suggested that HIV infection of terminally differentiated cells is not dependent on any of the karyophilic properties of viral proteins (Fouchier et al. 1997; Yamashita and Emerman 2005). Therefore, a well-defined role for the biological function of Vpr remains to be elucidated. In contrast, the ability of Vpx in promoting infection of non-dividing cells is emerging partly due to the efforts in delineating the biological function of Vpr.

3.3 Vpr Induces Cell Cycle Arrest

In particular, investigations focused on understanding the underlying mechanism of Vpr-mediated cell cycle arrest were instrumental in gaining a better understanding of the functional importance of Vpx. Vpr-induced cell cycle arrest was found to be dependent on its interaction with a cellular host factor, identified as DDB1-cullin 4-associated factor 1 (DCAF1). DCAF1 is identical to a cellular protein, previously reported to bind Vpr with high affinity and aptly named Vpr-binding protein (Zhang et al. 2001). Vpr recruits an E3 ubiquitin ligase complex through its interaction with DCAF1 (Fig. 2) and induces the ubiquitinylation and proteasomal degradation of an unknown host factor that is required for entry into mitosis (Belzile et al. 2007; DeHart et al. 2007; Hrecka et al. 2007; Schrofelbauer et al. 2007). Although it has been proposed that arresting cells in G2 would provide a significant boost in virus production from short-lived infected cells (Goh et al. 1998), G2 arrest may be induced as a coincidental consequence of a viral strategy to eliminate an intrinsic antiviral factor whose expression is linked to the cell cycle. Vpr appears to enhance virus replication under certain conditions, but it is unlikely to be linked to cell cycle arrest in macrophages, since these cells do not divide. A better understanding of the biologically significant role of Vpr in the lentiviral life cycle awaits identification of the host cell factor that is targeted for degradation by Vpr.

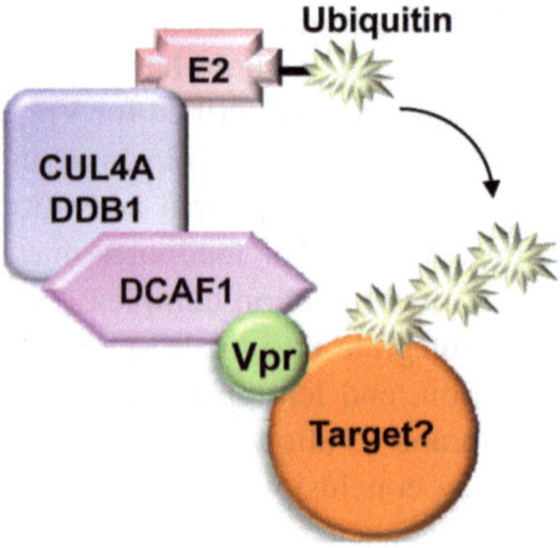

Fig. 2 Model for HIV-1 Vpr mechanism of action. HIV-1 Vpr recruits the CUL4A-DDB1 ubiquitin ligase through DCAF1 binding, which leads to the ubiquitination and inactivation of an unknown cellular target required for entry into mitosis. Reprinted from Ayinde et al. (2010)

3.4 Early Functional Studies of Vpx

HIV-2 and some SIVs encode both of the genetically related, although functionally distinct, accessory genes designated Vpr and Vpx. Unlike HIV-1 Vpr, Vpx does not induce G2 cell cycle arrest in dividing cells. Early studies determined that Vpx functions to enhance nuclear import of the preintegration complex in non-dividing cells, while it was found to be dispensable in cycling cells (Belshan et al. 2006; Fletcher et al. 1996; Planelles et al. 1996). Considerable evidence supports a functional role for Vpx in promoting nuclear localization of the preintegration complex containing the viral genome. It is efficiently packaged into the virion and generally thought to facilitate an early step in the viral life cycle (Pancio and Ratner 1998). Vpx associates with the preintegration complex, and residues constituting nuclear import signals within the protein have been characterized (Belshan and Ratner 2003; Pancio et al. 2000) and found to be necessary for productive infection of non-dividing cells (Belshan et al. 2006; Pancio et al. 2000). Vpx nuclear localization is enhanced via its interaction with a member of the Hsp40/DnaJ family of cellular chaperones, and modulation of cellular Hsp40/DnaJB6 levels impacts transport of HIV-2 preintegration complexes (Cheng et al. 2008). Furthermore, Vpx is an important determinant for establishing SIV infection in pigtailed macaques (Belshan et al. 2012; Hirsch et al. 1998) and promoting virus dissemination and pathogenesis (Hirsch et al. 1998).

3.5 Vpx Enhances Reverse Transcription

In addition to promoting nuclear import of preintegration complexes, in vitro studies with myeloid lineage cells led to the unexpected conclusion that Vpx functions to enhance the efficiency of reverse transcription (Fujita et al. 2008; Goujon et al. 2007; Sharova et al. 2008; Srivastava et al. 2008). The Vpx-mediated increase in reverse transcripts is cell-type dependent, with dendritic cells being some of the most restrictive to retrovirus infection (Goujon et al. 2007). Treatment of dendritic cells with the proteasome inhibitor MG132 partially restores infectivity of a Vpx-defective SIV, suggesting that Vpx is needed to counteract a proteasome-dependent antiviral mechanism present in dendritic cells (Goujon et al. 2007). Perhaps not surprisingly, Vpx specifically binds to the same DCAF1 subunit as Vpr to assemble a CUL4A-DDB1 ubiquitin ligase complex to target a cellular protein for proteasomal destruction (Bergamaschi et al. 2009; Sharova et al. 2008; Srivastava et al. 2008). While the cellular binding partner of Vpr remains to be identified, two research groups have recently demonstrated that SAMHD1 is the antiviral protein expressed in monocytes, dendritic cells, and macrophages that Vpx targets for ubiquitinylation and rapid proteasomal degradation (Hrecka et al. 2011; Laguette et al. 2011).

4 Sterile Alpha Motif and Histidine–Aspartic Domain Protein 1

4.1 Identifying the Restriction Factor SAMHD1

Using affinity purification and mass spectrometry, two independent groups identified SAMHD1 as a cellular protein that strongly interacts with Vpx (Hrecka et al. 2011; Laguette et al. 2011). SAMHD1 also associates with the cullin 4-based E3 ubiquitin ligase complex, but only when Vpx is part of the complex (Hrecka et al. 2011). Vpx acts as an adaptor to load SAMHD1 onto the CUL4A/DDB1/DCAF1 E3 ubiquitin ligase complex, resulting in ubiquitin modification and ultimately proteasomal degradation (Hrecka et al. 2011; Laguette et al. 2011). RNAi-mediated silencing of SAMHD1 allows HIV-1 to infect macrophages and dendritic cells independently of Vpx (Hrecka et al. 2011; Laguette et al. 2011). A consistent inverse correlation exists between the expression of SAMHD1 and permissiveness to HIV-1 infection (Laguette et al. 2011). Permissive cells such as Jurkat and monocytic U937 cells do not express SAMHD1, but non-permissive cells, including THP-1, monocytes, macrophages, and dendritic cells, exhibit high expression levels (Baldauf et al. 2012; Laguette et al. 2011). Artificially elevating SAMHD1 levels in normally permissive cells render them resistant to infection, unless Vpx is also expressed (Laguette et al. 2011). This experimental evidence strongly supports the conclusion that SAMHD1, identified through its ability to interact with Vpx, is the novel restriction factor that limits viral replication in myeloid lineage cells.

4.2 SAMHD1 Properties

SAMHD1 is a 626 amino acid protein that localizes to the nucleus and is uniquely comprised of a sterile alpha motif (SAM) and a histidine–aspartic (HD) domain that occurs in tandem (Aravind and Koonin 1998; Qiao and Bowie 2005). SAM domains are abundant protein interaction modules involved in protein–protein and protein–nucleic acid interactions that mediate many cellular processes, such as signal transduction and transcriptional regulation (Qiao and Bowie 2005). HD domains contain highly conserved histidine and aspartic acid residues and are found in a superfamily of enzymes with a protein fold predicted to have phosphohydrolase activity (Aravind and Koonin 1998; Zimmerman et al. 2008). Previously characterized HD domain-containing enzymes utilize nucleotides as substrates (Zimmerman et al. 2008). Structural studies and analysis of enzymatic activity indicate that SAMHD1 cleaves dNTPs to produce deoxynucleosides and triphosphates (Goldstone et al. 2011; Powell et al. 2011). The predominant subcellular localization of SAMHD1 appears to be in the nucleus (Berger et al. 2012; Brandariz-Nunez et al. 2012; Hofmann et al. 2012), although conflicting data indicates it may also localize to the cytoplasm in some cell types (Baldauf et al. 2012). Analysis of the primary amino acid sequence of SAMHD1 identified residues 11–14 (KRPR) as a match to the classical monopartite consensus nuclear localization signal (Brandariz-Nunez et al. 2012; Hofmann et al. 2012). Mutagenesis of the signal residues demonstrates that colocalization of Vpx and SAMHD1 in the nucleus is functionally important. Inactivation of the NLS shifts accumulation of SAMHD1 to the cytoplasm, but the cytoplasmic protein retains catalytic activity and antiviral function (Brandariz-Nunez et al. 2012; Hofmann et al. 2012). Cytoplasmic SAMHD1 can interact with Vpx, but this interaction is not sufficient for inducing the degradation of SAMHD1. There is a strong dependence on the Vpx–SAMHD1 interaction to occur in the nuclear compartment to initiate the process that results in depletion of cellular SAMHD1 by the proteasome (Brandariz-Nunez et al. 2012; Hofmann et al. 2012).

The SAMHD1 gene was previously cloned from dendritic cells and reported to be the human homologue of an interferon-γ-inducible gene in the mouse (Li et al. 2000). Treatment of human dendritic cells with either interferon-α, interferon-β, or interferon-γ does not increase SAMHD1 protein levels despite a transient increase in mRNA abundance (St Gelais et al. 2012). The absence of a regulatory effect in response to interferon in this system would make SAMHD1 unique among most antiviral restriction factors which are typically responsive to type I interferon (Bieniasz 2004; Harris et al. 2012). In contrast, treatment of monocytes or monocyte-derived macrophages with interferon significantly upregulated SAMHD1 levels, although variability in baseline expression levels of SAMHD1 in the different systems may account for the observed discordance (Berger et al. 2011; Rice et al. 2009). Interferon induction of SAMHD1 expression is also suggested by studies in which Vpx greatly enhances HIV-1 infection in type I interferon–treated macrophages (Gramberg et al. 2010). Interestingly, the loss of SAMHD1 expression due to genetic abnormalities in individuals with Aicardi–Goutières syndrome causes an inflammatory autoimmune state associated with elevated expression of type I interferon in the brain (Crow and Rehwinkel 2009; Rice et al. 2009). Clearly, there is a linkage between SAMHD1 expression and the immune system that implicates SAMHD1 as a negative regulator of the innate immune response (Rice et al. 2009).

4.3 Defects in SAMHD1 Cause Aicardi–Goutières Syndrome

Aicardi–Goutières syndrome is an autosomal neurological disorder that results in elevated type I interferon production phenotypically similar to that observed in congenital viral infections (Rice et al. 2009). Mutations in the 3′ repair exonuclease 1 (TREX1) or either subunit of the RNase H2 complex is causatively linked to AGS. TREX1 is a nuclease that preferentially degrades ssDNA in the cytosol to prevent the accumulation of DNA from endogenous retroelements or reverse-transcribed viral genomes that could otherwise trigger an autoimmune response (Stetson et al. 2008; Yan et al. 2010). Accumulation of cytosolic nucleic acid is detected by cellular receptors which trigger immune activation through a TLR-independent antiviral pathway (Stetson et al. 2008). HIV is normally able to evade this response, but excess HIV DNA accumulates in TREX1-deficient cells, resulting in detection and signaling to induce type 1 interferon (Yan et al. 2010). The RNase H2 complex also functions as a nuclease to remove misincorporated ribonucleosides from products of DNA replication, providing a mechanism to maintain chromosome integrity by avoiding elevated responses to DNA damage (Hiller et al. 2012). Like TREX1, this enzyme complex appears to facilitate HIV-1 infection, although its mechanism of action is unclear (Genovesio et al. 2011).

Prior to being identified as a lentiviral restriction factor, SAMHD1 was also linked to AGS and known to account for approximately ten percent of cases (Crow and Rehwinkel 2009). The loss of SAMHD1 expression results in the accumulation of cytosolic DNA and the associated increase in production of type I interferon (Rice et al. 2009). Like TREX1 and the RNase H2 complexes, SAMHD1 is also a modulator of the HIV-1 life cycle, but it functions to restrict rather than promote infection. Monocytes from SAMHD1-deficient AGS patients are highly susceptible to HIV-1 infection, while HIV-1 infection of healthy donor monocytes is very inefficient (Berger et al. 2011). In vitro, SAMHD1 does not degrade single-stranded DNA, double-stranded DNA, or RNA (Goldstone et al. 2011), so the mechanism by which abnormalities in SAMHD1 expression lead to the development of AGS differs from those of TREX1 and the RNase H2 complexes. Rather than degradation, the loss of SAMHD1 expression eliminates a cellular mechanism to inhibit the synthesis of cytosolic nucleic acid (Goldstone et al. 2011). Thus, there is an intricate relationship between nucleic acid metabolism, innate immune response, and susceptibility to lentiviral infection.

4.4 SAMHD1 Modulates Nucleotide Pools

SAMHD1 is a deoxyribonucleoside triphosphate triphosphohydrolase that converts dNTPs to deoxynucleosides and inorganic phosphate (Goldstone et al. 2011; Powell et al. 2011). Recombinant SAMHD1 rapidly hydrolyzes dGTP in vitro, while dATP, dCTP, and dTTP are not substrates when assayed individually.

However, dNTPs in combination with dGTP result in efficient hydrolysis of all four dNTPs (Goldstone et al. 2011; Powell et al. 2011). The crystal structure of the HD domain and molecular modeling indicate that dGTP binding mediates the allosteric activation of SAMHD1 to increase hydrolysis of all four dNTP substrates (Goldstone et al. 2011). Thus, dGTP is a preferred substrate and an enhancer of SAMHD1 enzymatic activity that modulates the intracellular levels of the four dNTPs.

Cellular dNTP levels are tightly regulated in response to expression of cell cycle–specific biosynthetic enzymes, such as ribonucleotide reductase (Bjursell and Skoog 1980). An optimized amount of the four dNTPs ensures fidelity during DNA replication and repair and minimizes negative consequences of inappropriate dNTP concentrations (Hofer et al. 2012; Mathews 2006; Niida et al. 2010). Accordingly, cells progressing through G1/S and S phase have relatively high concentrations of dNTPs (Bjursell and Skoog 1980). Conversely, in terminally differentiated non-cycling cells, such as macrophages, dNTP levels are relatively low (Diamond et al. 2004). Differential expression of SAMHD1 results in large cell-dependent differences in dNTP concentrations that highly correlate with antiviral activity (Table 1).

Levels of dNTPs in activated primary $CD4^+$ T cells are approximately 100 times higher than in monocyte-derived macrophages (Diamond et al. 2004). Reducing cellular dNTP concentrations to levels below that which is required by reverse transcription for efficient cDNA synthesis results in potent inhibition of retroviral replication (Diamond et al. 2004; Kim et al. 2012; Lahouassa et al. 2012). This antiviral mechanism is supported by studies in which RNAi-mediated depletion of SAMHD1 increases dNTP pools and renders the cells sensitive to HIV-1 infection (Kim et al. 2012; Lahouassa et al. 2012). Furthermore, addition of exogenous deoxynucleosides to culture medium increases intracellular dNTP levels and sensitivity to HIV-1 infection (Lahouassa et al. 2012). Recent studies have addressed whether a similar antiviral mechanism accounts for the differences in permissivity to HIV-1 infection of activated versus resting $CD4^+$ T cells.

Table 1 dNTP levels are variable in cells targeted by HIV-1

	dNTPs (uM)	Reverse transcription efficiency	Level of HIV-1 infection
Activated T cells	3–30	Very high	Very high
Resting T cells	0.3–5	Low	Very low
Macrophages	~0.05	Low	Low
Macrophages + Vpx	~1	High	High

The efficiency of reverse transcription is dependent on the concentration of dNTPs within cells infected by HIV-1. Suboptimal dNTP concentrations result in incomplete reverse transcription, abortive infection, and a block to productive infection. Vpx delivered to macrophages depletes SAMHD1, resulting in efficient reverse transcription and high levels of HIV-1 production. Adapted from Ayinde et al. (2012).

5 SAMHD1 Restricts HIV-1 Infection of Quiescent CD4$^+$ T Cells

Activated CD4$^+$ T cells are highly susceptible to productive HIV-1 infection, but infection of quiescent CD4$^+$ T cells is inefficient due to various post-entry blocks to replication (Pierson et al. 2002; Stevenson et al. 1990; Swiggard et al. 2004; Zack et al. 1990). Similar to infection of myeloid cells, reverse transcription in resting lymphocytes is inefficient (Plesa et al. 2007; Zack et al. 1990). Based on earlier work, it was unclear whether SAMHD1 contributed to the restriction to infection of resting lymphocytes, because dNTP levels are higher than in myeloid cells (Gao et al. 1993; Lahouassa et al. 2012) and SAMHD1 expression is not detected in T cell lines (Laguette et al. 2011). Analysis has revealed that SAMHD1 is highly expressed in resting CD4$^+$ T cells isolated from peripheral blood and lymphoid tissue (Baldauf et al. 2012; Descours et al. 2012). Expression levels are similar to those detected in myeloid cells, and analysis of sorted cells indicates that the various subsets of peripheral blood CD4$^+$ T lymphocytes express similarly high levels of SAMHD1 (Descours et al. 2012). Elevated expression of SAMHD1 and the profound inhibition of reverse transcription in quiescent lymphocytes raise the possibility that the mechanism of restriction detected in myeloid cells also functions in CD4$^+$ T cells. In support of this, delivery of Vpx into resting target cells or supplying exogenous nucleosides to boost cellular dNTP levels promotes efficient DNA synthesis in the absence of cellular activation (Baldauf et al. 2012; Descours et al. 2012). Resting CD4$^+$ T cells isolated from individuals with Aicardi–Goutières syndrome harboring a defective SAMHD1 gene are also susceptible to HIV-1 infection (Baldauf et al. 2012; Descours et al. 2012). Notably, SAMHD1 degradation was not sufficient for virus production, indicating that additional blocks in the viral life cycle are not bypassed by elevating dNTP levels in resting T cells (Baldauf et al. 2012). In addition, SAMHD1 expression levels are similar in both quiescent and activated CD4$^+$ T cells, suggesting that additional regulatory mechanisms influence dNTP levels in activated cells that are needed for DNA synthesis during proliferation (Descours et al. 2012).

6 SAMHD1 Influences Innate Immunity

Expression levels of SAMHD1 in cells of the immune system can have a major impact on the ability to sense retroviral infection and mount an antiviral immune response. In the context of HIV-1 infection, immune cells differ in their capacity to respond to infection. Plasmacytoid dendritic cells detect viral nucleic acid through their expression of Toll-like receptors 7 and 9 to activate TLR signaling (Gilliet et al. 2008). They are a main component of the innate antiviral immune system and respond rapidly to produce a large amount of type I interferon, as well as other antiviral cytokines, and play an important role in linking innate and adaptive immune responses (Gilliet et al. 2008). In contrast, monocyte-derived dendritic cells are largely resistant to HIV-1 infection and produce very little type I interferon when exposed to HIV (Fonteneau et al. 2004; Manel et al. 2010). Monocyte-derived macrophages are naturally infected by HIV-1, but the inefficiency of infection is likely to allow these cells to evade detection and escape an effective antiviral immune response (Hrecka et al. 2011; Laguette et al. 2011). Vpx enhances infection of myeloid cells resulting in increased sensitivity of detecting HIV that triggers the innate antiviral response and may contribute to development of more effective adaptive immunity (Manel et al. 2010). Vpx may be required by HIV-2 to infect cells with low dNTP pools in order to compensate for a less efficient reverse transcriptase (Fujita et al. 2012). HIV-2 infection is much less pathogenic than HIV-1 infection as evidenced by lower viral loads, slower rates of $CD4^+$ T cell decline, and delayed progression to clinical symptoms of immunodeficiency (Andersson et al. 2000; Marlink et al. 1994). Indeed, the vast majority of HIV-2 infected individuals are categorized as long-term non-progressors, who control viral loads in the absence of chemotherapy (de Silva et al. 2008; van der Loeff et al. 2010). It is likely that the tropism mediated by Vpx accounts substantially for the differences in the clinical outcomes of HIV-1 versus HIV-2 infection. Presumably, HIV-1 has not developed a strategy to counteract the antiviral activity of SAMHD1 because it is advantageous. Overall, the interplay between Vpx and SAMHD1 suggests that limiting HIV-1 infection of myeloid cells promotes viral persistence and escape from antiviral immune responses. Strategies designed to inhibit SAMHD1 expression and allow infection and activation of dendritic cells may increase HIV-1 detection and potentiate virus-specific adaptive immunity.

7 Conclusions

The accessory genes encoded by lentiviral genomes represent evolutionary adaptations that the virus relies on to enhance replication, and the study of accessory gene functions has revealed the complex interplay between viral and host factors that support or inhibit viral infection. Efforts to characterize the functional roles of lentiviral Vpr and Vpx have identified SAMHD1 as a host cell restriction factor that potently inhibits virus replication in non-cycling myeloid cells and quiescent $CD4^+$ T cells. The enzymatic activity of SAMHD1 regulates cellular dNTP levels, such that key substrates for reverse transcription are below levels needed for efficient conversion of viral RNA to genomic DNA. Vpx counteracts the SAMHD1-mediated restriction by recruiting the protein to the CUL4-based E3 ubiquitin ligase complex, resulting in its degradation by the proteasome. Depletion of SAMHD1 by Vpx, or the absence of SAMHD1 expression in individuals with Aicardi–Goutières syndrome, allows accumulation of cytosolic DNA that triggers an innate immune response with elevated production of interferon. HIV-1 has lost the ability to antagonize SAMHD1, and the low level of myeloid cell infection may be a viral strategy to evade immune surveillance to promote virus persistence and dissemination. A better understanding of SAMHD1 and innate immune responses during HIV infection may drive the development of novel therapeutic interventions and more effective vaccine strategies.

References

Accola MA, Bukovsky AA, Jones MS, Gottlinger HG (1999) A conserved dileucine-containing motif in p6 (gag) governs the particle association of Vpx and Vpr of simian immunodeficiency viruses SIV(mac) and SIV(agm). J Virol 73:9992–9999

Andersson S, Norrgren H, da Silva Z, Biague A, Bamba S, Kwok S, Christopherson C, Biberfeld G, Albert J (2000) Plasma viral load in HIV-1 and HIV-2 singly and dually infected individuals in Guinea-Bissau, West Africa: significantly lower plasma virus set point in HIV-2 infection than in HIV-1 infection. Arch Intern Med 160:3286–3293

Andrew A, Strebel K (2010) HIV-1 Vpu targets cell surface markers CD4 and BST-2 through distinct mechanisms. Mol Aspects Med 31:407–417

Aravind L, Koonin EV (1998) The HD domain defines a new superfamily of metal-dependent phosphohydrolases. Trends Biochem Sci 23:469–472

Ayinde D, Casartelli N, Schwartz O (2012) Restricting HIV the SAMHD1 way: through nucleotide starvation. Nat Rev Microbiol 10:675–680

Ayinde D, Maudet C, Transy C, Margottin-Goguet F (2010) Limelight on two HIV/SIV accessory proteins in macrophage infection: is Vpx overshadowing Vpr? Retrovirology 7:35

Bachand F, Yao XJ, Hrimech M, Rougeau N, Cohen EA (1999) Incorporation of Vpr into human immunodeficiency virus type 1 requires a direct interaction with the p6 domain of the p55 gag precursor. J Biol Chem 274:9083–9091

Baldauf HM, Pan X, Erikson E, Schmidt S, Daddacha W, Burggraf M, Schenkova K, Ambiel I, Wabnitz G, Gramberg T, Panitz S, Flory E, Landau NR, Sertel S, Rutsch F, Lasitschka F, Kim B, Konig R, Fackler OT, Keppler OT (2012) SAMHD1 restricts HIV-1 infection in resting CD4(+) T cells. Nat Med 18:1682–1687

Balliet JW, Kolson DL, Eiger G, Kim FM, McGann KA, Srinivasan A, Collman R (1994) Distinct effects in primary macrophages and lymphocytes of the human immunodeficiency virus type 1 accessory genes Vpr, Vpu, and Nef: mutational analysis of a primary HIV-1 isolate. Virology 200:623–631

Beer BE, Foley BT, Kuiken CL, Tooze Z, Goeken RM, Brown CR, Hu J, St Claire M, Korber BT, Hirsch VM (2001) Characterization of novel simian immunodeficiency viruses from red-capped mangabeys from Nigeria (SIVrcmNG409 and -NG411). J Virol 75:12014–12027

Belshan M, Kimata JT, Brown C, Cheng X, McCulley A, Larsen A, Thippeshappa R, Hodara V, Giavedoni L, Hirsch V, Ratner L (2012) Vpx is critical for SIVmne infection of pigtail macaques. Retrovirology 9:32

Belshan M, Mahnke LA, Ratner L (2006) Conserved amino acids of the human immunodeficiency virus type 2 Vpx nuclear localization signal are critical for nuclear targeting of the viral preintegration complex in non-dividing cells. Virology 346:118–126

Belshan M, Ratner L (2003) Identification of the nuclear localization signal of human immunodeficiency virus type 2 Vpx. Virology 311:7–15

Belzile JP, Duisit G, Rougeau N, Mercier J, Finzi A, Cohen EA (2007) HIV-1 Vpr-mediated G2 arrest involves the DDB1-CUL4AVPRBP E3 ubiquitin ligase. PLoS Pathog 3:e85

Bergamaschi A, Ayinde D, David A, Le Rouzic E, Morel M, Collin G, Descamps D, Damond F, Brun-Vezinet F, Nisole S, Margottin-Goguet F, Pancino G, Transy C (2009) The human immunodeficiency virus type 2 Vpx protein usurps the CUL4A-DDB1 DCAF1 ubiquitin ligase to overcome a postentry block in macrophage infection. J Virol 83:4854–4860

Bergamaschi A, Pancino G (2010) Host hindrance to HIV-1 replication in monocytes and macrophages. Retrovirology 7:31

Berger A, Sommer AF, Zwarg J, Hamdorf M, Welzel K, Esly N, Panitz S, Reuter A, Ramos I, Jatiani A, Mulder LC, Fernandez-Sesma A, Rutsch F, Simon V, Konig R, Flory E (2011) SAMHD1-deficient CD14+ cells from individuals with Aicardi-Goutieres syndrome are highly susceptible to HIV-1 infection. PLoS Pathog 7:e1002425

Berger G, Turpin J, Cordeil S, Tartour K, Nguyen XN, Mahieux R, Cimarelli A (2012) Functional analysis of the relationship between Vpx and the restriction factor SAMHD1. J Biol Chem 287:41210–41217

Bieniasz PD (2004) Intrinsic immunity: a front-line defense against viral attack. Nat Immunol 5:1109–1115

Bjursell G, Skoog L (1980) Control of nucleotide pools in mammalian cells. Antibiot Chemother 28:78–85

Brandariz-Nunez A, Valle-Casuso JC, White TE, Laguette N, Benkirane M, Brojatsch J, Diaz-Griffero F (2012) Role of SAMHD1 nuclear localization in restriction of HIV-1 and SIVmac. Retrovirology 9:49

Bukrinsky MI, Haggerty S, Dempsey MP, Sharova N, Adzhubel A, Spitz L, Lewis P, Goldfarb D, Emerman M, Stevenson M (1993) A nuclear localization signal within HIV-1 matrix protein that governs infection of non-dividing cells. Nature 365:666–669

Cheng X, Belshan M, Ratner L (2008) Hsp40 facilitates nuclear import of the human immunodeficiency virus type 2 Vpx-mediated preintegration complex. J Virol 82:1229–1237

Cohen EA, Terwilliger EF, Jalinoos Y, Proulx J, Sodroski JG, Haseltine WA (1990) Identification of HIV-1 Vpr product and function. J Acquir Immune Defic Syndr 3:11–18

Crow YJ, Rehwinkel J (2009) Aicardi-Goutieres syndrome and related phenotypes: linking nucleic acid metabolism with autoimmunity. Hum Mol Genet 18:R130–R136

de Silva TI, Cotten M, Rowland-Jones SL (2008) HIV-2: the forgotten AIDS virus. Trends Microbiol 16:588–595

DeHart JL, Zimmerman ES, Ardon O, Monteiro-Filho CM, Arganaraz ER, Planelles V (2007) HIV-1 Vpr activates the G2 checkpoint through manipulation of the ubiquitin proteasome system. Virol J 4:57

Descours B, Cribier A, Chable-Bessia C, Ayinde D, Rice G, Crow Y, Yatim A, Schwartz O, Laguette N, Benkirane M (2012) SAMHD1 restricts HIV-1 reverse transcription in quiescent CD4(+) T-cells. Retrovirology 9:87

Diamond TL, Roshal M, Jamburuthugoda VK, Reynolds HM, Merriam AR, Lee KY, Balakrishnan M, Bambara RA, Planelles V, Dewhurst S, Kim B (2004) Macrophage tropism of HIV-1 depends on efficient cellular dNTP utilization by reverse transcriptase. J Biol Chem 279:51545–51553

Dube M, Bego MG, Paquay C, Cohen EA (2010) Modulation of HIV-1-host interaction: role of the Vpu accessory protein. Retrovirology 7:114

Ellery PJ, Tippett E, Chiu YL, Paukovics G, Cameron PU, Solomon A, Lewin SR, Gorry PR, Jaworowski A, Greene WC, Sonza S, Crowe SM (2007) The CD16+ monocyte subset is more permissive to infection and preferentially harbors HIV-1 in vivo. J Immunol 178:6581–6589

Fassati A (2006) HIV infection of non-dividing cells: a divisive problem. Retrovirology 3:74

Fletcher TM 3rd, Brichacek B, Sharova N, Newman MA, Stivahtis G, Sharp PM, Emerman M, Hahn BH, Stevenson M (1996) Nuclear import and cell cycle arrest functions of the HIV-1 Vpr protein are encoded by two separate genes in HIV-2/SIV(SM). EMBO J 15:6155–6165

Fonteneau JF, Larsson M, Beignon AS, McKenna K, Dasilva I, Amara A, Liu YJ, Lifson JD, Littman DR, Bhardwaj N (2004) Human immunodeficiency virus type 1 activates plasmacytoid dendritic cells and concomitantly induces the bystander maturation of myeloid dendritic cells. J Virol 78:5223–5232

Fouchier RA, Meyer BE, Simon JH, Fischer U, Malim MH (1997) HIV-1 infection of non-dividing cells: evidence that the amino-terminal basic region of the viral matrix protein is important for gag processing but not for post-entry nuclear import. EMBO J 16:4531–4539

Fujita M, Nomaguchi M, Adachi A, Otsuka M (2012) SAMHD1-dependent and -independent functions of HIV-2/SIV Vpx protein. Front Microbiol 3:297

Fujita M, Otsuka M, Miyoshi M, Khamsri B, Nomaguchi M, Adachi A (2008) Vpx is critical for reverse transcription of the human immunodeficiency virus type 2 genome in macrophages. J Virol 82:7752–7756

Gao WY, Cara A, Gallo RC, Lori F (1993) Low levels of deoxynucleotides in peripheral blood lymphocytes: a strategy to inhibit human immunodeficiency virus type 1 replication. Proc Nat Acad Sci USA 90:8925–8928

Genovesio A, Kwon YJ, Windisch MP, Kim NY, Choi SY, Kim HC, Jung S, Mammano F, Perrin V, Boese AS, Casartelli N, Schwartz O, Nehrbass U, Emans N (2011) Automated genome-wide visual profiling of cellular proteins involved in HIV infection. J Biomol Screen 16:945–958

Gibbs JS, Lackner AA, Lang SM, Simon MA, Sehgal PK, Daniel MD, Desrosiers RC (1995) Progression to AIDS in the absence of a gene for Vpr or Vpx. J Virol 69:2378–2383

Gillick K, Pollpeter D, Phalora P, Kim EY, Wolinsky SM, Malim MH (2013) Suppression of HIV-1 Infection by APOBEC3 proteins in primary human CD4(+) T Cells is associated with inhibition of processive reverse transcription as well as excessive cytidine deamination. J Virol 87:1508–1517

Gilliet M, Cao W, Liu YJ (2008) Plasmacytoid dendritic cells: sensing nucleic acids in viral infection and autoimmune diseases. Nat Rev Immunol 8:594–606

Goh WC, Rogel ME, Kinsey CM, Michael SF, Fultz PN, Nowak MA, Hahn BH, Emerman M (1998) HIV-1 Vpr increases viral expression by manipulation of the cell cycle: a mechanism for selection of Vpr in vivo. Nat Med 4:65–71

Goldstone DC, Ennis-Adeniran V, Hedden JJ, Groom HC, Rice GI, Christodoulou E, Walker PA, Kelly G, Haire LF, Yap MW, de Carvalho LP, Stoye JP, Crow YJ, Taylor IA, Webb M (2011) HIV-1 restriction factor SAMHD1 is a deoxynucleoside triphosphate triphosphohydrolase. Nature 480:379–382

Goujon C, Riviere L, Jarrosson-Wuilleme L, Bernaud J, Rigal D, Darlix JL, Cimarelli A (2007) SIVSM/HIV-2 Vpx proteins promote retroviral escape from a proteasome-dependent restriction pathway present in human dendritic cells. Retrovirology 4:2

Gramberg T, Sunseri N, Landau NR (2010) Evidence for an activation domain at the amino terminus of simian immunodeficiency virus Vpx. J Virol 84:1387–1396

Gustin JK, Moses AV, Fruh K, Douglas JL (2011) Viral takeover of the host ubiquitin system. Front Microbiol 2:161

Harris RS, Hultquist JF, Evans DT (2012) The restriction factors of human immunodeficiency virus. J Biol Chem 287:40875–40883

Heinzinger NK, Bukinsky MI, Haggerty SA, Ragland AM, Kewalramani V, Lee MA, Gendelman HE, Ratner L, Stevenson M, Emerman M (1994) The Vpr protein of human immunodeficiency virus type 1 influences nuclear localization of viral nucleic acids in nondividing host cells. Proc Natl Acad Sci USA 91:7311–7315

Henderson LE, Sowder RC, Copeland TD, Benveniste RE, Oroszlan S (1988) Isolation and characterization of a novel protein (X-ORF product) from SIV and HIV-2. Science 241:199–201

Hiller B, Achleitner M, Glage S, Naumann R, Behrendt R, Roers A (2012) Mammalian RNase H2 removes ribonucleotides from DNA to maintain genome integrity. J Exp Med 209:1419–1426

Hirsch VM, Sharkey ME, Brown CR, Brichacek B, Goldstein S, Wakefield J, Byrum R, Elkins WR, Hahn BH, Lifson JD, Stevenson M (1998) Vpx is required for dissemination and pathogenesis of SIV(SM) PBj: evidence of macrophage-dependent viral amplification. Nat Med 4:1401–1408

Hofer A, Crona M, Logan DT, Sjoberg BM (2012) DNA building blocks: keeping control of manufacture. Crit Rev Biochem Mol Biol 47:50–63

Hofmann H, Logue EC, Bloch N, Daddacha W, Polsky SB, Schultz ML, Kim B, Landau NR (2012) The Vpx lentiviral accessory protein targets SAMHD1 for degradation in the nucleus. J Virol 86:12552–12560

Hrecka K, Gierszewska M, Srivastava S, Kozaczkiewicz L, Swanson SK, Florens L, Washburn MP, Skowronski J (2007) Lentiviral Vpr usurps CUL4-DDB1[VprBP] E3 ubiquitin ligase to modulate cell cycle. Proc Natl Acad Sci USA 104:11778–11783

Hrecka K, Hao C, Gierszewska M, Swanson SK, Kesik-Brodacka M, Srivastava S, Florens L, Washburn MP, Skowronski J (2011) Vpx relieves inhibition of HIV-1 infection of macrophages mediated by the SAMHD1 protein. Nature 474:658–661
Hu J, Switzer WM, Foley BT, Robertson DL, Goeken RM, Korber BT, Hirsch VM, Beer BE (2003) Characterization and comparison of recombinant simian immunodeficiency virus from drill (Mandrillus leucophaeus) and mandrill (Mandrillus sphinx) isolates. J Virol 77:4867–4880
Kappes JC, Morrow CD, Lee SW, Jameson BA, Kent SB, Hood LE, Shaw GM, Hahn BH (1988) Identification of a novel retroviral gene unique to human immunodeficiency virus type 2 and simian immunodeficiency virus SIVMAC. J Virol 62:3501–3505
Kaushik R, Zhu X, Stranska R, Wu Y, Stevenson M (2009) A cellular restriction dictates the permissivity of nondividing monocytes/macrophages to lentivirus and gammaretrovirus infection. Cell Host Microbe 6:68–80
Kewalramani VN, Emerman M (1996) Vpx association with mature core structures of HIV-2. Virology 218:159–168
Kim B, Nguyen LA, Daddacha W, Hollenbaugh JA (2012) Tight interplay among SAMHD1 protein level, cellular dNTP levels, and HIV-1 proviral DNA synthesis kinetics in human primary monocyte-derived macrophages. J Biol Chem 287:21570–21574
Laguette N, Sobhian B, Casartelli N, Ringeard M, Chable-Bessia C, Segeral E, Yatim A, Emiliani S, Schwartz O, Benkirane M (2011) SAMHD1 is the dendritic and myeloid-cell-specific HIV-1 restriction factor counteracted by Vpx. Nature 474:654–657
Lahouassa H, Daddacha W, Hofmann H, Ayinde D, Logue EC, Dragin L, Bloch N, Maudet C, Bertrand M, Gramberg T, Pancino G, Priet S, Canard B, Laguette N, Benkirane M, Transy C, Landau NR, Kim B, Margottin-Goguet F (2012) SAMHD1 restricts the replication of human immunodeficiency virus type 1 by depleting the intracellular pool of deoxynucleoside triphosphates. Nat Immunol 13:223–228
Li N, Zhang W, Cao X (2000) Identification of human homologue of mouse IFN-gamma induced protein from human dendritic cells. Immunol Lett 74:221–224
Manel N, Hogstad B, Wang Y, Levy DE, Unutmaz D, Littman DR (2010) A cryptic sensor for HIV-1 activates antiviral innate immunity in dendritic cells. Nature 467:214–217
Marlink R, Kanki P, Thior I, Travers K, Eisen G, Siby T, Traore I, Hsieh CC, Dia MC, Gueye EH et al (1994) Reduced rate of disease development after HIV-2 infection as compared to HIV-1. Science 265:1587–1590
Mathews CK (2006) DNA precursor metabolism and genomic stability. FASEB J Official Publ Fed Am Soc Exp Biol 20:1300–1314
Mehle A, Goncalves J, Santa-Marta M, McPike M, Gabuzda D (2004) Phosphorylation of a novel SOCS-box regulates assembly of the HIV-1 Vif-Cul5 complex that promotes APOBEC3G degradation. Genes Dev 18:2861–2866
Negri DR, Rossi A, Blasi M, Michelini Z, Leone P, Chiantore MV, Baroncelli S, Perretta G, Cimarelli A, Klotman ME, Cara A (2012) Simian immunodeficiency virus-Vpx for improving integrase defective lentiviral vector-based vaccines. Retrovirology 9:69
Neil S, Martin F, Ikeda Y, Collins M (2001) Post entry restriction to human immunodeficiency virus-based vector transduction in human monocytes. J Virol 75:5448–5456
Niida H, Shimada M, Murakami H, Nakanishi M (2010) Mechanisms of dNTP supply that play an essential role in maintaining genome integrity in eukaryotic cells. Cancer Sci 101:2505–2509
O'Brien WA, Namazi A, Kalhor H, Mao SH, Zack JA, Chen IS (1994) Kinetics of human immunodeficiency virus type 1 reverse transcription in blood mononuclear phagocytes are slowed by limitations of nucleotide precursors. J Virol 68:1258–1263
Pancio HA, Ratner L (1998) Human immunodeficiency virus type 2 Vpx-gag interaction. J Virol 72:5271–5275
Pancio HA, Vander Heyden N, Ratner L (2000) The C-terminal proline-rich tail of human immunodeficiency virus type 2 Vpx is necessary for nuclear localization of the viral preintegration complex in nondividing cells. J Virol 74:6162–6167

Peng G, Greenwell-Wild T, Nares S, Jin W, Lei KJ, Rangel ZG, Munson PJ, Wahl SM (2007) Myeloid differentiation and susceptibility to HIV-1 are linked to APOBEC3 expression. Blood 110:393–400
Pierson TC, Zhou Y, Kieffer TL, Ruff CT, Buck C, Siliciano RF (2002) Molecular characterization of preintegration latency in human immunodeficiency virus type 1 infection. J Virol 76:8518–8531
Pion M, Granelli-Piperno A, Mangeat B, Stalder R, Correa R, Steinman RM, Piguet V (2006) APOBEC3G/3F mediates intrinsic resistance of monocyte-derived dendritic cells to HIV-1 infection. J Exp Med 203:2887–2893
Planelles V, Benichou S (2009) Vpr and its interactions with cellular proteins. Curr Top Microbiol Immunol 339:177–200
Planelles V, Jowett JB, Li QX, Xie Y, Hahn B, Chen IS (1996) Vpr-induced cell cycle arrest is conserved among primate lentiviruses. J Virol 70:2516–2524
Plesa G, Dai J, Baytop C, Riley JL, June CH, O'Doherty U (2007) Addition of deoxynucleosides enhances human immunodeficiency virus type 1 integration and 2LTR formation in resting CD4(+) T cells. J Virol 81:13938–13942
Popov S, Rexach M, Zybarth G, Reiling N, Lee MA, Ratner L, Lane CM, Moore MS, Blobel G, Bukrinsky M (1998) Viral protein R regulates nuclear import of the HIV-1 pre-integration complex. EMBO J 17:909–917
Powell RD, Holland PJ, Hollis T, Perrino FW (2011) Aicardi-Goutieres syndrome gene and HIV-1 restriction factor SAMHD1 is a dGTP-regulated deoxynucleotide triphosphohydrolase. J Biol Chem 286:43596–43600
Qiao F, Bowie JU (2005) The many faces of SAM. Sci Sig Transduct knowl Environ: re7
Rice GI, Bond J, Asipu A, Brunette RL, Manfield IW, Carr IM, Fuller JC, Jackson RM, Lamb T, Briggs TA, Ali M, Gornall H, Couthard LR, Aeby A, Attard-Montalto SP, Bertini E, Bodemer C, Brockmann K, Brueton LA, Corry PC, Desguerre I, Fazzi E, Cazorla AG, Gener B, Hamel BC, Heiberg A, Hunter M, van der Knaap MS, Kumar R, Lagae L, Landrieu PG, Lourenco CM, Marom D, McDermott MF, van der Merwe W, Orcesi S, Prendiville JS, Rasmussen M, Shalev SA, Soler DM, Shinawi M, Spiegel R, Tan TY, Vanderver A, Wakeling EL, Wassmer E, Whittaker E, Lebon P, Stetson DB, Bonthron DT, Crow YJ (2009) Mutations involved in Aicardi-Goutieres syndrome implicate SAMHD1 as regulator of the innate immune response. Nat Genet 41:829–832
Rich EA, Chen IS, Zack JA, Leonard ML, O'Brien WA (1992) Increased susceptibility of differentiated mononuclear phagocytes to productive infection with human immunodeficiency virus-1 (HIV-1). J Clin Investig 89:176–183
Schrofelbauer B, Hakata Y, Landau NR (2007) HIV-1 Vpr function is mediated by interaction with the damage-specific DNA-binding protein DDB1. Proc Nat Acad Sci USA 104:4130–4135
Schule S, Kloke BP, Kaiser JK, Heidmeier S, Panitz S, Wolfrum N, Cichutek K, Schweizer M (2009) Restriction of HIV-1 replication in monocytes is abolished by Vpx of SIVsmmPBj. PLoS ONE 4:e7098
Sharova N, Wu Y, Zhu X, Stranska R, Kaushik R, Sharkey M, Stevenson M (2008) Primate lentiviral Vpx commandeers DDB1 to counteract a macrophage restriction. PLoS Pathog 4:e1000057
Sharp PM, Bailes E, Stevenson M, Emerman M, Hahn BH (1996) Gene acquisition in HIV and SIV. Nature 383:586–587
Sonza S, Maerz A, Deacon N, Meanger J, Mills J, Crowe S (1996) Human immunodeficiency virus type 1 replication is blocked prior to reverse transcription and integration in freshly isolated peripheral blood monocytes. J Virol 70:3863–3869
Srivastava S, Swanson SK, Manel N, Florens L, Washburn MP, Skowronski J (2008) Lentiviral Vpx accessory factor targets VprBP/DCAF1 substrate adaptor for cullin 4 E3 ubiquitin ligase to enable macrophage infection. PLoS Pathog 4:e1000059
St Gelais C, de Silva S, Amie SM, Coleman CM, Hoy H, Hollenbaugh JA, Kim B, Wu L (2012) SAMHD1 restricts HIV-1 infection in dendritic cells (DCs) by dNTP depletion, but its

expression in DCs and primary CD4(+) T-lymphocytes cannot be upregulated by interferons. Retrovirology 9:105
Stetson DB, Ko JS, Heidmann T, Medzhitov R (2008) Trex1 prevents cell-intrinsic initiation of autoimmunity. Cell 134:587–598
Stevenson M, Stanwick TL, Dempsey MP, Lamonica CA (1990) HIV-1 replication is controlled at the level of T cell activation and proviral integration. EMBO J 9:1551–1560
Swiggard WJ, O'Doherty U, McGain D, Jeyakumar D, Malim MH (2004) Long HIV type 1 reverse transcripts can accumulate stably within resting CD4(+) T cells while short ones are degraded. AIDS Res Hum Retroviruses 20:285–295
Triques K, Stevenson M (2004) Characterization of restrictions to human immunodeficiency virus type 1 infection of monocytes. J Virol 78:5523–5527
Tristem M, Purvis A, Quicke DL (1998) Complex evolutionary history of primate lentiviral Vpr genes. Virology 240:232–237
van der Loeff MF, Larke N, Kaye S, Berry N, Ariyoshi K, Alabi A, van Tienen C, Leligdowicz A, Sarge-Njie R, da Silva Z, Jaye A, Ricard D, Vincent T, Jones SR, Aaby P, Jaffar S, Whittle H (2010) Undetectable plasma viral load predicts normal survival in HIV-2-infected people in a West African village. Retrovirology 7:46
von Schwedler U, Kornbluth RS, Trono D (1994) The nuclear localization signal of the matrix protein of human immunodeficiency virus type 1 allows the establishment of infection in macrophages and quiescent T lymphocytes. Proc Nat Acad Sci USA 91:6992–6996
Wu X, Conway JA, Kim J, Kappes JC (1994) Localization of the Vpx packaging signal within the C terminus of the human immunodeficiency virus type 2 gag precursor protein. J Virol 68:6161–6169
Yamashita M, Emerman M (2005) The cell cycle independence of HIV infections is not determined by known karyophilic viral elements. PLoS Pathog 1:e18
Yan N, Regalado-Magdos AD, Stiggelbout B, Lee-Kirsch MA, Lieberman J (2010) The cytosolic exonuclease TREX1 inhibits the innate immune response to human immunodeficiency virus type 1. Nat Immunol 11:1005–1013
Yu X, Yu Y, Liu B, Luo K, Kong W, Mao P, Yu XF (2003) Induction of APOBEC3G ubiquitination and degradation by an HIV-1 Vif-Cul5-SCF complex. Science 302:1056–1060
Zack JA, Arrigo SJ, Weitsman SR, Go AS, Haislip A, Chen IS (1990) HIV-1 entry into quiescent primary lymphocytes: molecular analysis reveals a labile, latent viral structure. Cell 61:213–222
Zhang C, de Silva S, Wang JH, Wu L (2012) Co-evolution of primate SAMHD1 and lentivirus Vpx leads to the loss of the Vpx gene in HIV-1 ancestor. PLoS One 7:e37477
Zhang S, Feng Y, Narayan O, Zhao LJ (2001) Cytoplasmic retention of HIV-1 regulatory protein Vpr by protein-protein interaction with a novel human cytoplasmic protein VprBP. Gene 263:131–140
Zimmerman MD, Proudfoot M, Yakunin A, Minor W (2008) Structural insight into the mechanism of substrate specificity and catalytic activity of an HD-domain phosphohydrolase: the 5′-deoxyribonucleotidase YfbR from Escherichia coli. J Mol Biol 378:215–226

Rapid Adversarial Co-Evolution of Viruses and Cellular Restriction Factors

Welkin E. Johnson

Abstract Since the discovery of viruses over a century ago, virologists have recognized that host genetics plays a major role in viral tropism and the distribution of viruses in nature. Traditionally, studies of tropism have centered on identification of cellular factors required for viral replication, such as cell-surface entry receptors. However, over the past 20 years, there has been a steady increase in the identification and characterization of restriction factors (RFs), here defined as dominant cellular factors that have evolved specifically to interfere with viral replication. Genetic studies suggest that restriction factors impose significant barriers to interspecies movement of viruses and are therefore critical determinants of viral tropism. Furthermore, the scope of the ever-expanding list of restriction factors, and the variety of antiviral mechanisms they represent, testifies to the extraordinary impact viruses have had on organismal evolution—an impact hitherto under-appreciated by evolutionary biologists and virologists alike. Recent studies of RF-encoding genes that combine molecular evolutionary analysis with functional assays illustrate the potential for asking questions about virus–host interactions as they play out in natural populations and across evolutionary timescales. Most notably, it has become common to apply tests of positive selection to RF genes and couple these analyses with virological assays, to reveal evidence for antagonistic virus–host co-evolution. Herein, I summarize recent work on the evolutionary genetics of mammalian RFs, particularly those of humans, non-human primates, and model organisms, and how RFs can reveal the influence of virus–host interactions on organismal evolution. Because intensive investigation of RF evolution is fairly new (and because there is still much to learn), the discussion is organized around five broad, outstanding questions that will need to be answered before we can fully appreciate the evolutionary biology of restriction.

W. E. Johnson (✉)
Biology Department, Boston College, Chestnut Hill, MA 02467, USA
e-mail: welkin.johnson@bc.edu

B. R. Cullen (ed.), *Intrinsic Immunity*, Current Topics in Microbiology and Immunology 371, DOI: 10.1007/978-3-642-37765-5_5,

Contents

1 An Evolutionary Perspective on Restriction

It is difficult, if not impossible, to settle on a simple, universal description of what comprises a typical cellular restriction factor (RF). The list of RF loci in Table 1 is by no means comprehensive, but helps to illustrate the nature of the problem. Among this sampling of well-studied RFs, there is no unifying element of structure, mechanism, gene ontology, or viral target that is shared by all or even a majority of the factors (Table 1). Thus, there are RFs that target viral nucleic acids, such as the DNA-editing enzymes encoded by the *APOBEC3* cluster, or the ADARs, the restriction nucleases and CRISPRS of bacteria, or the RNAi-mediated mechanisms of plants, fungi, and invertebrates; there are cytosolic proteins, such as Fv1 in mice and mammalian TRIM5, that target viral capsids; there are membrane-associated proteins that target replication, including ER-associated factors (such as Flv in mice) and plasma membrane proteins (Fv4, IFITMs, tetherin). Moreover, for every stage in a canonical viral replication cycle (attachment and entry, replication, expression, assembly and egress), at least one host-encoded barrier has been identified (and in most cases, several examples are known). For example, the murine Fv4 protein and the mammalian IFITM proteins interfere with attachment or entry; Fv1 and TRIM5 interfere with early post-entry of retroviruses; mammalian APOBEC3s and the murine Flv interfere with viral genome replication of reverse-transcribing viruses and flaviviruses, respectively; ovine enJS56A5 interferes with assembly of exogenous JSRV; and *mammalian tetherin* interferes with release of a multitude of enveloped viruses.

Table 1 Known and proposed antiviral intrinsic restriction factors

Gene	Organism(s)	Viral stages targeted	Viruses	Antagonist examples	Pos. selection
TRIM5	Mammals	Post-entry	Retroviruses	Unk	Johnson and Sawyer 2009; Sawyer et al. 2005; Song et al. 2005
BST2/tetherin	Mammals	Release	Enveloped viruses	Vpu, Nef, K5, viral glycoproteins	McNatt et al. 2009
APOBEC3	Mammals	RT/integration	Reverse-transcribing viruses	Vif, Bet	Sawyer et al. 2004; Zhang and Webb 2004
SAMHD1	Mammals	Genome replication	Lentiviruses (others?)	Vpx, Vpr	Laguette et al. 2012; Lim et al. 2012a
Schlafen 11	Mammals	Translation	HIV, others?	unk	
Flv	Mice	Replication	Flaviviruses	unk	
viperin	Mammals	?	Various	unk	Lim et al. 2012b
MxA	Mammals	Viral proteins; Various?	Various	Various	Mitchell et al. 2012
Mov10	Mice	RNA	Retroviruses, retrotransposons	unk	
IFITMs	Mammals	Fusion/entry?		unk	
Fv1	Mice	Post-entry	Retroviruses	unk	Qi et al. 1998; Yan et al. 2009
Fv4/Akvr	Mice	Entry	MLV	unk	
enJSRV	Sheep	Assembly?	JSRV, ENTV	unk	
Restriction/methylation	Bacteria	DNA	DNA phage	unk	
ev-3, ev-6, ev-9	Chickens	Entry	Avian leukemia virus	unk	
CRISPRS	Bacteria	DNA?	Bacteriophage, T.E.s	unk	
RNAi	Plants, fungi, invertebrates	RNA	Various	Norovirus VP1; Cricket paralysis virus protein 1A	

T.E. - Transposable elements; unk - unknown

Alternatively, RF can be parsed according to the breadth of viral targets they encompass. Thus, there are those that have broad targeting properties, thereby encompassing many unrelated viruses (such as tetherin, which can restrict a variety of enveloped viruses as they bud from the plasma membrane); specialized factors being those that appear to be viral genus or viral species specific (e.g., Fv4 and enJSRV) and those that fall somewhere in between, demonstrating activity either for many viruses within a particular family, such as Flv (flaviviruses) and TRIM5 (retroviruses), or for distantly related viruses that retain a shared replication mechanism, such as APOBEC3G (which has been shown to restrict retroviruses, hepadnaviruses, and retrotransposons).

Above all else, there is the fact that the RF genes are not related to one another in terms of primary sequence, tertiary structure, or mechanism of action. However, there is at least one defining feature that consistently applies to most or all cellular restriction factors (RFs)—they bear distinctive hallmarks of natural selection and in particular positive selection. Thus, we can define cellular RFs as *adaptations of the host, naturally selected for the ability to interfere directly with viral replication at the level of the cell.*

Although it is conceivable that the RF genes have additional functions unrelated to inhibiting viral infection, for many of them, there are compelling reasons to believe that the primary or only function (that is, the evolutionarily selected function) is to defend the host cell from infection with pathogenic viruses. As will be discussed below, many RF genes display signatures of strong, positive selection, including greater-than-expected levels of non-synonymous divergence (between species) and elevated polymorphism within populations, consistent with genes directly involved in evolutionary conflict (as is expected of immune effectors that defend against viruses). Some RF genes also show considerable structural variation between species, as is the case with the IFITM, TRIM5, and APOBEC3 genes (Münk et al. 2012; Sawyer et al. 2007; Zhang et al. 2012). In some cases, there are species in which RF functions have been lost, for example, TRIM5 appears to be inactive in both canines and felines (McEwan et al. 2009; Sawyer et al. 2007). Owl monkeys retain a TRIM5 gene, but the entire SPRY domain has been replaced with a cyclophilin-A-like domain (Nisole et al. 2004; Sayah et al. 2004). Some species of Asian macaques also carry a TRIMCyp allele (Brennan et al. 2008; Liao et al. (2007); Newman et al. 2008; Virgen et al. 2008; Wilson et al. 2008b), and recent evidence suggests that previous TRIMCyp fusions arose and were subsequently lost during evolution of old-world primates (Malfavon-Borja et al. 2013). A significant number of humans are homozygous for an inactivating mutation in the TRIM5 RING domain (Sawyer et al. 2006). At least two subspecies of mouse lack a complete Fv1 gene (Qi et al. 1998), and some humans have a large deletion, resulting in fusion of the *APOBEC3A–APOBEC3B* genes (Kidd et al. 2007). While all of these patterns are fairly typical of genes encoding immune-related functions, such variation is unusual among proteins subject to purifying selection, such as those with essential or conserved functions.

2 How do Cellular Restriction Factors Evolve?

There is an abundance of evidence indicating that genes encoding many of the cellular RFs are subject to strong positive selection, a property they share with many other immunity-related genes, as well as genes involved in sexual reproduction. Such genes are often described as being engaged in evolutionary "arms races," a scenario in which genes co-evolve as a result of direct evolutionary competition. Thus, virus–host co-evolution unfolds as a series of adaptations (e.g., in host RFs) and counteradaptations (in viral proteins that interact with host RFs) (Meyerson and Sawyer 2011). In the case of RF genes, the selective advantage conferred on the host is resistance to viral infection and the deleterious effects of pathogenesis on fitness. For viruses, because existence depends absolutely on uninterrupted access to host cells, only those that effectively negotiate the barriers imposed by RFs will survive.

Evolutionary arms races are often depicted as a cyclical, back-and-forth process of evolution involving two co-evolving lineages—for example, two interacting genes involved in sexual reproduction, or a host defense gene and the viral gene whose protein it targets. Such models should not be taken literally, however, for two reasons. First, the rates at which virus and host populations change are drastically different. Relative to most hosts, viruses have very large effective population sizes, very short generation times, and in many cases highly error-prone genome replication, allowing viral lineages to adapt to host genetic barriers within the course of a single-host generation. Second, host evolution with regard to any one locus is slower by several orders of magnitude, and host adaptation to viruses is much more likely to involve many different kinds of loci, including those affecting behavior, physiology, and immunity. Thus, host RF genes that are subject to positive selection over long evolutionary periods (hundreds of thousands or millions of years) may in fact reflect selection not by a single pathogen (and its direct descendants), but by the whole spectrum of viral pathogens encountered by the host lineage across many generations.

As a consequence of this "antagonistic" mode of evolution, the most consistently observed property of RF genes is the enrichment of positively selected sites in the protein-coding sequence. Indeed, it has become increasingly common to analyze RF genes for tell-tale signs of positive selection in parallel with structural and functional characterization of the proteins they encode (specific examples from the literature are listed in the last column of Table 1). Because scanning sequences for evidence of positive selection is increasingly common in the literature on RFs, I will begin with a very brief overview of the methods involved.

2.1 Measuring Rates of Non-synonymous Substitution

The most common means for determining whether RF genes have evolved under positive selection is to ask whether fixation of non-synonymous mutations exceeds expected values. This can be done most easily for nucleotide substitution data, because the rate of synonymous variation can be taken as a measure of the background or neutral rate of substitution. If the rate at which non-synonymous substitutions are fixed in a population exceeds the expectation based on the fixation of synonymous changes (assumed to be neutral or nearly neutral), this is taken as presumptive evidence for positive selection. Thus, dN > dS (or dN/dS > 1) reflects positive selection (selection for change); dN = dS (or dN/dS = 1) reflects little or no selection; and dN < dS (or dN/dS < 1) reflects purifying selection (selection to eliminate deleterious changes). Such measurements are frequently referred to as dN/dS values (or sometimes as Ka/Ks) (Hurst 2002).

Traditional methods for estimating dN/dS were based on averaging the number of changes in each category across the length of divergent genes. Because even genes under strong positive selection will be composed of a mix of residues under purifying selection and residues under positive selection, and because the former tend to outnumber the latter (even in RF genes), detecting positive selection by methods that average along the entire protein sequence is very conservative. More recent methods developed for identifying individual residues under positive selection are much more sensitive (Yang and Bielawski 2000). Codon-based methods require two inputs: (1) a sequence alignment comprised of multiple orthologs of the RF gene of interest and (2) a phylogenetic tree describing the evolutionary relationship between the taxa represented in the alignment. While the alignment indicates sites of variation, the phylogenetic tree serves to determine the time and direction of sequence change—as a simple example, if an alignment has a mix of lysine (AAA) and arginine (AGA) codons at a particular position, the tree provides evidence for the most probable direction of change (was it AAA → AGA or AGA → AAA?).

Together, these inputs represent three inherent assumptions of the analysis: (1) that the sequences in the alignment are true orthologs; (2) that homologous codons have been correctly aligned to one another; and (3) that the tree represents an accurate depiction of the evolutionary relationships between the organismal taxa being investigated. Violations of these assumptions, including alignment of non-orthologous sequences, misalignment of codons, and incorrect phylogenies, can result in the appearance of increased variation and therefore false-positive identification of residues under selection. Most RF genes, by virtue of evolving under positive selection, are inherently variable, increasing the likelihood that mistakes in alignment or tree inference will confound the analysis. A significant number of RF genes also belong to larger gene families, and in several cases, individual loci have sometimes undergone lineage-specific expansions or contractions in copy number (Münk et al. 2012; Nisole et al. 2004; Sawyer et al. 2007; Sayah et al. 2004; Zhang et al. 2012). Thus, the assumption of orthology can also be tricky when analyzing RF genes for positive selection.

The dN/dS ratio is calculated by counting the number of observed non-synonymous and synonymous differences in a codon, then dividing by the number of non-synonymous and synonymous sites in that same codon. Here, the term "sites" refers to mutational opportunity. For example, there are nine single-nucleotide substitutions that will change the sequence of a given triplet codon; in the case of the lysine codon AAA, there is one synonymous substitution (changes the codon from Lys AAA to Lys AAG), whereas the other eight substitutions result in a different amino acid. Thus, there are $3 \times 1/9 = 1/3$ synonymous sites and $3 \times 8/9 = 8/3$ non-synonymous sites in the Lys (AAA) codon. Key modifications of the procedure are then incorporated to account for differences in rates of different kinds of substitutions (e.g., transitions versus transversions) and to correct for multiple hits (Emerman and Malik 2010; Hurst 2002; Yang 2006; Yang 2005; Yang and Bielawski 2000).

The application of dN/dS to RFs is especially informative when coupled with molecular assays to confirm the functional consequences of variation at the identified sites. A striking observation that has emerged from many such studies is that positively selected sites often define physically interacting domains between the RF and its viral-binding partners. In other words, selection is most apparent for those residues that are involved in recognition of viral targets or that themselves are bound by virus-encoded antagonists. For example, in the case of primate TRIM5α, sites with dN/dS > 1 cluster strongly in the V1 region of the SPRY domain, and experimentally introduced substitutions in and around these sites clearly alter target specificity (Sawyer et al. 2005; Yang and Bielawski 2000). Similarly, of six codons with dN/dS > 1 in murine Fv1, which restricts replication of murine leukemia viruses (MLVs) in a manner similar to TRIM5α, variation at two is known to be involved in discriminating between closely related strains of MLV (Yan et al. 2009).

Similar approaches can also help pinpoint potential sites of interaction between cellular RFs and virus-encoded antagonists. For example, molecular evolutionary analysis of the interferon-inducible MxA gene revealed a cluster of positively selected sites and helped identify a specific region (loop 4) as a primary virus interaction domain (Mitchell et al. 2012). Interestingly, in the case of SAMHD1 antagonism by lentiviral Vpx proteins, independent studies identified a different set of positively selected sites, in the N-terminal domain of SAMHD1 in one case and the C-terminal domain in the other (Laguette et al. 2012; Lim et al. 2012a). Specifically, Lim et al. found that at least two residues with dN/dS > 1 in the N-terminus affect degradation by lentiviral Vpx, while Laguette et al. identified a separate set of sites in the C-terminal domain that affected binding and degradation. The conflicting results could mean that the independent sites form part of a single binding domain in the folded protein, or that the Vpx homologues tested, despite being closely related, have evolved to target different regions of the protein. Importantly, the fact that the two studies did not identify the same constellation of positively selected sites (despite using very similar approaches) speaks to the need for experimental confirmation, whenever possible, of sites identified by in silico analysis.

One of the most striking examples is tetherin (also known as BST2) (Neil et al. 2008; Van Damme et al. 2008). In the laboratory, tetherin has demonstrable restriction activity against a broad range of unrelated enveloped viruses (Evans et al. 2010). As a consequence, it is not surprising that scans of the tetherin-coding sequence reveal multiple discrete patches of positive selection (McNatt et al. 2009).

In fact, several unrelated tetherin antagonists have been identified, including cytosolic proteins (e.g., the K5 protein of KSHV and the Nef proteins of certain primate lentiviruses) and membrane proteins (e.g., the Vpu protein of HIV-1 and the HIV-2 and Ebola virus glycoproteins) (Evans et al. 2010; Hauser et al. 2011; Le Tortorec and Neil 2009; Neil and Bieniasz 2009). Thus, the abundance and variety of tetherin antagonists is consistent with the distribution of positive selection across the protein sequence.

At least one study indicates that the tetherin molecule can tolerate an extraordinarily high level of primary sequence variation and still preserve function, as long as secondary structural features are maintained (Perez-Caballero et al. 2009). This result raises the possibility that the tetherin sequence is less constrained than other RFs. Thus, in the case of tetherin, it may be especially important to distinguish the effects of positive selection from the effects of relaxed purifying selection (both of which can lead to increased fixation of non-synonymous substitutions). This is a case for which experimental corroboration was therefore essential in identifying specific, positively selected sites involved in viral antagonism.

2.2 Limitations of dN/dS Analysis

Detection is more difficult when positive selection results in fixation of other classes of sequence change, such as insertions, deletions, recombination events, structural variation, or mutations in non-coding regions. Statistical methods for analyzing these kinds of changes are still being developed (for example, see (Haygood et al. 2010; Wong and Nielsen 2004) and have not been widely applied to the analysis of cellular RFs. A few examples of potentially altering non-coding sequences have appeared in the literature on RFs. For example, Diehl et al. surveyed insertions in the introns of primate TRIM5 and found an elevated rate of fixation of inserted LTR elements in the first and fourth introns, suggesting that these insertions may have been selected for altered regulation or transcription of the TRIM5 gene (Diehl et al. 2013). As another example, the novel TRIMCyp fusion protein found in several species of Asian macaques (*Macaca sp*) represents the *de novo* evolution of a new exon, a process that required two steps that were almost certainly selectively advantageous: the line-mediated insertion of a CypA reading frame downstream of the TRIM5 gene and elimination via a single-point mutation of the 3′ splice acceptor site for exon 7 of TRIM5 (Brennan et al. 2008; Brennan et al. 2007; Newman et al. 2008). The former created the coding sequence for a new exon, while the latter presumably permitted splicing to bypass exon 7 and exon 8 (which has subsequently accumulated missense mutations) and splice in-frame to the CypA insertion. Together, these mutations (one insertional and one a non-coding substitution) were likely subject to strong positive selection.

The SPRY domain of some TRIM5α orthologs in primates has differences in length due to expansions and contractions of short segments of sequence (Song et al. 2005). These regions overlap with clusters of positive selection, and site-directed mutagenesis indicates that determinants of target specificity can be found within the expanded sequences (Sawyer et al. 2005; Song et al. 2005; Wilson et al. 2008a). Thus, it is reasonable to assume that positive selection has at times favored changes in sequence due to insertions/duplications of sequence. Importantly, length variation and overlapping non-synonymous substitution make these regions especially prone to errors in sequence alignment and could artificially enhance dN/dS values for codons within these segments.

Claims of positive selection based on dN/dS of whole genes or individual codons implicitly assume that evolution proceeds in successive waves of mutation, selection, and fixation. A caveat to many studies is that each taxon (or species) is typically represented by a single allele, either cloned by RT-PCR or recreated based on a genome assembly. RF genes, like many immune system effectors, tend to be variable within species (polymorphism), a consequence of the same selective forces that cause them to be among the most variable genes between species (divergence). Much of the RF literature refers to "species-specific" restriction profiles, often based on testing of a single, representative ortholog. However, in cases where more extensive within-species sampling has occurred, it is not unusual to find significant non-synonymous polymorphism. For example, non-synonymous polymorphism in RFs is found in rabbits (TRIM5), mice (Fv1, APOBEC3), humans (TRIM5, APOBEC3G), and old-world monkeys (TRIM5, APOBEC3G, tetherin)—in most of these cases, allelic differences have been associated with functional differences in target specificity (Goldschmidt et al. 2006; Newman et al. 2006; Sawyer et al. 2006; Speelmon et al. 2006; Yan et al. 2009). For example, a single amino acid difference between the $Fv1^{n}$ and $Fv1^{b}$ alleles determines patterns of restriction of MLV strains, and the TRIM5; likewise, APOBEC3G loci of rhesus macaques are both highly polymorphic, and specific polymorphisms can be associated with differences in targeting (in the case of TRIM5) and susceptibility to Vif (in the case of APOBEC3G) (Bishop et al. 2001; Kirmaier et al. 2010; Weiler et al. 2006; Krupp et al. unpub).

Interestingly, the distribution of polymorphic sites in an RF gene can also be indicative of a special form of positive selection called balancing selection (Charlesworth 2006). In the case of the rhesus macaque TRIM5 locus, clustering of non-synonymous polymorphisms within the coiled-coil and SPRY domains overlapped with sites of $dN/dS > 1$ was tightly linked to silent polymorphisms and included trans-species polymorphisms (cases in which both alleles are also present in another, related species) (Newman et al. 2006). These patterns—clustering of high-frequency non-synonymous and synonymous differences and sharing of alleles between divergent species—are expected of loci under long-term balancing selection, which occurs when a combination of selective forces acts to simultaneously preserve multiple alleles in a population over long periods of time, for example by virtue of heterozygote advantage (Charlesworth 2006). The presence of these patterns, together with the discovery of a TRIMCyp fusion allele in multiple species of macaque, raises the possibility that the ancestors of extant macaques were subject to frequent, possibly concurrent, episodes of positive selection incurred by unknown retroviruses (Newman et al. 2006, 2008).

3 What are the Evolutionary Origins of Restriction Factors?

The collection of genes that qualify as cellular, antiviral RFs is diverse, and aside from positive selection, no single defining property applies to all RFs. However, signatures of positive selection alone do not immediately discriminate between RFs and other classes of genes that are also engaged in so-called antagonistic evolutionary arms races (e.g., genes involved in sexual reproduction), and additional lines of evidence, such as laboratory demonstration of antiviral activity, are required to identify a *bona fide* RF. Which raises the question of whether there are properties, or combinations of properties, that could guide the discovery of new RF genes or, at a minimum, provide a list of likely candidates? Perhaps more importantly, can patterns be discerned among the known RFs that can lead to theories about where and how new RFs evolve?

As more RFs are discovered and characterized, it may be that some properties are frequently associated with restriction. For example, a significant number of restrictions involve targeting of viral nucleic acid, a feature shared with many innate pattern recognition receptors. Examples include the proteins of the APOBEC3 cluster, the bacterial endonucleases, CRISPRS, AGO, ADARS, and MOV10. Not surprisingly, many RFs are also interferon-stimulated genes (ISGs), including SAMHD1, tetherin, MxA, viperin, and the IFITMS (Neil and Bieniasz 2009; Schoggins et al. 2011); some reports suggest that IFN treatment also induces expression of TRIM5 and APOBEC3 proteins. The overlap with properties of traditional innate immune factors, including detection of nucleic acid and IFN regulation, suggests that this pool of genes may be an evolutionary source of new RFs or that selection acts to ensure that RFs often become embedded in existing immune regulatory networks.

Are some gene families or superfamilies more likely than others to produce RFs? Most of the genes of the APOBEC3 cluster have now been shown to have antiviral activity, and it may be that similar genes will eventually yield new RFs. Certainly, the TRIM family of genes (tripartite motif family) has yielded a surprising number of antiviral genes, including genes with direct effector function (e.g., TRIM5, TRIMCypA1, TRIMCypA2, TRIM28) and genes that involve either detecting of PAMPs or activation of antiviral responses (TRIM25). Many of the TRIM family members have no reported functions, and there may be additional RFs among them. The modularity of TRIM genes is striking: While they share the N-terminal TRIM motif (an RING domain, one or two B-boxes, and a coiled-coil), the C-terminal domains include a diverse array of unrelated protein structures. This may reflect evolutionary adaptability (the replacement of the TRIM5α SPRY domain by CypA domains is a striking example) and possibly structural properties that are especially useful for restriction.

By far, the most common source of cellular RFs is endogenous viral sequences that have been co-opted by evolution (exaptation) to defend against exogenous infection. At least three stages of the retroviral replication cycle are now known to be restricted by such elements: entry, early post-entry, and assembly. Among the best characterized of these is the Fv1 locus of mice. Fv1 was first defined genetically in murine models of retrovirus-induced cancer (Steeves and Lilly 1977). Mapping and subsequent cloning of Fv1 revealed it to be the remnants of an ancient retroviral insertion, encoding a portion of what was once the structural gene (*gag*) of a retrovirus (Best et al. 1996). Similar to TRIM5α-mediated restriction, expression of Fv1 in the cytoplasm interferes with infection by murine leukemia viruses at an early post-entry step in the retroviral replication cycle. Resistance to Fv1-mediated restriction maps to specific amino acids in the MLV Gag protein—suggesting that Fv1 inhibits viral replication by interacting directly with the capsid core of the virus. Surveys of all four subgenera of mice (genus *Mus*) revealed that the Fv1 insertion may date back as far as 7 million years, close to the estimated time of divergence between the four subgenera (Yan et al. 2009).

A second endogenous viral sequence mediates the Fv4 restriction of mice. In this case, the Fv4 gene (also known as *Akvr-1*) encodes the remnants of an MLV-related envelope glycoprotein (Dandekar et al. 1987; Gardner et al. 1991; Ikeda and Sugimura 1989). Expression of exogenous retroviral Env proteins often results in superinfection interference, a means of preventing reinfection of the same cell (Coffin 1997). Fv4 may work by the same mechanism, setting up a dominant interference that prevents exogenous viruses that utilize the same receptor from infecting. At least three related endogenous retroviral loci of domestic chickens, *ev-3*, *ev-6*, and *ev-9* have been reported to interfere with exogenous avian leukosis virus (ALV) infection through expression of viral envelope glycoprotein (Robinson et al. 1981), in a manner analogous to Fv4-mediated inhibition of MLV infection in wild mice. More recently, studies of the pathogenic jaagsiekte sheep retrovirus (JSRV) led to the discovery of endogenous JSRV loci in sheep that interfere with replication of exogenous JSRV by multiple mechanisms, either by blocking entry (similar to murine Fv4 and chicken *ev-3, ev-6,* and *ev-9*) or by dominant negative interference with virion assembly (Murcia et al. 2007; Mura et al. 2004; Arnaud et al. 2008).

What is the likelihood that other ERV loci function as host restrictions? Endogenous retroviral sequences abound in the genomes of all vertebrate organisms; the human genome and the genomes of related, non-human primates contain hundreds of thousands of ERV-related loci, and most mammalian species have numbers of similar magnitude (Gifford and Tristem 2003; Jern and Coffin 2008). The ERVs that serve as RFs of chickens, mice, and sheep were discovered fortuitously in the course of studying exogenous viruses in specific animal models of disease; it seems improbable that these will be the only examples, and it may be just a matter of time (and effort) before additional examples emerge. Conceivably, additional examples could be discovered based on a priori assumptions—maintenance of an intact reading frame for a viral gene and evidence for natural selection (either purifying or diversifying selection).

While the human genome is not known to encode completely intact proviruses, members of some human ERV families still have viral genes with intact reading frames. In most cases, it is not clear if these provide a cellular function. It may also be that the most likely source of ERV that still retains function (in humans or any species) will be found among more recent integrations. As with Fv4, however, the younger an insertion is, the less likely it is to be fixed in a population. Therefore, very recent insertions may not be well represented in whole-genome assemblies, particularly from genome projects based on a single individual or type specimen (humans are an exception).

In addition to the ERVs, it is now appreciated that many non-retroviral virus lineages are represented by genes or gene fragments resident in host germline DNA, although to a lesser extent than retroviruses (Belyi et al. 2010; Feschotte and Gilbert 2012; Gilbert and Feschotte 2010; Johnson 2010; Katzourakis and Gifford 2010; Liu et al. 2011). Among these, it is possible to identify some that have maintained intact or partially intact reading frames over long stretches of evolutionary time, raising the intriguing possibility that some of these may encode (or at one time encoded) cellular restrictions (Horie et al. 2010; Katzourakis and Gifford 2010; Kobayashi et al. 2011).

4 How do Viruses Adapt to Genetic Barriers Imposed by Restriction?

Mirroring the great diversity of forms and functions among the cellular RFs, the viral mechanisms for counteracting cellular restrictions are also numerous. Viral strategies for evading restriction can be placed into two different categories: (1) passive evasion, achieved through mutation and selection of resistant alleles in target molecules (analogous to the role of antigenic variation in immune escape) and (2) active evasion, achieved through expression of virus-encoded factors that inhibit RF activity.

4.1 Passive Evasion

The underlying mechanisms for evading restriction via sequence evolution are essentially the same as well-established mechanisms for immune evasion and drug resistance, the product of errors in genome replication, large effective population sizes (relative to hosts), and relatively short generation times (on the order of hours to days for most viruses). The topic of viral evolution and escape has been extensively reviewed and is beyond the scope of this chapter (for recent, comprehensive treatments see Domingo et al. 2012 and Domingo and Holland 2008 and references therein.

Viral escape from restriction through selection of mutations in the target can be informative, potentially aiding in the identification of viral targets of restriction and underscoring the significance of restriction in vivo. For example, a virally encoded antagonist of TRIM5α has not been reported, and patterns of retroviral susceptibility and resistance to TRIM5α proteins instead result from amino acid differences in the N-terminal domain of the retroviral capsid protein. Escape from TRIM5α-mediated restriction has been observed in vivo (Kirmaier et al. 2010) and during directed evolution in tissue culture (Ohkura et al. 2011; Pacheco et al. 2010), with changes seen in multiple, distinct domains of the viral capsids. Thus, it is not clear which structural elements of the capsid, or even which form of the CA protein (e.g., monomers or multimers), comprise the molecular pattern(s) targeted by TRIM5 proteins. Nonetheless, the amazing breadth of TRIM5α restriction (in some cases, individual orthologs have demonstrated activity against retroviruses from three or more genera) suggests that the unique fold of the retroviral capsid is somehow exploited as a particularly effective target. The evolutionary convergence of multiple proteins on this target, including Fv1, TRIM5α, TRIMCypA1, and TRIMCypA2, is a persuasive argument for both the conservation of the target (whatever that may be) and the fitness costs that must be associated with changes in retrovirus–capsid architecture.

4.2 Active Antagonism

Viral antagonists have been identified for many RFs (Table 1). Not surprisingly, RFs with broad activity against multiple viruses, such as tetherin or RNAi, are targeted by diverse antagonists (Evans et al. 2010; van Mierlo et al. 2012). The genesis of viral accessory genes, including those that function to antagonize RF function, occurs by a variety of mechanisms. In many cases, viral accessory genes are functionally modified homologues of host genes, presumably derived by capture of sequences from a host cell—this is particularly evident for large DNA viruses, such as herpesviruses and poxviruses (Swain and Coffin 1992). Mechanisms of gene capture can vary, but probably include recombination, transposition, and, in the case of retroviruses, possibly capture of read-through mRNA from proviruses. These genes frequently display evolutionary plasticity, in many cases taking on additional functions or switching functions (neofunctionalization). For example, the envelope proteins of HIV-2 and Ebola virus, in addition to mediating viral entry, have anti-tetherin activity (Evans et al. 2010; Hauser et al. 2011). De novo evolution of anti-tetherin activity has been observed in vivo—in this case, replication of an *Nef*-deleted SIV in rhesus macaques selected for changes in the cytoplasmic tail of envelope that rescued the Nef defect by conferring the ability to antagonize host tetherin (Serra-Moreno et al. 2011).

New virally encoded antagonists can also arise by gene duplication followed by functional diversification, the *Vpr* and *Vpx* genes of primate lentiviruses being a prominent example. In the case of large DNA viruses, where the mutation rate is usually significantly lower than in RNA viruses, gene copy-number variation may in fact be a significant contributor to evolutionary plasticity, including adaptation to overcome cellular restrictions. Recently, mechanistic proof-of-principle for a gene expansion mechanism in poxviruses was demonstrated by passaging an attenuated vaccinia virus strain under conditions that selected for expansion and modification of the viral K3L gene (Elde et al. 2012).

5 Can Restriction Factor Loci Reveal the Impact of Viruses on Host Evolution?

The list of RFs in Table 1 is not meant to be comprehensive, but rather to convey the fact that the phenomenon of cellular restriction occurs throughout all domains of life. This is not surprising, given that modern methods of discovery have isolated or detected viruses using samples taken from virtually every kind of environment on the planet in which cellular organisms are found (Witzany 2012). That life evolved in concert with viruses is also apparent from organismal genome assemblies, where endogenized viral sequences, the remnants of ancient epidemics, and outbreaks are found by the thousands (Feschotte and Gilbert 2012; Gifford and Tristem 2003; Jern and Coffin 2008; Johnson 2010). It is difficult to imagine that viruses have not had a major influence on organismal evolution across all domains and their constituent taxa. It is conceivable, by combining molecular virology, evolutionary biology, and population genetics, that we can now begin to ask specific questions about when, where, and how viruses have had a specific role in our own evolution and the evolution of our non-human relatives. Because RF gene sequences are exquisitely sensitive to selective pressures imposed by viruses, their analysis is highly relevant to such studies. While the idea is fairly new, some examples will help illustrate the potential of RFs to be exploited for this purpose.

5.1 Positive Selection as Evidence for Specific Virus–Host Interactions in the Past

When positively selected sites correlate with targeting by a specific viral antagonist, it is common to conclude that past selection on the sites was driven by a related virus (Patel et al. 2012; Compton and Emerman 2013; Lim et al. 2010a, 2012b; Patel et al. 2011). For example, although the MAVS protein is not considered an RF per se, its central role in the innate effector response to viral infection makes it a likely target for viral antagonism. Indeed, Patel et al. found that the target of the hepatitis C virus NS3/4a protease in MAVS is also under strong positive selection in primates and that naturally occurring variation at these sites gives differential sensitivity to HCV NS3/4A and to homologous proteases from other hepaciviruses (Patel et al. 2012). Such results can be viewed as evidence that hepaciviruses were present during evolution of modern old-world primates.

While such hypotheses are reasonable, the possibility remains that suitable surfaces for antagonistic interactions may be limited to certain structures or by conformational accessibility and that unrelated viral antagonists may therefore share similar or overlapping binding sites as a result of convergent evolution. For example, given the clear convergence of multiple antagonists on specific regions of tetherin (Evans et al. 2010), it cannot be formally proven that other, unknown viruses may have exploited the same or closely overlapping targets. Likewise, while positive selection itself can operate directly on contact residues, it can also work by allosteric mechanisms. In the latter case, the same sites may be selected in response to different antagonists, even if their contact residues or binding sites differ. A striking example, in which positive selection worked on spatially distinct (yet interacting) residues, comes from a structure–function analysis of the interaction between cellular PKR and the vaccinia virus antagonist K3L (Elde et al. 2008). If convergence turns out to be common, correlations between positively selected sites and activity of a particular antagonist cannot alone prove that a particular kind of virus was responsible for selection. Structural data for RF antagonist interactions, as they become available, will hopefully provide added resolution to evolutionary reconstructions.

5.2 Connecting RF Evolution with Specific Viral Families

Based on current understanding of certain RFs, and in some cases, the corresponding viral antagonists, it may be possible to infer a role for specific types of viruses in the evolution of host lineages. Retroviruses are especially attractive for these kinds of studies for at least two reasons. First, due to the use of retroviruses as model systems for the past century, a great deal is known about the genetics of susceptibility to infection, and multiple cellular RFs have been identified and characterized (Table 1). Second, the presence of thousands of endogenized retroviruses in mammalian genomes can provide a second, corroborating source of evidence (Emerman and Malik 2010; Feschotte and Gilbert 2012; Patel et al. 2011). Some examples drawn from the literature on retroviral restriction serve to illustrate the potential of RF as markers of past virus–host interactions.

In the case of retroviral restriction, evolution has produced RFs that target at least four stages of the replication cycle, including entry (Fv4 in mice; *ev-3, ev-6, and ev-9* of chickens; *enJS loci* of sheep); post-entry (Fv1, TRIM5, TRIMCypA1, TRIMCypA2); reverse transcription (APOBEC3); and assembly (enJS56A1). Notably, the early post-entry stage of the retroviral replication cycle is the target of three completely unrelated protein domains: the SPRY domain of mammalian TRIM5α proteins; the murine Fv1 protein; and in a spectacular instance of convergent evolution, the CypA domains of the independently formed owl monkey and macaque TRIMCypA fusion proteins (TRIMCypA1 and TRIMCypA2, respectively). It is also noteworthy that these factors are widely distributed in terms of age and taxa—TRIM5 and APOBEC3 probably trace back to a common ancestor of mammals, and evidence suggests that both have functioned as RFs more or less continuously throughout mammalian evolution. Fv1 is found in at least three subgenera of mice (genus *Mus*) and probably originated by retroviral insertion < 7 million years ago. Fv4, the ev loci of chickens, and the enJSRV loci of sheep are all much more recent, species-specific RFs. The two TRIMCyp genes evolved independently, one in new-world primates (*Aotus* sp) and the other in Asian macaques (*Macaca* sp). From these observations, we can glean two points of significance: First, resistance to retroviruses is a frequently selected for phenotype in evolution and second, the initial stages of replication (entry and early post-entry) may be particularly effective as targets for restriction.

5.3 Extinct Viruses

Several studies have used endogenous retrovirus sequences as "representatives" of ancient exogenous infections, in some cases even resurrecting ancient viral proteins in the laboratory, to ask whether a particular extinct virus has tropism for cells of extant species or to infer a role for an RF in protecting ancestral populations from the virus (Dewannieux et al. 2006; Jern et al. 2007; Kaiser et al. 2007; Lee and Bieniasz 2007; Perez-Caballero et al. 2008). For example, viruses bearing resurrected capsid proteins from extinct primate retroviruses have been tested for restriction by human TRIM5α (Kaiser et al. 2007; Perez-Caballero et al. 2008). Similarly, reconstructed ERV envelope proteins have been used to identify and study the cellular receptors of extinct primate retroviruses (Soll et al. 2010). An important caveat to such studies is the degree to which ERVs are faithful representatives of past infections. Given the stochastic nature of ERV formation, and the likelihood that selection and random genetic drift eliminate the vast majority of integrations before they can become fixed elements in the genome, it is probably the case that ERVs are a highly skewed, underrepresentative sampling of all past exogenous infections.

Nevertheless, the presence of ERVs constitutes definitive evidence that a related, exogenous virus was able to infect an ancestral organism, and the sequences of such loci can sometimes reveal patterns consistent with susceptibility to restriction. For example, inspection of endogenous sequences in the mouse genome related to a particular class of murine leukemia virus revealed patterns highly consistent with APOBEC3-mediated hypermutation, including an excess of G-to-A mutations and a gradient of mutation from the 5′ to the 3′ end of the genome, strong indications that the infectious, replicating virus that generated the ERV insertions was sensitive to APOBEC3 or an APOBEC3-like factor (Jern et al. 2007). Similar conclusions were obtained by inspection of a subset of endogenous, gammaretrovirus-like capsid sequences from old-world primates (Perez-Caballero et al. 2008)

5.4 Evidence for Recent, Virus-Driven Evolution of Modern Species

While codon-based methods for detecting positive selection are powerful, there are limitations. In particular, such methods are not readily applied to the detection of recent episodes of natural selection, for example, selection occurring within a single population or among closely related populations. DN/dS analysis, for example, is best suited for identifying positive selection representing evolution on the scale of millions of years and thus is not as useful for identifying genes at the center of more recent episodes of positive selection (for example, events well within the last 1 million years). There are now a variety of approaches being developed for this purpose, most of which are based on distribution of allele frequencies or the predicted effects of selection on neighboring, linked loci (Andersen et al. 2012; Grossman et al. 2010; Sabeti et al. 2002; Zhao et al. 2012). Although these have not yet been widely applied to the study of cellular RFs, there are studies that have successfully applied such approaches to other kinds of loci involved in susceptibility to infectious diseases, including HIV-1 (Sabeti et al. 2005), malaria (Sabeti et al. 2002), and Lassa Fever (Andersen et al. 2012). Such methods could be useful for de novo discovery of candidate RFs or for assessing the impact of selection on evolution of an RF locus and linked genes.

In cases where a localized viral outbreak is well documented and intergenerational samples from the host population are available, it may be possible to use simple genotype frequencies as an indicator of selection (Sullivan et al. 2001). Although not a viral disease, historical outbreaks of Kuru in Papua New Guinea provide a very interesting case in point. Kuru is a transmissible prion disease, and among the Fore people, transmission was correlated with the custom of feasting on brain of deceased relatives. Heterozygosity at position 129 (M/V) in the human prion protein gene (PRNP) has been associated with resistance to prion diseases (Collinge et al. 1991; Palmer et al. 1991). Mead and colleagues discovered that heterozygosity at position 129 in elderly members of the Fore, who would have been alive before the custom of mortuary feasts ended, deviates significantly from Hardy–Weinberg equilibrium (HWE), whereas younger generations of the Fore, and other populations that did not suffer epidemics of Kuru, do not deviate from expected values (based on allele frequencies) (Mead et al. 2008).

There are some reports of recent, retrovirus-driven selection operating on human RFs. For example, Sawyer et al. reported that the human *TRIM5* locus has a lower-than-expected level of polymorphism in the C-terminal portion of the gene (encoding the viral-targeting domain) suggestive of a selective sweep in recent human history (<500,000 years); the authors suggest that this event could have contributed to the modern HIV-1 epidemic by virtue of limiting functional diversity of human TRIM5 (Sawyer et al. 2006). The application of more sophisticated methods (Andersen et al. 2012; Grossman et al. 2010; Sabeti et al. 2002; Zhao et al. 2012), combined with increased understanding of RF biology, should allow more precise evaluation of the impact of viruses on extant host populations, including humans.

6 What is the Biological Significance of Viral Restriction in Nature?

Demonstrable antiviral activity in vitro or in model organisms, taken together with phylogenetic evidence of long-term evolutionary conflict, can be taken as strong evidence that RFs function primarily to suppress or prevent the trafficking of viruses between hosts. Despite the extensive work and associated publications focusing on individual RFs and their mechanisms of action, surprisingly little attention has been paid to the role(s) RFs may (or may not) play in outbred populations with regard to the spectrum of viruses that organisms encountered in their natural ecological environments. Nonetheless, we can consider three situations in nature where RFs might be expected to exert significant influence:

1. *Persistence of endemic viruses in their natural hosts.*
2. *Cross-species transmission and emergence in new host populations.*
3. *Expression and reinsertion of endogenous viruses in the host genome.*

These are not mutually exclusive, and it is reasonable to ask whether any given RF might play a role in some or all of these scenarios. While experimental evidence exists in support of all three situations, when the question is framed in the larger context of host evolution, the most compelling case can be made for the second scenario. The first is also likely, but much more difficult to detect. For a variety of reasons (discussed below), the third seems unlikely as a general role for RFs, although it is possible that specific types of RF may serve this function under the right circumstances.

6.1 Endemic Viruses and Persistence in Natural Hosts

The extent to which a given RF could influence spread of a pathogen once it has become established in a population is not clear and may be difficult to prove. In the case of positive interactions with cellular co-factors, purifying selection acts on the virus population to preserve sequences involved in maintaining the interaction. Similarly, in the case of a virus that is well adapted to its host population, purifying selection should work to eliminate changes that result in gain-of-sensitivity to RFs. In fact, candidate gene studies in human HIV/AIDS cohorts have found only weak or no evidence of a correlation between variation in human RF genes, including TRIM5 and APOBEC3G, and set-point viral replication levels in patient plasma (e.g., An et al. 2004; Goldschmidt et al. 2006; Javanbakht et al. 2006; Speelmon et al. 2006).

Like HIV-1 in humans, SIVmac strains are well adapted for replication in rhesus macaques. The most widely used strains, SIVmac239 and SIVmac251, give a highly reproducible infection in experimentally infected animals, and a majority of animals progress to AIDS in less than 2 years (Desrosiers 1990; Johnson et al. 2003). Functional variation in rhesus macaque immune effectors, including APOBEC3 and TRIM5, has very weak impact on SIVmac viremia (Lim et al. 2010b; Weiler et al. 2006). In contrast, the related SIVs of sooty mangabeys (SIVsm), which have little or no history of adaptation to macaques, are sensitive to multiple macaque RFs (described below). These observations are consistent with the idea that once a virus has adapted to the host population, the impact of restriction may be minimal.

Surprisingly, restriction-sensitive mutations in HIV-1 have been detected in vivo, including APOBEC3-sensitive mutations in Vif and TRIM5-sensitive mutations in capsid (Battivelli et al. 2011; Simon et al. 2005). It is possible that such mutations may increase in frequency, if they are selected in response to other pressures, such as the need to escape from virus-specific antibodies or cytotoxic T lymphocytes. For example, Battivelli et al. found that naturally occurring CTL escape mutations also affect sensitivity to human TRIM5α in vitro (Battivelli et al. 2011). Simon et al. found significant numbers of Vif-defective variants in infected patients, raising the possibility that incomplete antagonism of APOBEC3 proteins might be of benefit to the virus (for example, by enhancing the overall mutation rate) (Simon et al. 2005). Interestingly, although replication of SIVmac239 Nef-deleted mutants is attenuated in vivo, viral replication is sufficient to generate tetherin-resistant mutations in the viral *env* gene (Serra-Moreno et al. 2011), and mice expressing restrictive APOBEC3 alleles can nonetheless be infected with APOBEC3-sensitive viruses (Okeoma et al. 2007; Ross 2009; Santiago et al. 2008). Together, such studies suggest that sensitivity to a single restriction is not necessarily an absolute barrier to viral replication, a situation that is most likely to occur in the case of viruses that are otherwise well adapted to a host population. In the case of APOBEC3G, it has even been hypothesized that weak sensitivity could play in favor of the virus by allowing small increases in mutation rate (Malim 2009; Simon et al. 2005). Hypothetically, selection for weak-to-moderate restriction could be beneficial to a virus, if it moderates virulence and increases the probability of viral transmission. Further studies, of human cohorts or animal models, are still needed to establish the extent to which any given RF is biologically significant in the context of viruses that are well established in a host population.

6.2 Cross-species transmission and emergence of novel pathogens in a new host population

Although a role for RFs in suppressing viral replication during cross-species transmission can be inferred from laboratory experiments and molecular evolutionary analyses, the influence of RFs on emergence has not been well studied, particularly in outbred populations where genetic variation would be expected to play a role.

Perhaps, the most direct data come from studies of rhesus macaques involved in preclinical AIDS research using the pathogenic AIDS virus, SIVmac. In the 1970s, SIVmac emerged in captive colonies of rhesus macaques in the United States. The origin of the virus was discovered to be endemic SIV of sooty mangabeys (SIVsm) that was accidentally introduced into macaques, probably sometime in the 1960s or early 1970s (Apetrei et al. 2004; Marx et al. 2004). Isolation and cloning of SIVmac in the 1980s lead to development of the SIV/macaque model for studying AIDS and AIDS vaccines (Desrosiers 1990). Importantly, SIVsm and SIVmac infections in rhesus macaques can also serve as a model system for understanding the impact of restriction on interspecies infection, adaptation, and emergence of a viral pathogen.

Recently, genotyping of rhesus macaque colonies revealed that this species is polymorphic in multiple RF loci, including TRIM5, APOBEC3G, and *tetherin* (McNatt et al. 2009; Newman et al. 2006; Weiler et al. 2006). Kirmaier et al. reported that SIVsm strains are sensitive to restriction by the most common alleles of rhesus TRIM5 and that emergence of SIVmac in these animals therefore required evolution of TRIM5-resistant capsid proteins (Kirmaier et al. 2010). Interestingly, these authors also reported a strong correlation between TRIM5 genotype and viral replication levels in vivo and documented successful adaptation of SIVsm by retrospective analysis of samples from experimentally infected rhesus macaques. A similar situation is found with the APOBEC3G locus of rhesus macaques; in this case, high-frequency alleles of rhesus A3G are resistant to degradation by the Vif protein of SIVsm, whereas the Vif protein of SIVmac has evolved compensating mutations that confer the ability to degrade rhesus APOBEC3G alleles (Krupp et al. unpub). Taken together, these results indicate that the emergence of SIVsm as SIVmac in macaque colonies was coupled to adaptive evolution to overcome barriers presented by at least two RFs, TRIM5, and APOBEC3G.

To date, there is still no tractable animal model for HIV-1 infection and AIDS that recapitulates AIDS in all its parameters. The finding that HIV-1 is sensitive to the rhesus macaque homologues of APOBEC3G, TRIM5α, and tetherin has spurred attempts to generate rationally designed, simian-tropic HIV (stHIV) that circumvents these genetic barriers (Ambrose et al. 2007). Compared to the apparent ease with which SIVsm and SIVmac, including attenuated strains such as SIVmac239deltaNef, adapt in rhesus macaques, the difficulties in generating macaque models of HIV-1 infection indicate that the presence of multiple restrictions, perhaps acting in concert, can significantly increase the barrier to successful cross-species transmission and emergence. In fact, the degree to which a virus must be "preadapted" to the genetic environment of a new host in order to successfully emerge is an important, unanswered question in viral ecology (Holmes and Drummond 2007), and the specific influence of host-encoded restrictions on this process has not been well studied.

The capacity of a virus to spread in a new population of hosts can be represented as a function of the reproductive ratio (R_0), being the average number of new secondary infections that result from a single infected individual (Heffernan et al. 2005; Nowak and May 2000). Typically, it is assumed that viruses with $R_0 < 1$ cannot successfully emerge in the new host population. Antia et al. modeled initial transmission events and concluded that even viruses with $R_0 < 1$ should be able to initiate successful emergence, but the frequency of emergence is sensitive to stochastic factors that influence ability to adapt during the initial stages of infection and spread and depends on mutation rate (μ) to the power of the number of evolutionary steps required to achieve $R_0 > 1$ (Antia et al. 2003). The effect is predicted to be most pronounced either if adaptation is multistep (requires multiple simultaneous or near-simultaneous mutations) or if the pathogen must pass through a fitness valley in transition from $R_0 < 1$ to $R_0 > 1$ (Antia et al. 2003). Although speculative, this suggests a potential role for RFs at the interface between natural hosts (reservoirs) and emergent hosts (new hosts), whereby frequent viral incursions into a new population could provide recurring selection, including positive selection on RF genes. This could take place, for example, when hosts sharing overlapping ranges come into frequent physical contact, which occurs during fights over territory (leading to blood-borne exposure); when one host regularly preys on a reservoir species (and is thus exposed to virus endemic to the prey); or in situations where ecological change leads to an increase in the incidence of cross-species exposures. This would involve those RFs, acting alone or in combinations, that decrease the ability of the nascent pathogen to maintain uninterrupted chains-of-transmission in the new host, perhaps by increasing the number of mutational steps required by the virus. Interestingly, Zhao et al. have surveyed human populations in central West Africa, where zoonotic exposure to simian immunodeficiency viruses is probably common, and found signs of recent positive selection operating on several genes, including Cul5 (a binding partner of APOBEC3G) and TRIM5 (Zhao et al. 2012).

6.3 Amplification of Endogenous Viruses in the Host Genome

Several authors have suggested that RFs may serve to suppress transposition or replication of endogenous viral elements already present in the host genome (Sawyer et al. 2005; Schumacher et al. 2008). Demonstration of inhibitory activity in tissue culture systems would seem to support this hypothesis; however, while the measured activity is real, the relevant question is whether such interactions occur in nature, and if they do, are they sufficient to explain adaptive evolution of the RF? As an extreme example, human APOBEC3G can inhibit replication of Ty1 elements in yeast (Schumacher et al. 2005), but it is unlikely that humans maintain a functional APOBEC3G as a defense against fungal retrotransposons—in other words, observed activity is not sufficient to identify an evolutionary cause.

It is helpful to first consider the stages involved in genesis of an endogenous virus in the germline from a population–genetics perspective. The initial invasion of the germline with an endogenous virus requires insertion of a provirus into germline DNA (by infection of germline tissue) (Lock et al. 1988), which may be in and of itself a low-probability event (and certainly not given for any particular retrovirus). Thus, a great many ancient retroviruses may not even be represented among genomic ERV loci. At this point, the provirus represents a single, insertional mutation, subject to the same range of stochastic factors that determine whether any new mutation is lost or fixed in a population. For neutral insertions, loss is far more likely than fixation. If the insertion is deleterious to the host, loss may be hastened by negative selection. Thus, beginning at the moment of provirus formation, it is likely that there is strong selection in favor of insertions that are already defective or minimally deleterious to the organism. At this point, the provirus is still a minor allele in the host population (the uninterrupted chromosomal site being the major allele) and therefore subject to the vagaries of random genetic drift—a process that more often than not will lead to eventual loss of the provirus in most cases. So, only after many generations (possibly hundreds or thousands), will a small minority of proviruses actually become fixed in the population.

Throughout this period, the proviruses are subject to the same random mutational events that act on all genes, including nucleotide substitutions and recombinational events—and in the case of ERV, recombination between the LTRs often leads to loss of the intervening viral sequences (Jern and Coffin 2008; Johnson and Coffin 1999). Given all these immediate effects of selection operating directly on retroviral sequences throughout the long process of endogenization, it is difficult to imagine that ERVs as a group exert stronger selective pressure on RF genes than the intensity of selection imposed by exogenous, pathogenic viruses. This is especially true when the slow, stochastic process of ERV formation is contrasted with the relatively immediate effects of exogenous viral outbreaks, which will influence large fractions of a host population more or less simultaneously (even within a single generation) and give rise to repeated cycles of infection in a relatively short evolutionary time span—in some cases, may be even imposing a selective sweep (Sawyer et al. 2006). Finally, RF may play an important role during the period of exogenous spread that precedes endogenization, such that those ERV sequences that ultimately take up residence in the germline may have already passed through the selective filters established by host-encoded restrictions—in which case, phenotypic properties of the ERV may be quite different from those of the original, exogenously replicating virus. Thus, there is a good chance that the demonstrated activity of individual RFs against endogenous viruses is a holdover from a time when the virus circulated in exogenous form (i.e., not because the activity has present-day function against ERV) or that the activity simply reflects chance similarities between the endogenous viral targets being tested and the exogenous targets that drive evolution of the RF in nature.

7 Summary

The fact that so many diverse genes are dedicated to defending the host cell against viruses is compelling evidence that viruses have had a major impact on the course of organismal evolution. By virtue of evolving at the physical interface between viruses and hosts, cellular RFs are exquisitely sensitive indicators of virus–host co-evolution. Pioneering work thus far has focused largely on reconstructing deep evolutionary history of RFs, on the order of millions of years, to generate testable hypotheses that can be addressed, at least in part, using laboratory experiments. Importantly, the ability to study virus–host interactions as they unfolded over long stretches of evolutionary time and in outbred populations will help us better predict how modern epidemics will ultimately play out. An important direction for the study of RFs is to also understand how they influence the trafficking and evolution of modern viruses—that is, how do they govern such processes as interspecies transmission, adaptation, and emergence of new infectious diseases? Going forward, the identification and study of cellular RFs, as well as other related components of intrinsic and innate immunity, will necessarily involve the intersection of molecular virology, evolutionary biology, and immunology and should be solidly grounded in the framework of population–genetics theory.

7.1 Some Unanswered Questions

- Do specific patterns of evolution of RF genes correlate temporally with major phylogeographic transitions or phylogenetic events, such as major migrations, formation of land bridges, exchange of fauna, and speciation events?
- What are the respective contributions of positive acting co-factors and inhibitory restriction factors to cross-species transmission and emergence of viruses?
- When and how do cellular RFs originate?
- Does positive selection influence regulation of RF genes?
- How do cellular RFs become embedded in existing gene regulatory networks?
- Are there criteria that can be used for identifying new, candidate RFs in the absence of virological data?
- Where do viral antagonists come from (particularly genes like Nef, Vpr, Vpx, Vif, with no obvious homologues in nature)?
- Do accessory genes of viruses evolve more rapidly than other parts of the genome? Does the rate of accessory gene evolution differ between natural reservoirs and new, emergent hosts?
- Given that many viruses actively antagonize innate effectors, do accessory genes of unrelated viruses exhibit "cross talk," complementing one another during co-infection?

References

Ambrose Z, KewalRamani VN, Bieniasz PD, Hatziioannou T (2007) HIV/AIDS: in search of an animal model. Trends Biotechnol 25:333–337. doi:10.1016/j.tibtech.2007.05.004
An P, Bleiber G, Duggal P et al (2004) APOBEC3G genetic variants and their influence on the progression to AIDS. J Virol 78:11070–11076. doi:10.1128/JVI.78.20.11070-11076.2004

Andersen KG, Shylakhter I, Tabrizi S et al (2012) Genome-wide scans provide evidence for positive selection of genes implicated in Lassa fever. Philos Trans R Soc Lond B Biol Sci 367:868–877. doi:10.1098/rstb.2011.0299
Antia R, Regoes RR, Koella JC, Bergstrom CT (2003) The role of evolution in the emergence of infectious diseases. Nature 426:658–661. doi:10.1038/nature02104
Apetrei C, Robertson DL, Marx PA (2004) The history of SIVS and AIDS: epidemiology, phylogeny and biology of isolates from naturally SIV infected non-human primates (NHP) in Africa. Front Biosci 9:225–254
Arnaud F, Varela M, Spencer TE, Palmarini M (2008) Endogenous retroviruses. Cell Mol Life Sci 65:3422–3432. doi:10.1007/s00018-008-8500-9
Battivelli E, Migraine J, Lecossier D et al (2011) Gag cytotoxic T lymphocyte escape mutations can increase sensitivity of HIV-1 to human TRIM5alpha, linking intrinsic and acquired immunity. J Virol 85:11846–11854. doi:10.1128/JVI.05201-11
Belyi VA, Levine AJ, Skalka AM (2010) Unexpected inheritance: multiple integrations of ancient bornavirus and ebolavirus/marburgvirus sequences in *Vertebrate genomes*. PLoS Pathog 6:e1001030. doi:10.1371/journal.ppat.1001030.t006
Best S, Tissier PL, Towers G, Stoye JP (1996) Positional cloning of the mouse retrovirus restriction gene Fvl. Nature 382:826–829. doi:10.1038/382826a0
Bishop KN, Bock M, Towers G, Stoye JP (2001) Identification of the regions of Fv1 necessary for murine leukemia virus restriction. J Virol 75:5182–5188 doi: 10.1128/JVI.75.IL518296-5188
Brennan G, Kozyrev Y, Hu SL (2008) From the cover: TRIMCyp expression in old world primates *Macaca nemestrina* and *Macaca fascicularis*. Proc Natl Acad Sci 105:3569–3574. doi:10.1073/pnas.0709511105
Brennan G, Kozyrev Y, Kodama T, Hu S-L (2007) Novel TRIM5 isoforms expressed by *Macaca nemestrina*. J Virol 81:12210–12217
Charlesworth D (2006) Balancing selection and its effects on sequences in nearby genome regions. PLoS Genet 2:e64
Coffin JM (1997) Retroviruses. CSHL Press, New York
Collinge J, Palmer MS, Dryden AJ (1991) Genetic predisposition to iatrogenic Creutzfeldt-Jakob disease. Lancet 337:1441–1442
Compton AA, Emerman M (2013) Convergence and divergence in the evolution of the APOBEC3G-Vif interaction reveal ancient origins of simian immunodeficiency viruses. PLoS Pathog 9:e1003135. doi:10.1371/journal.ppat.1003135
Dandekar S, Rossitto P, Pickett S et al (1987) Molecular characterization of the Akvr-1 restriction gene: a defective endogenous retrovirus-borne gene identical to Fv-4r. J Virol 61:308–314
Desrosiers R (1990) The simian immunodeficiency viruses. Annu Rev Immunol 8:557–578
Dewannieux M, Harper F, Richaud A et al (2006) Identification of an infectious progenitor for the multiple-copy HERV-K human endogenous retroelements. Genome Res 16:1548–1556. doi:10.1101/gr.5565706
Diehl WE, Johnson WE, Hunter E (2013) Elevated rate of fixation of endogenous retroviral elements in Haplorhini *TRIM5* and *TRIM22* genomic sequences: impact on transcriptional regulation. PLoS One 8(3):e58532
Domingo E, Holland JJ (2008) Origin and evolution of viruses. Academic Press, New York
Domingo E, Sheldon J, Perales C (2012) Viral quasispecies evolution. Microbiol Mol Biol Rev 76:159–216. doi:10.1128/MMBR.05023-11
Elde NC, Child SJ, Eickbush MT et al (2012) Poxviruses deploy genomic accordions to adapt rapidly against host antiviral defenses. Cell 150:831–841. doi:10.1016/j.cell.2012.05.049
Elde NC, Child SJ, Geballe AP, Malik HS (2008) Protein kinase R reveals an evolutionary model for defeating viral mimicry. Nature 457:485–489. doi:10.1038/nature07529
Emerman M, Malik HS (2010) Paleovirology–modern consequences of ancient viruses. PLoS Biol 8:e1000301. doi:10.1371/journal.pbio.1000301
Evans DT, Serra-Moreno R, Singh RK, Guatelli JC (2010) BST-2/tetherin: a new component of the innate immune response to enveloped viruses. Trends Microbiol 18:388–396. doi:10.1016/j.tim.2010.06.010

Feschotte C, Gilbert C (2012) Endogenous viruses: insights into viral evolution and impact on host biology. Nat Rev Genet 13:283–296. doi:10.1038/nrg3199

Gardner MB, Kozak CA, O'Brien SJ (1991) The Lake Casitas wild mouse: evolving genetic resistance to retroviral disease. Trends Genet 7:22–27

Gifford R, Tristem M (2003) The evolution, distribution and diversity of endogenous retroviruses. Virus Genes 26:291–315

Gilbert C, Feschotte C (2010) Genomic fossils calibrate the long-term evolution of hepadnaviruses. PLoS Biol. doi:10.1371/journal.pbio.1000495

Goldschmidt V, Bleiber G, May M et al (2006) Role of common human TRIM5alpha variants in HIV-1 disease progression. Retrovirology 3:54

Grossman SR, Shlyakhter I, Shylakhter I et al (2010) A composite of multiple signals distinguishes causal variants in regions of positive selection. Science 327:883–886. doi:10.1126/science.1183863

Hauser H, Lopez LA, Yang SJ et al (2011) HIV-1 Vpu and HIV-2 Env counteract BST-2/tetherin by sequestration in a perinuclear compartment. Retrovirology 8:85. doi:10.1186/1742-4690-8-85

Haygood R, Babbitt CC, Fedrigo O, Wray GA (2010) Contrasts between adaptive coding and noncoding changes during human evolution. Proc Natl Acad Sci 107:7853–7857. doi:10.1073/pnas.0911249107

Heffernan JM, Smith RJ, Wahl LM (2005) Perspectives on the basic reproductive ratio. J R Soc Interface 2:281–293. doi:10.1098/rsif.2005.0042

Holmes EC, Drummond AJ (2007) The evolutionary genetics of viral emergence. Curr Top Microbiol Immunol 315:51–66

Horie M, Honda T, Suzuki Y et al (2010) Endogenous non-retroviral RNA virus elements in *Mammalian genomes*. Nature 463:84–87. doi:10.1038/nature08695

Hurst LD (2002) The Ka/Ks ratio: diagnosing the form of sequence evolution. Trends Genet 18:486

Ikeda H, Sugimura H (1989) Fv-4 resistance gene: a truncated endogenous *Murine leukemia* virus with ecotropic interference properties. J Virol 63:5405–5412

Javanbakht H, An P, Gold B et al (2006) Effects of human TRIM5alpha polymorphisms on antiretroviral function and susceptibility to human immunodeficiency virus infection. Virology 354:15–27

Jern P, Coffin JM (2008) Effects of retroviruses on host genome function. Annu Rev Genet 42:709–732. doi:10.1146/annurev.genet.42.110807.091501

Jern P, Stoye JP, Coffin JM (2007) Role of APOBEC3 in genetic diversity among endogenous *Murine leukemia* viruses. PLoS Genet 3:2014–2022. doi:10.1371/journal.pgen.0030183

Johnson WE (2010) Endless forms most viral. PLoS Genet 6:e1001210. doi:10.1371/journal.pgen.1001210

Johnson WE, Coffin JM (1999) Constructing primate phylogenies from ancient retrovirus sequences. Proc Natl Acad Sci USA 96:10254–10260

Johnson WE, Lifson JD, Lang SM et al (2003) Importance of B-cell responses for immunological control of variant strains of simian immunodeficiency virus. J Virol 77:375–381

Johnson WE, Sawyer SL (2009) Molecular evolution of the antiretroviral TRIM5 gene. Immunogenetics 61:163–176. doi:10.1007/s00251-009-0358-y

Kaiser SM, Malik HS, Emerman M (2007) Restriction of an extinct retrovirus by the human TRIM5 antiviral protein. Science 316:1756–1758. doi:10.1126/science.1140579

Katzourakis A, Gifford RJ (2010) Endogenous viral elements in animal genomes. PLoS Genet 6:e1001191. doi:10.1371/journal.pgen.1001191

Kidd JM, Newman TL, Tuzun E et al (2007) Population stratification of a common APOBEC gene deletion polymorphism. PLoS Genet 3:e63. doi:10.1371/journal.pgen.0030063

Kirmaier A, Wu F, Newman RM et al (2010) TRIM5 suppresses cross-species transmission of a primate immunodeficiency virus and selects for emergence of resistant variants in the new species. PLoS Biol. doi:10.1371/journal.pbio.1000462

Kobayashi Y, Horie M, Tomonaga K, Suzuki Y (2011) No evidence for natural selection on endogenous borna-like nucleoprotein elements after the divergence of old world and new world monkeys. PLoS ONE 6:e24403. doi:10.1371/journal.pone.0024403

Laguette N, Rahm N, Sobhian B et al (2012) Evolutionary and functional analyses of the interaction between the myeloid restriction factor SAMHD1 and the lentiviral Vpx protein. Cell Host Microbe 11:205–217. doi:10.1016/j.chom.2012.01.007

Le Tortorec A, Neil SJD (2009) Antagonism to and intracellular sequestration of human tetherin by the human immunodeficiency virus type 2 envelope glycoprotein. J Virol 83:11966–11978. doi:10.1128/JVI.01515-09

Lee Y, Bieniasz PD (2007) Reconstitution of an infectious human endogenous retrovirus. PLoS Pathog 3:e10. doi:10.1371/journal.ppat.0030010

Liao C, Kuang Y, Liu HL et al (2007) A novel fusion gene mnTRIMCyp in pig-tailed macaques determines its susceptibility to HIV-1 infection. Unpublished

Lim ES, Fregoso OI, McCoy CO et al (2012a) The ability of primate lentiviruses to degrade the monocyte restriction factor SAMHD1 preceded the birth of the viral accessory protein Vpx. Cell Host Microbe 11:194–204. doi:10.1016/j.chom.2012.01.004

Lim ES, Malik HS, Emerman M (2010a) Ancient adaptive evolution of tetherin shaped the functions of Vpu and Nef in human immunodeficiency virus and primate lentiviruses. J Virol 84:7124–7134. doi:10.1128/JVI.00468-10

Lim ES, Wu LI, Malik HS, Emerman M (2012b) The function and evolution of the restriction factor viperin in primates was not driven by lentiviruses. Retrovirology 9:55. doi:10.1186/1742-4690-9-55

Lim S-Y, Rogers T, Chan T et al (2010b) TRIM5α modulates immunodeficiency virus control in *Rhesus* monkeys. PLoS Pathog 6:e1000738. doi:10.1371/journal.ppat.1000738.t002

Liu H, Fu Y, Li B et al (2011) Widespread horizontal gene transfer from circular single-stranded DNA viruses to eukaryotic genomes. BMC Evol Biol 11:276. doi:10.1186/1471-2148-11-276

Lock LF, Keshet E, Gilbert DJ et al (1988) Studies of the mechanism of spontaneous germline ecotropic provirus acquisition in mice. EMBO J 7:4169–4177

Malfavon-Borja R, Wu LI, Emerman M, Malik HS (2013) Birth, decay, and reconstruction of an ancient TRIMCyp gene fusion in primate genomes. Proc Natl Acad Sci. doi:10.1073/pnas.1216542110

Malim MH (2009) APOBEC proteins and intrinsic resistance to HIV-1 infection. Philos Trans R Soc Lond B Biol Sci 364:675–687. doi:10.1098/rstb.2008.0185

Marx PA, Apetrei C, Drucker E (2004) AIDS as a zoonosis? Confusion over the origin of the virus and the origin of the epidemics. J Med Primatol 33:220–226. doi:10.1111/j.1600-0684.2004.00078.x

McEwan WA, Schaller T, Ylinen LM et al (2009) Truncation of TRIM5 in the *Feliformia* explains the absence of retroviral restriction in cells of the domestic cat. J Virol 83:8270–8275. doi:10.1128/JVI.00670-09

McNatt MW, Zang T, Hatziioannou T et al (2009) Species-specific activity of HIV-1 Vpu and positive selection of tetherin transmembrane domain variants. PLoS Pathog 5:e1000300. doi:10.1371/journal.ppat.1000300

Mead S, Whitfield J, Poulter M et al (2008) Genetic susceptibility, evolution and the kuru epidemic. Philos Trans R Soc Lond B Biol Sci 363:3741–3746. doi:10.1098/rstb.2008.0087

Meyerson NR, Sawyer SL (2011) Two-stepping through time: mammals and viruses. Trends Microbiol 19:286–294. doi:10.1016/j.tim.2011.03.006

Mitchell PS, Patzina C, Emerman M et al (2012) Evolution-guided identification of antiviral specificity determinants in the broadly acting interferon-induced innate immunity factor MxA. Cell Host Microbe 12:598–604. doi:10.1016/j.chom.2012.09.005

Mura M, Murcia P, Caporale M et al (2004) Late viral interference induced by transdominant gag of an endogenous retrovirus. Proc Natl Acad Sci USA 101:11117–11122. doi:10.1073/pnas.0402877101

Murcia PR, Arnaud F, Palmarini M (2007) The transdominant endogenous retrovirus enJS56A1 associates with and blocks intracellular trafficking of Jaagsiekte sheep retrovirus gag. J Virol 81:1762–1772. doi:10.1128/JVI.01859-06

Münk C, Willemsen A, Bravo IG (2012) An ancient history of gene duplications, fusions and losses in the evolution of APOBEC3 mutators in mammals. BMC Evol Biol 12:71. doi:10.1186/1471-2148-12-71

Neil S, Bieniasz P (2009) Human immunodeficiency virus, restriction factors, and interferon. J Interferon Cytokine Res 29:569–580. doi:10.1089/jir.2009.0077

Neil SJD, Zang T, Bieniasz PD (2008) tetherin inhibits retrovirus release and is antagonized by HIV-1 Vpu. Nature 451:425–430. doi:10.1038/nature06553

Newman RM, Hall L, Connole M et al (2006) Balancing selection and the evolution of functional polymorphism in old world monkey TRIM5 {alpha}. Proc Natl Acad Sci USA 103:19134–19139

Newman RM, Hall L, Kirmaier A et al (2008) Evolution of a TRIM5-CypA splice isoform in old world monkeys. PLoS Pathog 4:e1000003. doi:10.1371/journal.ppat.1000003

Nisole S, Lynch C, Stoye JP, Yap MW (2004) A TRIM5-cyclophilin a fusion protein found in owl monkey kidney cells can restrict HIV-1. Proc Natl Acad Sci USA 101:13324–13328. doi:10.1073/pnas.0404640101

Nowak M, May RM (2000) Virus dynamics. Oxford University Press, Oxford

Ohkura S, Goldstone DC, Yap MW et al (2011) Novel escape mutants suggest an extensive TRIM5α binding site spanning the entire outer surface of the *Murine leukemia* virus capsid protein. PLoS Pathog 7:e1002011. doi:10.1371/journal.ppat.1002011

Okeoma CM, Lovsin N, Peterlin BM, Ross SR (2007) APOBEC3 inhibits mouse mammary tumour virus replication in vivo. Nature 445:927–930. doi:10.1038/nature05540

Pacheco B, Finzi A, Stremlau M, Sodroski J (2010) Adaptation of HIV-1 to cells expressing *Rhesus* monkey TRIM5α. Virology 408:204–212. doi:10.1016/j.virol.2010.09.019

Palmer MS, Dryden AJ, Hughes JT, Collinge J (1991) Homozygous prion protein genotype predisposes to sporadic Creutzfeldt-Jakob disease. Nature 352:340–342. doi:10.1038/352340a0

Patel MR, Emerman M, Malik HS (2011) Paleovirology—ghosts and gifts of viruses past. Curr Opin Virol 1:304–309. doi:10.1016/j.coviro.2011.06.007

Patel MR, Loo Y-M, Horner SM et al (2012) Convergent evolution of escape from hepaciviral antagonism in primates. PLoS Biol 10:e1001282. doi:10.1371/journal.pbio.1001282

Perez-Caballero D, Soll SJ, Bieniasz PD (2008) Evidence for restriction of ancient primate gammaretroviruses by APOBEC3 but not TRIM5alpha proteins. PLoS Pathog 4:e1000181. doi:10.1371/journal.ppat.1000181

Perez-Caballero D, Zang T, Ebrahimi A et al (2009) tetherin inhibits HIV-1 release by directly tethering virions to cells. Cell 139:499–511. doi:10.1016/j.cell.2009.08.039

Qi CF, Bonhomme F, Buckler-White A et al (1998) Molecular phylogeny of Fv1. Mamm Genome 9:1049–1055

Robinson HL, Astrin SM, Senior AM, Salazar FH (1981) Host susceptibility to endogenous viruses: defective, glycoprotein-expressing proviruses interfere with infections. J Virol 40:745–751

Ross SR (2009) Are viruses inhibited by APOBEC3 molecules from their host species? PLoS Pathog 5:e1000347. doi:10.1371/journal.ppat.1000347

Sabeti PC, Reich DE, Higgins JM et al (2002) Detecting recent positive selection in the human genome from haplotype structure. Nature 419:832–837. doi:10.1038/nature01140

Sabeti PC, Walsh E, Schaffner SF et al (2005) The case for selection at CCR5-Δ32. PLoS Biol 3:e378. doi:10.1371/journal.pbio.0030378.st005

Santiago ML, Montano M, Benitez R et al (2008) APOBEC3 encodes Rfv3, a gene influencing neutralizing antibody control of retrovirus infection. Science 321:1343–1346. doi:10.1126/science.1161121

Sawyer SL, Emerman M, Malik HS (2007) Discordant evolution of the adjacent antiretroviral genes TRIM22 and TRIM5 in mammals. PLoS Pathog 3:e197. doi:10.1371/journal.ppat.0030197

Sawyer SL, Emerman M, Malik HS (2004) Ancient adaptive evolution of the primate antiviral DNA-editing enzyme APOBEC3G. PLoS Biol 2:E275. doi:10.1371/journal.pbio.0020275

Sawyer SL, Wu LI, Akey JM et al (2006) High-frequency persistence of an impaired allele of the retroviral defense gene TRIM5alpha in humans. Curr Biol 16:95–100. doi:10.1016/j.cub.2005.11.045

Sawyer SL, Wu LI, Emerman M, Malik HS (2005) Positive selection of primate TRIM5alpha identifies a critical species-specific retroviral restriction domain. Proc Natl Acad Sci USA 102:2832–2837. doi:10.1073/pnas.0409853102

Sayah D, Sokolskaja E, Berthoux L, Luban J (2004) Cyclophilin a retrotransposition into TRIM5 explains owl monkey resistance to HIV-1. Nature 430:569–573

Schoggins JW, Wilson SJ, Panis M et al (2011) A diverse range of gene products are effectors of the type I interferon antiviral response. Nature 472:481–485. doi:10.1038/nature09907

Schumacher AJ, Haché G, Macduff DA et al (2008) The DNA deaminase activity of human APOBEC3G is required for Ty1, MusD, and human immunodeficiency virus type 1 restriction. J Virol 82:2652–2660. doi:10.1128/JVI.02391-07

Schumacher AJ, Nissley DV, Harris RS (2005) APOBEC3G hypermutates genomic DNA and inhibits Ty1 retrotransposition in yeast. Proc Natl Acad Sci USA 102:9854–9859. doi:10.1073/pnas.0501694102

Serra-Moreno R, Jia B, Breed M et al (2011) Compensatory changes in the cytoplasmic tail of gp41 confer resistance to tetherin/BST-2 in a pathogenic nef-deleted SIV. Cell Host Microbe 9:46–57. doi:10.1016/j.chom.2010.12.005

Simon V, Zennou V, Murray D et al (2005) Natural variation in Vif: differential impact on APOBEC3G/3F and a potential role in HIV-1 diversification. PLoS Pathog 1:e6. doi:10.1371/journal.ppat.0010006

Soll SJ, Neil SJD, Bieniasz PD (2010) Identification of a receptor for an extinct virus. Proc Natl Acad Sci 107:19496–19501. doi:10.1073/pnas.1012344107

Song B, Gold B, Ohuigin C et al (2005) The B30.2 (SPRY) domain of the retroviral restriction factor TRIM5alpha exhibits lineage-specific length and sequence variation in primates. J Virol 79:6111–6121

Speelmon E, Livingston-Rosanoff D, Li S et al (2006) Genetic association of the antiviral restriction factor TRIM5alpha with human immunodeficiency virus type 1 infection. J Virol 80:2463–2471

Steeves R, Lilly F (1977) Interactions between host and viral genomes in *Mouse leukemia*. Annu Rev Genet 11:277–296. doi:10.1146/annurev.ge.11.120177.001425

Sullivan AD, Wigginton J, Kirschner D (2001) The coreceptor mutation CCR5Delta32 influences the dynamics of HIV epidemics and is selected for by HIV. Proc Natl Acad Sci USA 98:10214–10219. doi:10.1073/pnas.181325198

Swain A, Coffin J (1992) Mechanism of transduction by retroviruses. Science 255:841–845

Van Damme N, Goff D, Katsura C et al (2008) The interferon-induced protein BST-2 restricts HIV-1 release and is downregulated from the cell surface by the viral Vpu protein. Cell Host Microbe 3:245–252. doi:10.1016/j.chom.2008.03.001

van Mierlo JT, Bronkhorst AW, Overheul GJ et al (2012) Convergent evolution of argonaute-2 slicer antagonism in two distinct insect RNA viruses. PLoS Pathog 8:e1002872. doi:10.1371/journal.ppat.1002872

Virgen CA, Kratovac Z, Bieniasz PD, Hatziioannou T (2008) From the cover: independent genesis of chimeric TRIM5-cyclophilin proteins in two primate species. Proc Natl Acad Sci 105:3563–3568. doi:10.1073/pnas.0709258105

Weiler A, May G, Qi Y et al (2006) Polymorphisms in eight host genes associated with control of HIV replication do not mediate elite control of viral replication in SIV-infected Indian *Rhesus macaques*. Immunogenetics 58:1003–1009. doi:10.1007/s00251-006-0166-6

Wilson SJ, Webb BLJ, Maplanka C et al (2008a) *Rhesus macaque* TRIM5 alleles have divergent antiretroviral specificities. J Virol 82:7243–7247. doi:10.1128/JVI.00307-08

Wilson SJ, Webb BLJ, Ylinen LMJ et al (2008b) From the cover: independent evolution of an antiviral TRIMCyp in *Rhesus macaques*. Proc Natl Acad Sci 105:3557–3562. doi:10.1073/pnas.0709003105

Witzany G (2012) Viruses: essential agents of life. Springer, Berlin

Wong WSW, Nielsen R (2004) Detecting selection in noncoding regions of nucleotide sequences. Genetics 167:949–958. doi:10.1534/genetics.102.010959

Yan Y, Buckler-White A, Wollenberg K, Kozak CA (2009) Origin, antiviral function and evidence for positive selection of the gammaretrovirus restriction gene Fv1 in the genus Mus. Proc Natl Acad Sci 106:3259–3263. doi:10.1073/pnas.0900181106

Yang Z (2005) The power of phylogenetic comparison in revealing protein function. Proc Natl Acad Sci USA 102:3179–3180. doi:10.1073/pnas.0500371102
Yang Z (2006) Computational molecular evolution. Oxford University Press, Oxford
Yang Z, Bielawski J (2000) Statistical methods for detecting molecular adaptation. Trends Ecol Evol 15:496–503. doi:10.1016/S0169-5347(00)01994-7
Zhang J, Webb D (2004) Rapid evolution of primate antiviral enzyme APOBEC3G. Hum Mol Genet 13:1785–1791
Zhang Z, Liu J, Li M et al (2012) Evolutionary dynamics of the interferon-induced transmembrane gene family in vertebrates. PLoS ONE 7:e49265. doi:10.1371/journal.pone.0049265
Zhao K, Ishida Y, Oleksyk TK et al (2012) Evidence for selection at HIV host susceptibility genes in a west central African human population. BMC Evol Biol 12:237. doi:10.1186/1471-2148-12-237

RNA Interference-Mediated Intrinsic Antiviral Immunity in Plants

György Szittya and József Burgyán

Abstract RNA interference (RNAi) is an evolutionarily conserved, sequence-specific gene-inactivation system that plays an essential role in many biological processes, such as genome defense against mobile DNA elements or regulation of factors involved in plant and animal development. In higher plants and invertebrates, it also functions as a powerful antiviral mechanism. To overcome antiviral RNAi, viruses have evolved suppressor proteins which counteract host RNAi-based antiviral processes and target one or more key points in the RNAi machinery. Here, we review recent progress in our understanding of the mechanism and function of antiviral RNAi in plants and on the viral responses through the expression of silencing suppressor proteins. As a counter-attack RNAi may also regulate innate immunity in plants and contribute to a novel layer of defense against pathogen attack. We also discuss emerging evidence that viruses use RNAi to manipulate host gene expression to modify the cellular environment for the benefit of invading viruses.

Contents

G. Szittya (✉) · J. Burgyán (✉)
Agricultural Biotechnology Center, Gödöllő, Hungary
e-mail: szittya@abc.hu

J. Burgyán
e-mail: burgyan@abc.hu

B. R. Cullen (ed.), *Intrinsic Immunity*, Current Topics in Microbiology and Immunology 371, DOI: 10.1007/978-3-642-37765-5_6,

1 Introduction

Viruses are obligate intracellular pathogens that manipulate and exploit the molecular mechanisms of the host to survive in a hostile cellular environment. The presence of viruses and their propagation induces diverse mechanisms in the host for combating viral infection at both the single-cell and the whole-organism levels. Viruses are obligate intracellular pathogens infecting almost all living organisms. During the replication of viruses at a given point of their life cycle, they reach the stage of ssRNA or dsRNA, which can trigger host defense responses. These mechanisms range from RNA interference (RNAi) a mechanism mainly found in plants and lower eukaryotes, to the sophisticated interferon-regulated gene response of higher animals (Ding 2010; Csorba et al. 2009; Ding and Voinnet 2007).

RNAi plays an essential role in many plant biological processes, including developmental timing and patterning, transposon control, DNA methylation, and chromatin modification as well as antiviral defense. Although it operates through multiple pathways, RNAi relies on a set of core processes triggered by double-stranded (ds) or self-complementary foldback RNAs that are processed into 21–24-nt-short small-interfering (si)RNA or micro (mi)RNA duplexes by the RNase III-type DICER enzymes (Bartel 2004; Bernstein et al. 2001; Baulcombe 2004). These miRNAs and siRNAs guide a multiprotein effector complex, the RNA-induced silencing complex (RISC) (Hammond et al. 2000; Tomari and Zamore 2005), of which Argonaute protein (AGO) is the slicer component showing similarity to RNase H (Song et al. 2004; Liu et al. 2004; Tomari and Zamore 2005). RISC is the executioner of RNAi, inhibiting target RNA expression. The specific recognition of target sequences is guided by the si/miRNAs through a base-pairing mechanism, whereas the slicing of target RNA is carried out by the AGO proteins at the post-transcriptional level (Fig. 1) (Bartel 2004; Almeida and Allshire 2005; Brodersen et al. 2008; Eamens et al. 2008).

The model plant *Arabidopsis thaliana* encodes four DICER paralogues with specialized functions (Baulcombe 2004). Dicer-like (DCL)-1 mainly contributes to the production of miRNAs from non-coding, imperfect stem-loop precursor RNAs (Voinnet 2009), whereas populations of 21, 22 and 24-nt-long short-interfering (si)RNAs are synthesized from long, perfectly or near-perfectly base-paired dsRNAs through the action of DCL4, DCL2, and DCL3, respectively (Brodersen and Voinnet 2006; Vazquez 2006; Chapman and Carrington 2007).Upon processing,

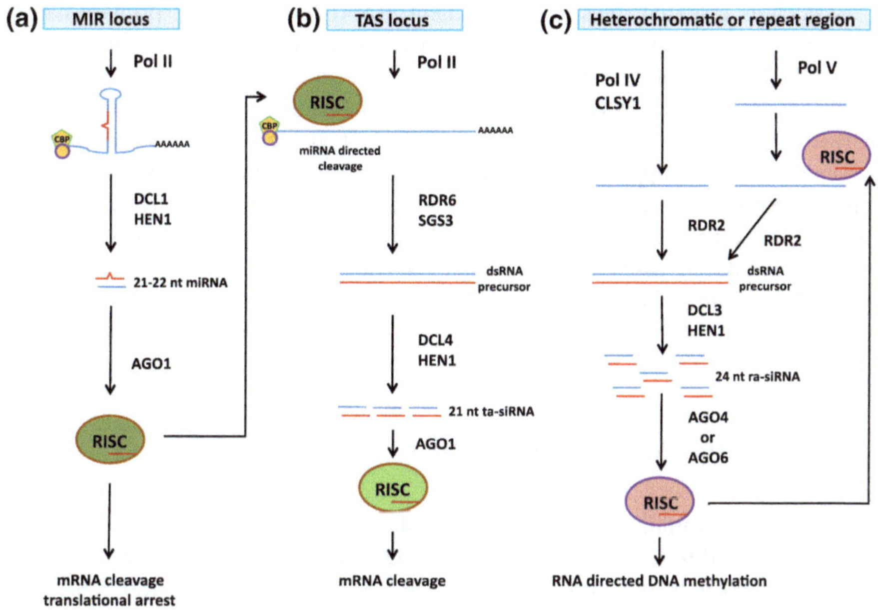

Fig. 1 Endogenous small RNA pathways in plants. **a** Plant miRNA genes (MIRs) are transcribed into primary miRNA transcripts by RNA polymerase II (Pol II) and the transcript contains a cap and polyA tail. This primary transcript forms a hairpin structure that is cleaved by DCL1 to produce the miRNA/miRNA* duplex. The duplex is methylated by HEN1. The mature miRNA duplex is bound by AGO 1, and the miRNA* strand is eliminated. Plant miRNAs have near-perfect complementarity to their target sites and they bind to these sites to guide the cleavage of target mRNAs, although there are examples where the translation of the mRNA is suppressed without cleavage. The cap-binding complex (CBP) promotes miRNA biogenesis but is not essential. **b** transacting siRNAs belong to a special class of endogenous siRNAs. Ta-siRNAs require components of both the miRNA and siRNA pathways for their biogenesis. They are produced from long non-coding transcripts made by Pol II that are converted into dsRNA by an RNA-dependent RNA polymerase 6 (RDR6) and SGS3. This long dsRNA is then cleaved in a phased dicing reaction by DCL4 to generate 21-nt ta-siRNAs which are methylated by HEN 1. The phase of the dicing reaction is determined by the initial miRNA cleavage site. Ta-siRNAs target mRNAs in a similar way as miRNAs. **c** The vast majority of endogenous siRNAs in plants form a complex population of more than 100,000 different siRNAs transcribed from thousands of loci. The biogenesis of the heterochromatic siRNAs requires RNA polymerase IV (PolIV), which is unique to the plant kingdom and homologues to DNA-dependent RNA polymerase II. Pol IV produces a transcript that is converted to a dsRNAs precursor by RDR2 and subsequently cleaved by DCL3 to produce 24-nt ra-siRNAs that are methylated by HEN 1 and are bound by AGO 4 or AGO 6. AGO-bound ra-siRNAs can induce heterochromatin formation or cleavage of Pol V (another plant-specific RNA polymerase)-produced transcripts, which serves as an amplification loop to produce more ra-siRNAs. CLASSY 1 (CLSY1) is a putative chromatin-remodeling protein that acts together with Pol IV

small RNAs (sRNA) are incorporated into RISC containing one of the ten AGO proteins that are involved in RNAi in *Arabidopsis,* thereby controlling the expression of genes involved in various pathways.

Besides its essential roles in development, RNAi constitutes the primary natural plant immune system against viruses. Several lines of evidence underline the antiviral role of RNAi; first, viral siRNAs are invariably detected during viral infection, which is a hallmark of the activation of antiviral RNAi (Hamilton and Baulcombe 1999; Szittya et al. 2002). Second, most if not all plant viruses have evolved virulence factors called viral suppressors of RNAi (VSRs) to overcome the RNAi-based host defense. (Silhavy et al. 2002; Ding and Voinnet 2007; Burgyan and Havelda 2011). Furthermore, plants infected by viruses defective in VSR expression rapidly recover from virus-induced diseases and symptoms (Ratcliff et al. 1997, 1999; Silhavy et al. 2002). Third, plants carrying mutations in components of the RNAi pathway are more susceptible to some, but not all, virus infections (Ding and Voinnet 2007).

2 RNAi-Based Antiviral Immunity

Viruses are potentially both initiators and targets of antiviral plant response (Ratcliff et al. 1997), and it has been also demonstrated that these responses are RNA-mediated and that viral siRNAs (vsiRNAs), derived from viral RNA templates, represent the specificity determinant of this antiviral response. The main steps in the formation of an effective antiviral RNAi are (1) activation of RNAi in the cell by the incoming viral RNA, where structured or double-stranded RNAs are recognized by plant Dicer-like (DCL) enzymes, producing vsiRNAs; (2) the protection of vsiRNAs by 2′-O methylation by HEN1; (3) antiviral RISC assembly, in which vsiRNAs are recruited by AGO proteins; (4) targeting cognate viral RNAs by vsiRNA-guided RISC. Alternatively, vsiRNAs or aberrant viral RNAs can enter the plant RNA-dependent RNA polymerase (RDR)-mediated amplification cycle to enhance the antiviral silencing response (Fig. 2).

2.1 Activation of RNAi and Production of vsiRNAs

The majority of known plant viruses have RNA genomes and replicate via double-stranded RNA replication intermediates, and it is widely assumed that these dsRNA molecules are the main trigger of antiviral RNAi. However, it turned out that induction of the antiviral response is much more complex. The probability that viral RNAs are present in a naked form in the plant cell is very small. The majority of viral RNAs are in an encapsidated form or in other protein complexes for replication or movement. Moreover, viral replication usually takes place inside specialized replication compartments (Burgyan et al. 1996), and the viral dsRNA replication intermediates can immediately be unwound by viral or host RNA helicases (Ahlquist 2002). It is likely that highly structured single-stranded viral RNAs with stem-loop structures are also recognized by the silencing machinery and directly diced by plant DCLs into vsiRNAs (Fig. 2). The sequencing and experimental data of vsiRNAs strongly support this model since the resulting

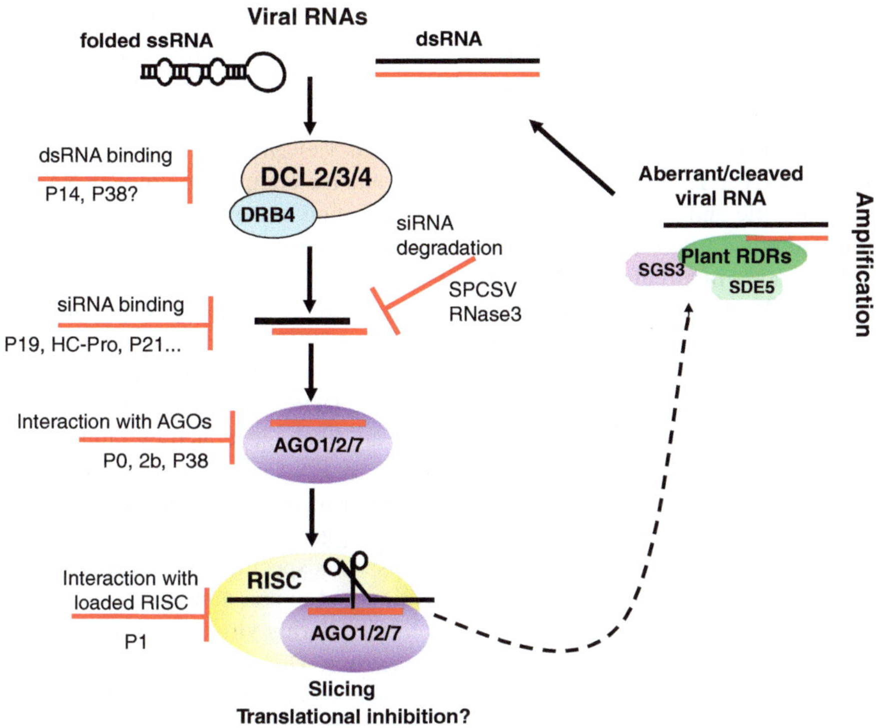

Fig. 2 Antiviral RNAi in plant and its suppression by virus-encoded silencing suppressors. RNAi is initiated by the recognition of viral dsRNAs or partially double-stranded hairpin RNAs, which are processed to viral siRNAs (vsiRNAs) by dsRNA-specific RNases called Dicer-like proteins (DCL2/3/4). In the next step, HSP90-activated AGO1/2/7 is loaded with vsiRNA, thereby forming large RISC, which has likely incorporated other proteins such as the GW motif-containing AGO interactors. Afterward, the vsiRNA-loaded RISC targets viral RNAs by slicing or translational arrest. Secondary vsiRNAs are produced in an amplification loop through the actions of RNA-dependent RNA polymerases (RDRs) and cofactors (SGS3 and SDE5). As indicated, viral silencing suppressors can disrupt these pathways at multiple points preventing the assembly of different effectors or inhibiting their actions

vsiRNA molecules are imperfect duplexes (Molnar et al. 2005) have a non-random distribution along the viral genome, and they map asymmetrically to the positive strand of the viral RNA (Molnar et al. 2005; Ho et al. 2006; Donaire et al. 2009; Qi et al. 2009; Pantaleo et al. 2010; Szittya et al. 2010). Similarly, in the case of *Cauliflower mosaic virus* (CaMV), the 35S polycistronic transcript of this dsDNA virus contains a highly structured 600-nt leader, which is the major vsiRNA source (Moissiard and Voinnet 2006). However, this massive production of leader-derived vsiRNAs does not restrict viral accumulation. They may serve as a decoy diverting the silencing machinery from viral transcripts (Blevins et al. 2011). In viroid-infected plants, the strandedness of viroid-specific siRNAs is also asymmetrical and they are preferentially derived from the highly structured plus sense viroid RNA sequence (Itaya et al. 2007), although recent deep sequencing data have shown a more symmetrical origin of viroid siRNAs (Navarro et al. 2009, 2012). Furthermore, in plants infected by the Potyvirus *Turnip mosaic virus* (TuMV), which has a positive ssRNA genome that expresses a polyprotein, the sequenced vsiRNAs showed similar amounts of (+) and (−) strand vsiRNAs (Ho et al. 2006). This result may suggest that the TuMV-derived vsiRNAs are processed mostly from dsRNA. It is likely that the strandedness and the composition of vsiRNAs are virus specific and strongly depend on the replication strategy, compartmentalization, and other different characteristics of the given virus.

In the case of circular ssDNA geminiviruses, early reports suggested that vsiRNAs are likely derived from dsRNAs formed by overlapping sense-antisense transcripts (Blevins et al. 2006; Ding and Voinnet 2007; Akbergenov et al. 2006). Indeed, it has been shown very recently that the majority of viral siRNAs accumulating during geminiviral infection are plant RDR-independent primary siRNAs, and dsRNA precursors of these siRNAs are likely generated by bidirectional readthrough transcription of circular viral DNA by RNA polymerase II. In contrast to transgenic mRNA, geminiviral mRNAs appear to be poor templates for RDR-dependent production of secondary siRNAs (Aregger et al. 2012). These findings demonstrate that the generation of vsiRNAs is virus specific, indicating that plant DCLs are adapted to different viral replication and expression strategies and are able to recognize different RNA structures formed during virus life cycles.

The *Arabidopsis thaliana* genome encodes four DCLs for sRNA processing: DCL1 to DCL4. Specific DCLs have major functions in specific silencing pathways but functional redundancy exists between members: DCL1 contributes to miRNA production and has no or little role in the antiviral response. DCL2, DCL3, and DCL4 are able to recognize viral structures and, respectively, generate vsiRNAs of 22, 24, and 21 nt in length (Deleris et al. 2006; Blevins et al. 2006; Aregger et al. 2012).

Biogenesis of vsiRNAs needs the coordinated and hierarchical action of DCL enzymes, and RNA virus infection is mainly affected by DCL4 and to a lesser extent by DCL2. Inactivation of DCL4 reveals the surrogate antiviral role of DCL2. Inactivation of both DCL2 and DCL4 was necessary and sufficient to restore systemic infection of a suppressor-deficient virus, indicating the crucial role of DCL4 and DCL2 in the antiviral response (Deleris et al. 2006; Bouche et al. 2006; Llave 2010). This finding is further supported by the observation that in an immune precipitation experiment, AGO1 was found loaded with 21- and 22-nt vsiRNAs in *Nicotiana benthamiana* plants infected with p19 mutant tombusvirus (Burgyan et al., unpublished result). Interestingly in the absence of DCL4, the resultant 22-nucleotide viral siRNAs alone did not guide efficient silencing in a CMV–*Arabidopsis* (virus–host) combination (Wang et al. 2011).

Upon DNA virus infection, the production of 24-nt vsiRNAs by DCL3 is also sufficient for virus-induced gene silencing (Blevins et al. 2006). DCL3-dependent 24-nt-long vsiRNAs have also been detected in *Tobacco rattle virus* (TRV) and CMV-infected wild-type (wt) plants or *Turnip crinkle virus* (TCV)-infected *dcl4/dcl2* double-mutant *Arabidopsis* plants (Deleris et al. 2006; Qu et al. 2008; Zhang et al. 2012).

The participation of DCL1 in the antiviral silencing induced by RNA viruses is very limited since DCL1-dependent vsiRNAs are hardly detected in the *dcl2/dcl3/dcl4* triple-mutant plants (Deleris et al. 2006; Bouche et al. 2006; Blevins et al. 2006). However, DCL1 promotes DCL3- and DCL4-derived vsiRNAs accumulation upon dsDNA (CaMV) or ssDNA (geminivirus) infection. Very likely, DCL1 excises the stem-loop structures of CaMV 35S leader transcripts, which are very similar to pri- or pre-miRNAs, and renders them more accessible to other DCLs (Moissiard and Voinnet 2006). Recent findings suggest that this leader region is transcribed into long sense- and antisense-RNAs, which serve as a template for a massive quantity of 21-, 22-, and 24-nt viral siRNAs (Blevins et al. 2011). An opposite effect of DCL1 was observed in plants infected with TCV: the disruption of DCL1 function led to higher expression of DCL4 and DCL3 and enhanced antiviral response, suggesting that these proteins are under DCL1-negative control (Qu et al. 2008).

In several virus–plant combinations, in addition to core dicing and slicing components of RNAi, RNAi-mediated viral immunity requires host RDRs to amplify the antiviral plant response and produce viral secondary siRNAs (Fig. 2). *Arabidopsis* encodes six identified RDRs (Wassenegger and Krczal 2006), and they play a role in the amplification of the silencing response by production of secondary vsiRNAs (Voinnet 2005; Diaz-Pendon et al. 2007; Llave 2010). The amplification and the high level of secondary vsiRNA in many virus infections depend on the combined activity of the host-encoded RDRs such as RDR1, RDR2, and RDR6, suggesting that aberrant viral single-stranded RNAs lacking quality control marks might be converted by RDR enzymes to dsRNAs, which could serve as a substrate for secondary vsiRNA production (Donaire et al. 2008; Diaz-Pendon et al. 2007; Bao et al. 2009; Garcia-Ruiz et al. 2010; Qu 2010). These siRNAs, upon incorporation into RISC, execute effector steps of RNAi and also direct further amplification rounds by releasing the cleaved target RNAs, thus providing additional templates for RDR enzymes (Voinnet 2005; Vaucheret 2006).

It has also been suggested that the RDR-dependent secondary vsiRNAs can drive a more effective antiviral response (Vaistij and Jones 2009; Wang et al. 2010). However, this is not a general phenomenon for all plant viruses since other studies have shown that loss-of-function mutations in RDR6 have no detectable impact on the production of vsiRNAs and virus accumulation in *Arabidopsis* plants infected with TRV, TCV, and cr-TMV (Dalmay et al. 2000, 2001; Blevins et al. 2006; Deleris et al. 2006; Szittya et al. 2010).

These findings indicate that although there are very conserved steps in the silencing-based antiviral response, plants are able to respond specifically to different viruses, demonstrating the versatility of this antiviral surveillance mechanism.

2.2 Assembly of Antiviral RNA-Induced Silencing Complexes

The *Arabidopsis* genome contains ten AGO proteins, AGO1 to AGO10, and they are the catalytic components of the effector complexes (RISC) of RNAi. AGO proteins interact with small RNAs to inhibit gene expression by target cleavage or translational arrest in all RNAi-related pathways known so far (Fig. 1). Endogenous or exogenous small RNAs are loaded into distinct AGO-containing effector complexes to guide them to their RNA target molecules (Ding and Voinnet 2007; Vaucheret 2008; Hutvagner and Simard 2008). In plants, the loading of siRNAs into a particular AGO complex is preferentially, but not exclusively, dictated by their 5′ terminal nucleotides (Mi et al. 2008; Takeda et al. 2008). It has been shown previously that both AGO1 and AGO7 function to ensure the efficient clearance of viral RNAs and that AGO7 appears to work as a surrogate slicer in the absence of AGO1 (Qu et al. 2008). Moreover, it is likely that AGO1 is capable of targeting viral RNAs with more compact structures, whereas AGO7 favors less structured RNA targets (Qu et al. 2008). Our knowledge is still limited about plant si/miRNA RISC assembly, although a recently developed in vitro system has provided further insights into plant RISC development and RNA targeting mediated by this effector (Iki et al. 2010, 2011). Accumulating evidence has shown that AGO2 is also involved in RNAi-based antiviral responses (Harvey et al. 2011; Scholthof et al. 2011; Wang et al. 2011; Zhang et al. 2011, 2012; Jaubert et al. 2011). Moreover, complementation assays in *ago2-1* plants with different AGO2 forms demonstrated the requirement for AGO2 catalytic residues for antiviral activity, emphasizing its role in antiviral defense in plants (Carbonell et al. 2012). Indeed a recent report has shown a direct evidence for AGO2 slicer activity (Schuck et al. 2013).

2.3 Viral RNA Targeting by Antiviral RISC

The DCL-mediated processing of viral dsRNA regions into vsiRNA in theory could be enough for viral RNA degradation. However, *dcl2/dcl3, dcl2/dcl4, and dcl3/dcl4* mutant plants infected with TRV had approximately equivalent levels of vsiRNAs but only *dcl2/dcl4* plants showed strong viral symptoms and high virus titer (Deleris et al. 2006). Strikingly, high levels of viral siRNAs in the absence of AGO1, AGO2, or both were insufficient to confer resistance, suggesting that AGO1 and AGO2 are essential for the antiviral activities of viral siRNAs downstream of DCL-mediated siRNA biogenesis (Wang et al. 2011).

AGO1 was suggested to be involved in antiviral silencing, as hypomorphic *ago1* mutants are hypersensitive to CMV infection (Morel et al. 2002). Pull-down experiments revealed that AGO1 recruits vsiRNAs and the AGO1-vsiRNA complex is a major player in antiviral defense (Zhang et al. 2006). In addition, it has been reported that both AGO2 and AGO5 can bind CMV-derived vsiRNAs, selecting for short RNAs having 5′ A and C nucleotides, respectively (Takeda et al. 2008; Mi et al. 2008).

More direct evidence for the existence of antiviral RISC comes from studies with the positive-strand RNA *Cymbidium ringspot virus* (CymRSV). Two vsiRNA-containing silencing complexes, which co-fractionated with miRNA-containing complexes, were detected in infected plants: the smaller one at approximately the AGO1-siRNA size (150 kDa), the so-called minimal RISC, and a high molecular weight (670 kDa) multiprotein complex (Pantaleo et al.

2007) probably analogous to animal RISC (Pham et al. 2004). A similar complex was isolated in separate experiments involving another tombusvirus. This complex contained vsiRNAs and exhibited *in vitro* nuclease activity, which preferentially targeted homologous viral sequences (Omarov et al. 2007). Strikingly, viral RNA was targeted in a non-random fashion in hot spots by the antiviral RISC in Cym19stop suppressor mutant virus-infected plants (Pantaleo et al. 2007). Those regions of viral RNA that show hot spots for vsiRNA generation probably form strong secondary structures, which are selectively recognized by DCLs. However, these hot spots are poor targets for RISC-mediated cleavage, since RNA sequences possessing strong secondary structures are not accessible to RISC (Szittya et al. 2002; Ameres et al. 2007; Pantaleo et al. 2007). The accessibility of the viral RNA is probably also influenced by encapsidation, formation of replication complexes containing host and viral proteins and compartmentalization of virus replication. It has been shown recently that there is asymmetry in the strandedness of virus-derived siRNAs, showing that the majority of viral siRNAs have plus-stranded viral sequences (Molnar et al. 2005; Ho et al. 2006; Donaire et al. 2009; Qi et al. 2009; Szittya et al. 2010). This finding suggests that viral siRNA-guided RISC should target more frequently the viral strand having negative polarity than the plus-stranded viral RNA. Indeed, in previous experiments, strand-specific sensors were used for sensing antiviral RISC-mediated cleavages, and the sensor RNAs carrying (−) strand sequences were better targets than the (+) strand sensors (Pantaleo et al. 2007; Szittya et al. 2010). It is worth noting that the level of negative-strand viral RNA is a rate-limiting factor for viral replication; thus preferential targeting of the negative viral strand could make the antiviral silencing response very efficient and very attractive for plant defense. The analysis of 5' RNA cleavage products of sensor RNAs and viral RNAs reveals the presence of non-templated U residues at the cleavage site (Pantaleo et al. 2007), this is the signature of RISC action (Shen and Goodman 2004), confirming the presence of RISC-mediated slicing.

According to the current model of virus-induced RNAi (Fig. 2), a large amount of vsiRNA originates from partially base-paired regions of plus-stranded viral RNAs (Molnar et al. 2005; Ding and Voinnet 2007; Szittya et al. 2010). Thus, plus-stranded vsiRNAs could also potentially target plus-stranded viral RNA through translational arrest. Indeed, recent findings suggest that translational arrest could also be a widespread way to inhibit gene expression by plant miRNAs and siRNAs (Brodersen et al. 2008; Lanet et al. 2009). Moreover, a novel role of AGO4 has been suggested for specific translational control of viral RNA (Bhattacharjee et al. 2009). AGO7 was also shown to function as a surrogate slicer in the absence of AGO1 in the clearance of viral RNA of TCV and favors less structured RNA targets (Qu et al. 2008). More recently, Scholthof and co-workers reported that a *N. benthamiana* AGO (NbAGO) with similarity to *Arabidopsis* AGO2 is involved in antiviral defense against TBSV. The activity of this NbAGO2 was shown to be directly associated with anti-TBSV RNAi, suggesting that NbAGO2 might primarily function in antiviral defense (Scholthof et al. 2011).

3 Strategies of Antiviral RNAi Suppression

The fact that most viruses have evolved viral suppressors of RNAi (VSRs) underlines the antiviral nature of RNAi and reveals a pathogen counter-defensive strategy leading to the active suppression of host surveillance (Voinnet et al. 1999; Silhavy and Burgyan 2004; Ding and Voinnet 2007; Burgyan and Havelda 2011). Indeed, plant viruses often cause severe symptoms and damage, demonstrating an efficient counter-defensive viral strategy. The first VSRs were identified more than a decade ago (Anandalakshmi et al. 1998; Brigneti et al. 1998; Kasschau and Carrington 1998), which was followed by the discovery of several additional VSRs (Silhavy and Burgyan 2004; Burgyan and Havelda 2011). In the past, more than 50 individual VSRs have been identified from almost all plant virus genera, underlining the need of their expression for successful virus infection (Ding and Voinnet 2007; Diaz-Pendon and Ding 2008; Csorba et al. 2009). Available data suggest that virtually all plant viruses encode at least one suppressor, but in many cases viruses encode more than one (e.g., carmo-, clostero-, crini- and begomoviruses) (Csorba et al. 2009). VSRs are surprisingly diverse within and across kingdoms, with no obvious sequence homology, and are therefore considered to be the outcome of recent evolutionary processes. The various VSRs are able to target all steps of the RNAi pathway, such as viral RNA recognition, dicing, RISC assembly, RNA targeting, and amplification (Fig. 2). Since the first discovery of VSRs, a lot of effort has gone into understanding the molecular basis of RNAi suppression. However, our knowledge of the molecular mechanisms of action of VSRs is still restricted to only a few cases, in large part because the majority of VSRs have essential viral functions besides being RNAi suppressors. This multifunctional nature of VSRs often causes serious difficulties in the exploration of their mechanisms, because the inactivation of VSRs often leads to a loss of viability of the given virus.

3.1 Viral Suppressors Interacting with RNAi-Related RNAs

The inhibition of viral RNA recognition and the subsequent dicing by plant DICER effectors is not a frequent strategy of known VSRs. Two viral proteins have been identified which inhibit the processing of dsRNA to siRNAs in agroinfiltration assays: P14 of *Pothos latent aureusvirus* (PoLV) and P38 of *Turnip crinkle virus* (TCV). In addition, P38 and P14 have been shown to bind dsRNA in a size-independent way (Merai et al. 2006, 2005). However, the dsRNA-binding properties of P38 have been questioned using transgenic plants (Schott et al. 2012). Instead, p38 was shown to inhibit antiviral RNAi through AGO1 binding and interference with the AGO1-dependent homeostatic network, which leads to the inhibition of *Arabidopsis* DCLs (Azevedo et al. 2010). Genetic evidence was also reported, that P38 specifically inhibits Dicer-like 4 (DCL4) activity, which has been shown to be the primary antiviral DICER in the *Arabidopsis* model plant

(Deleris et al. 2006). In addition, the P6 VSR of CaMV (Love et al. 2007) has been shown to interfere with vsiRNA processing. P6 was previously described as a viral translational transactivator protein essential for virus biology. Importantly, P6 has two importin-alpha-dependent nuclear localization signals, which are mandatory for CaMV infectivity. A recent discovery showed that one of the nuclear functions of P6 is to suppress RNAi by interacting with double-stranded RNA-binding protein 4 (DRB4), which is required for the function of DCL4 (Haas et al. 2008).

The most common suppression strategy, evolved by several viral genera, is duplex siRNA sequestration (Lakatos et al. 2006; Merai et al. 2006; Csorba et al. 2007, 2009; Ding and Voinnet 2007; Wu et al. 2010), which prevents assembly of the RISC effector complex. These siRNA-binding VSRs are unrelated proteins, although they share analogous biochemical properties, suggesting their independent evolution in different viruses. The P19 protein of tombusviruses, probably the best-known VSR so far, prevents RNAi by siRNA sequestration through binding duplex siRNAs with high affinity (Silhavy et al. 2002). Crystallographic studies have shown that P19 forms a tail-to-tail homodimer, which acts like a molecular caliper, measuring the length of siRNA duplexes and binding them in a sequence-independent manner, selecting for the 19-bp-long dsRNA region of the typical siRNA duplex (Vargason et al. 2003; Ye et al. 2003). Thus, the P19 VSR evolved to bind and inactivate vsiRNAs, which are the most conserved elements of the RNAi pathway. Indeed, the P19 VSR was also able to inhibit siRNA-directed RNA cleavage and to suppress the assembly of the RISC *in vitro* (Lakatos et al. 2006; Schuck et al. 2013) or *in vivo* systems (Dunoyer et al. 2004). In addition, recent findings also demonstrate that P19 inhibits the spread of the siRNA duplex identified as the signal of systemic RNAi (Dunoyer et al. 2010). In a very recent report, Schott and co-workers strongly criticized the reliability of *in vitro* siRNA-binding assays, which are generally used to identify the molecular mechanism of RNAi inhibition. Using transgenic *Arabidopsis* lines expressing four distinct VSRs (the P38, the P15 protein of *Peanut clump virus*, the Hc-Pro of TuMV, and the tombusviral P19), they concluded that the siRNA/miRNA sequestration strategy of P19 is an exception instead of representing a common suppression strategy of VSRs. In contrast, previous studies using virus-infected plants have generated direct *in vivo* evidence, using pull-down approaches, that P19 s of CymRSV and *Carnation Italian ringspot virus* (CIRV), Hc-Pro of *Tobacco etch virus* (TEV), and P21 of *Beet yellows virus* efficiently bind vsiRNAs in virus-infected plants (Lakatos et al. 2004, 2006; Lozsa et al. 2008). Importantly, virus-expressed Hc-Pro also bound plant miRNA duplexes (Lakatos et al. 2006). However, P19 expressed by CIRV efficiently bound vsiRNAs but failed to bind miRNA duplexes (Lozsa et al. 2008). These findings indicate that transgenically expressed VSRs may not really reflect their normal biological function. It has also been reported that transgenically expressed P19 and HC-Pro VSRs prevent the essential siRNA and miRNA 2′-O methylation steps (Yu et al. 2006; Ebhardt et al. 2005), which appears to protect small RNAs from oligouridinilation and subsequent degradation (Li et al. 2005). Thus, it seems that the inhibitory effect of VSRs may require the temporal and spatial co-expression of the suppressor and the endogenous siRNAs and miRNAs (Lozsa et al. 2008).

The 2b protein of *Cucumovirus Tomato aspermy virus* (TAV) and B2 of the insect-infecting *Flock House virus* also show siRNA-binding activity; however, studies have shown that the structures of silencing suppressor proteins and their mode of binding siRNAs do not share any similarity (Chen et al. 2008; Chao et al. 2005).

In sum, plants fail to confine the virus infection and spread, since vsiRNAs are sequestered by siRNA-binding VSRs before they can be incorporated into the RNAi based–antiviral effector complexes.

The P14 protein of PoLV binds long and short dsRNAs (including duplex siRNAs) in a size-independent way (Merai et al. 2005, 2006). P14 may interact with viral dsRNAs, inhibiting the RNA-silencing machinery on two levels: (1) by siRNA sequestration (Merai et al. 2006), and (2) by interfering with DCL4-mediated vsiRNA processing. The inhibition of DCL4 by P38 has been confirmed experimentally (Deleris et al. 2006). In contrast to dsRNA-binding VSRs, the AC4 suppressor of *African cassava mosaic virus* binds single-stranded small RNAs bound by AGOs and prevents holo-RISC assembly. AC4 also inhibits miRNA-mediated negative regulation of endogenous genes (Chellappan et al. 2005). *Rice stripe virus* NS3 and *Grapevine virus A* P10 proteins are also able to sequester ss-siRNA molecules (Xiong et al. 2009; Zhou et al. 2006) implying, in part at least, a similar strategy to AC4.

A very similar outcome is achieved by adopting a completely different strategy in the case of the Crinivirus *Sweet potato chlorotic stunt virus* (SPCSV). An SPCSV-encoded RNaseIII endonuclease cleaves the 21-, 22- and 24-vsiRNAs into 14-bp products, which are inactive in the RNAi pathways (Cuellar et al. 2009).

The V2 protein from *Tomato yellow leaf curl virus* has been shown to be an efficient suppressor of RNAi (Zrachya et al. 2007; Glick et al. 2008), and V2 was proposed to interact with the tomato protein SGS3 in infected plant cells (Glick et al. 2008). However, *in vitro* studies on V2 show that it outcompetes SGS3 protein for binding a dsRNA with 5′ ssRNA overhangs, whereas a V2 mutant lacking the suppressor function *in vivo* cannot efficiently overcome SGS3 binding (Fukunaga and Doudna 2009). These findings not only predict a new type of RNA-binding VSR but also may reveal a new RNA intermediate, which is essential for SDS3-/RDR6-dependent siRNA formation in the plant (Kumakura et al. 2009).

3.2 Viral Suppressors Interacting with RNAi-Related Host Proteins

The inhibiton of antiviral RNAi also happens through direct or indirect interactions between VSRs and the protein components of RISC. The 2b protein of CMV, one of the first described VSRs (Brigneti et al. 1998), prevents the spread of the long-range silencing signal, thus facilitating systemic virus infection (Guo and Ding 2002). Consistently, the 2b protein of Fny-CMV has been found to physically interact with the PAZ and part of the PIWI-domain of AGO1, inhibiting the slicing activity of AGO1 (Zhang et al. 2006). The Fny-CMV 2b protein was found to preferentially co-localize with the AGO1 protein in the cell's nucleus and also in cytoplasmic foci (Mayers et al. 2000). Fny-CMV 2b protein expression phenocopies the *ago1–27* mutant phenotype (Zhang et al. 2006). The VSR mutant CMV-Δ2b can be rescued by the *dcl2/dcl4* host double mutant, which is impaired in vsiRNA production, suggesting that 2b is dispensable for infection and spread in a host defective in small RNA-directed immunity (Diaz-Pendon et al. 2007). The 2b protein is also known to bind siRNAs and long dsRNA (Goto et al. 2007) and to inhibit the production of viral secondary siRNAs (Diaz-Pendon et al. 2007). Thus, cucumovirus 2b proteins have a dual mode of silencing inhibition, either by sequestering siRNAs or by interacting with AGO1 and preventing RISC assembly.

The P0 protein of the phloem-limited poleroviruses also targets the AGO1 protein, the core component of RISC for degradation (Pazhouhandeh et al. 2006; Bortolamiol et al. 2007; Baumberger et al. 2007). The P0 protein is indispensable for viral infection since null mutations of P0 in *Beet western yellows virus* (BWYV) and *Potato leafroll virus* strongly diminish or completely abolish virus accumulation (Mayo and Ziegler-Graff 1996). In contrast to the RNA-binding VSRs, P0 has no RNA-binding activity (Zhang et al. 2006; Csorba et al. 2010). Instead, it interacts with the SCF family of E3-ligase S-phase kinase-related protein1 (SKP1) components, orthologous to *Arabidopsis* ASK1 and ASK2, by means of its minimal F-box motif and promotes AGO1 degradation (Pazhouhandeh et al. 2006; Baumberger et al. 2007; Bortolamiol et al. 2007). The mutation of the F-box motif of P0 leads to the loss of silencing suppressor activity, suggesting the involvement of the proteasome pathway in AGO1 degradation. However, P0-mediated AGO1 degradation is insensitive to proteasome inhibitors, which is inconsistent with the idea that AGO1 is targeted by P0 for ubiquitination and proteasome-dependent degradation. Furthermore, P0 does not appear to interact directly with AGO1 (Baumberger et al. 2007; Bortolamiol et al. 2007; Csorba et al. 2010). Indeed, it has been found that P0 cannot interfere with the slicer activity of pre-programmed siRNA-/miRNA-containing AGO1, but it prevents the *de novo* formation of siRNA-/miRNA-loaded AGO1 (Csorba et al. 2010). The transgenic expression of P0 in *Arabidopsis* leads to severe developmental abnormalities, similar to those induced by mutants affecting the miRNA pathways. This effect of P0 is accompanied by AGO1 protein decay *in planta* and enhanced levels of several miRNA-target transcripts (Bortolamiol et al. 2007).

Two recent studies illustrated a new strategy for VSR binding to AGO whereby the viral proteins mimic the cellular glycine/tryptophan (GW) repetitive motif (Azevedo et al. 2010; Giner et al. 2010). The GW linear peptide motif has been identified in different silencing-related proteins, and it functions as an "Ago hook" to interact with AGO proteins (Till and Ladurner 2007; Karlowski et al. 2010). The AGO-interacting GW motif has been identified in the largest subunit, NRPD1b, of RNA polymerase V in plants (El-Shami et al. 2007), in human GW182 protein (Behm-Ansmant et al. 2006; Eulalio et al. 2008; Liu et al. 2005), and in the Tas3 homologue of the GW182 RITS complex component in *Schizosaccharomyces pombe* (Buhler et al. 2006; Verdel et al. 2004). In addition, the RdDM effector KTF1 contains several GW motifs and was identified as an AGO4-binding element (Bies-Etheve et al. 2009; He et al. 2009). It was shown that the P38 VSR protein contains two GW repeats and interacts with *A. thaliana* AGO1 but not AGO4 (Azevedo et al. 2010). Direct AGO1 binding to the individual N-terminal and C-terminal GW motifs of P38 *in vitro* was also detected, and the interaction was abolished by changing GW to GA. Point mutations in the P38 GW residues are sufficient to abolish TCV virulence, which is restored in the *A. thaliana* ago1 hypomorphic mutant, uncovering both physical and genetic interactions between the two proteins. It has been suggested that P38 compromises small RNA loading into AGO1 by interacting with non-loaded AGO1 (Azevedo et al. 2010). It was also shown that AGO1 inactivation by P38 profoundly impacts the cellular availability of the four *Arabidopsis* DCLs, uncovering an AGO1-dependent, homeostatic network that functionally connects these factors (Azevedo et al. 2010). Recently, it was shown that P38 also interacts with *Arabidopsis* AGO2; however, this interaction also depends on a region of P38, which is outside of the GW motif (Zhang et al. 2012). The P1 VSR of *Sweet potato mild mottle ipomovirus* (SPMMV) also uses a GW Ago hook to inhibit siRNA-/miRNA-loaded RISC activity (Giner et al. 2010). It has been shown that the interaction between P1 and small RNA-loaded AGO1 is specific and direct. The suppression activity mapped to the N-terminal part of P1, containing three GW motifs, and site-directed mutagenesis proved that these motifs are essential for both binding and suppression of AGO1 functions. An attractive model is therefore suggested for the mechanism of silencing suppression mediated by P1, that is, it can mimic essential AGO1 interactors and outcompete them for RISC binding. Indeed, these AGO1 interactors have recently been identified in *Arabidopsis* (Garcia et al. 2012). Alternatively, P1 might compromise base-pairing between the small RNA-loaded AGO1 and the target RNA (Giner et al. 2010).

The action of P38 and P1 VSRs indicates a new mechanism for the suppression of RNAi by mimicking the plant GW proteins that hook up to AGOs, and it might represent a widespread strategy by pathogens to counteract RNAi-based host defenses.

3.3 Other RNAi Suppressor Strategies

Recent work has revealed a new strategy for antiviral silencing suppression through the specific induction of the plant miRNA miR-168, which has an important function in AGO1 homeostasis (Vaucheret et al. 2006; Varallyay 2010). It has been observed that CymRSV-infection enhances the expression of AGO1 mRNA; however, this was not followed by increased AGO1 protein levels. It was also shown that the P19 protein specifically increased the expression of miR-168, which controls the AGO1 mRNA level. By contrast, p19 suppressor mutant virus infection did not result in miR-168 induction and enhanced accumulation of the

AGO1 protein, suggesting that the P19 VSR inhibits the translation of AGO1 mRNA by enhancing the endogenous miR-168 level to alleviate the antiviral function of the AGO1 protein (Varallyay et al. 2010). However, the mechanism behind the specific induction of miR-168 and the inhibition of AGO1 translation is not fully understood. Importantly, it has also been demonstrated that in virus-infected plants, P19 is not able to efficiently bind miRNA duplexes, including miR-168 (Lozsa et al. 2008) suggesting that the specific induction of miR-168 by P19 is not mediated through miR-168 binding. The down-regulation of AGO1 protein may have some collateral effects similar to the recently described AGO1 quenching by P38, which impacts the cellular availability of the four *Arabidopsis* DCLs, compromising the AGO1-dependent homeostatic network (Azevedo et al. 2010). An insufficient level of AGO1 protein in virus-infected plants can lead to the mis-regulation of miRNA targets, resulting in disordered host cell gene expression, which could facilitate the development of viral symptoms.

Host factors involved in both RNAi and viral replication have been proposed to play roles in RNAi suppression during infection by *Red clover necrotic mosaic virus* (RCNMV). The putative host factor involved in both processes could be the DCL1 protein, because miRNA biogenesis is inhibited by virus replication and *dcl1* mutant plants show reduced susceptibility to RCNMV infection (Takeda et al. 2005). In the suggested scenario, DCL1 and its homologues are recruited by the viral replication complex and are therefore depleted from the silencing pathways.

4 Effects of Viral Suppressors on Endogenous RNAi Pathways and Induced Viral Symptoms

The antiviral RNAi pathway and the components of the endogenous RNAi pathways overlap therefore VSRs may also have the capacity to interfere with many aspects of these endogenous pathways. This hypothesis has been supported by the observation that many VSRs are pathogenicity determinants and responsible for virus-induced symptoms (Voinnet 2005). Furthermore, in many cases, transgenic plants expressing VSRs constitutively show developmental abnormalities that mimic the phenotype of miRNA deficient hypomorphic *dcl1* mutant plants (Chapman et al. 2004; Dunoyer et al. 2004; Kasschau et al. 2003; Mallory et al. 2002; Jay et al. 2011).

4.1 Viral Suppressors Interaction with Endogenous Small (si-, mi- and hcRNAs) RNAs

siRNA-binding VSRs (e.g., P19, Hc-Pro, P122, and P21) can interact with endogenous siRNA and miRNA biogenesis and can result in the overexpression of siRNA-regulated genes. Indeed, *Arabidopsis* plants stably expressing the

P19 protein altered the function of miRNAs and induced developmental defects by preventing the activity of the mature miRNA strand through the sequestering miRNA/miRNA* duplexes (Dunoyer et al. 2004). Furthermore, transgenic *Arabidopsis* lines expressing P38, P15, Hc-Pro, and P19 were also found to be compromised for the loading of siRNAs into AGO1. However, in the case of P38, the inhibition of siRNA loading was mainly the consequence of its direct binding to AGO1. Surprisingly, only P19 was found to efficiently prevent both siRNA and miRNA loading into AGO1. Interestingly, all VSR-transgenic plants efficiently inhibited miRNA-mediated target regulation (Schott et al. 2012) independently of miRNA sequestration. The use of these transgenic systems provides a tool to study the components of the RNAi pathways. However, they do not necessarily reproduce faithfully the molecular effects of viral infections on the RNAi pathways. It is important to note that during infections, the expression of VSRs is restricted to virus-infected tissues and compartments and is also limited by time.

Other VSRs target different stages of the RNAi pathway. The V2 VSR of TYLCV interacts directly with the tomato SGS3 protein. Since SGS3 is also required for the endogenous tasiRNA pathway, the potential inactivation of SGS3 by V2 also may affect the development of TYLCV disease symptoms. Accordingly, both TYLCV-infected tomatoes and tomato plants carrying a mutation within the *sgs3* locus exhibit an aberrant leaf phenotype (Glick et al. 2008). Some of the VSRs can inhibit HEN1-mediated methylation (Lozsa et al. 2008). All endogenous classes of plant small RNAs are 2'-O methylated by HEN1, which protects them from degradation (Yu et al. 2005; Park et al. 2005; Li et al. 2005). In agreement with this, transgenic *Arabidopsis* plants expressing different VSR proteins, including HC-Pro, P21, and P19, inhibited HEN1-mediated methylation (Yu et al. 2006). However, in natural virus infection, it was found that the HEN1-dependent methylation of viral siRNAs was variable, depending on the virus species. It was shown that P19 had only a slight effect on viral siRNA 3′ modification; however, Hc-Pro significantly inhibited the 3′modification of vsiRNAs and miRNAs as well (Lozsa et al. 2008). These results further emphasize the intrinsic difference between the use of VSR-expressing transgenic plants and natural infections.

VSRs can also suppress RNAi-based antiviral immunity through protein–protein interaction. Several VSRs were shown to interact with AGO1, including P38 (Azevedo et al. 2010), P1 of SPMMV (Giner et al. 2010) and the 2b protein of the CMV Fny strain (Zhang et al. 2006). The poleroviral P0 and P0(PE) of the Enamovirus *Pea enation mosaic virus-1* VSR proteins were suggested to act as an F-Box protein targeting AGO proteins for degradation, thereby preventing RISC assembly (Baumberger et al. 2007; Bortolamiol et al. 2007; Csorba et al. 2010; Derrien et al. 2012; Fusaro et al. 2012). Inhibiting AGO1 function during virus infection can influence the regulation of miRNA-regulated endogenous genes. Accordingly, 2b, P0, and P1 are pathogenicity determinants (Pfeffer et al. 2002; Voinnet et al. 1999; Brigneti et al. 1998), and AGO1 suppression by P38 profoundly modifies a homeostatic interaction network linking the four *Arabidopsis* DCLs. This is thought to result from reduced levels of miR162-containing AGO1, which enhances DCL1 accumulation which in turn decreases DCL4 and DCL3 levels through an unidentified mechanism (Azevedo et al. 2010).

In *Arabidopsis,* the AGO1 level is maintained by an autoregulatory feedback loop by miR-168 (Vaucheret et al. 2006). The P19 VSR of CymRSV specifically increases the expression of miR-168 to alleviate plant antiviral function. Therefore, P19 can also inhibit the translation of AGO1 mRNA by enhancing endogenous miR-168 levels (Varallyay et al. 2010).

In plants, siRNAs can also target homologous genomic DNA sequences for cytosine methylation, a process known as RNA-directed DNA methylation (RdDM). This mechanism transcriptionally silences many endogenous loci, including transposons and DNA repeats, and can be utilized as a defense against DNA viruses (Matzke et al. 2009). As a consequence, several viruses encode proteins that can suppress RdDM including the P6 protein of CaMV (Haas et al. 2008) and the C2 protein (also known as AL2 and AC2) of geminiviruses (Wang et al. 2005; Voinnet et al. 1999; van Wezel et al. 2002; Buchmann et al. 2009; Raja et al. 2008; Yang et al. 2012). It was shown that geminivirus AL2 and L2 (AL2 related) proteins interact with and inactivate an adenosine kinase, which is required for efficient production of S-adenosyl methionine, an essential methyltransferase cofactor. This inhibition protects viral DNA from methylation (Raja et al. 2008). Indeed, transgenic expression of AL2 or L2 causes a global reduction in cytosine methylation and therefore geminivirus proteins can reverse RdDM by non-specifically inhibiting cellular transmethylation reactions (Buchmann et al. 2009).

4.2 The Effect of Viral siRNAs on Host Transcripts

During virus infection, DCL4 and DCL2 act in a hierarchical manner to produce 21- and 22-nt-long viral siRNAs (Deleris et al. 2006; Fusaro et al. 2006). These vsiRNAs can be incorporated into AGO complexes and can inhibit the expression of viral RNAs (Pantaleo et al. 2007; Omarov et al. 2007; Llave 2010). Furthermore, it was also demonstrated by several groups that recombinant viruses carrying endogenous sequences can guide AGO-containing effector complexes to endogenous mRNA transcripts (Becker and Lange 2010). These observations led to the hypothesis that vsiRNAs can also specifically regulate host gene expression if host genes share sequence complementarity with vsiRNA. Accordingly, bioinformatic analysis of small RNA deep–sequencing data from a crucifer-infecting strain of *Tobacco mosaic virus* (TMV-Cg)-infected *Arabidopsis* showed that a large set of host genes could be potentially targeted by TMV-Cg-derived vsiRNAs. A small subset of the predicted host targets were selected for experimental validation and two of them, which encode a cleavage and polyadenylation specificity factor (CPSF30, At1g30460) and an unknown protein similar to translocon-associated protein alpha (TRAP α, At2g16595), respectively, yielded a positive result upon cleavage validation by 5′-RACE assays (Qi et al. 2009). In another study, it was reported that vsiRNAs derived from the CaMV 35S leader sequence bear near-perfect sequence complementarity to *Arabidopsis* transcripts, and evidence was provided that at least one of the vsiRNAs specifically down-regulated the *Arabidopsis* At1g76950 mRNA (Moissiard and Voinnet 2006). Moreover, it was found in two independent studies that vsiRNAs from the Y-satellite of CMV specifically cleave the mRNA of *Nicotiana tabacum* magnesium protoporphyrin cheletase subunit I (CHLI, the key gene involved in chlorophyll synthesis), thus inducing the typical yellow symptoms in tobacco. It was demonstrated using 5'-RACE, GFP sensor experiment, and transgenic plants that vsiRNA-directed cleavage of the CHLI host gene is solely responsible for the Y-satellite-induced symptoms (Shimura et al. 2011; Smith et al. 2011). Similarly, viroid-derived siRNAs resulting from the host defensive response, via RNAi, can target plant cell mRNAs and trigger a signal cascade leading to symptoms such as severe leaf albinism (Navarro et al. 2012). Moreover, a recent study analyzing the data sets by integrating the vsiRNAs and degradom libraries from *Vitis vinifera* identified a pool of vsiRNAs, which are responsible for the cleavage of host mRNAs. Among them, several encode putative proteins involved in ribosome biogenesis and in biotic and abiotic stresses. The bioinformatic analysis of these data also showed that a large fraction of vsiRNAs did not induce host mRNA cleavage, suggesting that relatively few vsiRNAs are involved in the inhibition of host mRNAs (Miozzi et al. 2013).

vsiRNAs can also target homologous genomic DNA sequences through RdDM to interfere with gene expression at a transcriptional level. A PVX vector carrying the promoter sequence of the GFP transgene triggered transcriptional gene silencing through the sequence-specific epigenetic modification of DNA and chromatin (Jones et al. 1999, 2001). Furthermore, CaMV infection of a transgenic oilseed rape line containing an herbicide tolerance gene regulated by the 35S promoter, altered the expression of the herbicide tolerance gene such that plants became susceptible to the herbicide. Susceptibility to the herbicide was most likely the result of transcriptional gene silencing of the transgene (Al-Kaff et al. 1998).

These observations suggest that viruses might also be able to harness the RNAi-based antiviral plant response for the benefit of the virus and that vsiRNAs which create a cellular environment more suitable for viral proliferation (i.e., by targeting host defense factors) may be positively selected (Ruiz-Ferrer and Voinnet 2009).

5 RNAi may Regulate Innate Immunity in Plants

In plants, RNAi is considered as the primary antiviral defense system (Ding and Voinnet 2007). However, plants also respond to pathogen infection using innate immunity, which is also a conserved and evolutionary ancient mechanism (Ding and Voinnet 2007; Jones and Dangl 2006; Chisholm et al. 2006). To recognize invading pathogens, plant genomes contain two major classes of innate immune receptors. The primary immune response comprises the pattern recognition receptors (PRRs) that recognize conserved pathogen-associated molecular patterns (PAMPs) by the diverse transmembrane pattern recognition receptors (PRRs) and activate PAMP-triggered immunity (PTI). However, to counteract PTI and establish infection in susceptible host, pathogens can suppress PTI, usually through secreted proteins called effectors. To recognize effector proteins, plant genomes have also evolved large numbers of resistance (R) genes with nucleotide-binding (NB) and leucine-rich repeat (LRR) domains (Meyers et al. 2005), which function as a second line of defense that mediate intercellular recognition of specific pathogen effectors and initiate effector-triggered immunity (ETI). ETI is considered to be an accelerated and amplified version of the PTI response. The ETI signaling cascade often leads to disease resistance, usually through a hypersensitive cell death response (HR) at the infection site (Jones and Dangl 2006). There is no evidence for a viral PAMP so far. However, several viral proteins including VSRs (e.g., P126 of TMV, P38) can function as effector proteins that trigger R gene-dependent hypersensitive responses in specific hosts (Zvereva and Pooggin 2012). Although R-genes provide protection from diverse pathogens, their large numbers and cell death-triggering activity are potential threats to plant fitness (Tian et al. 2003). Despite the enormous agricultural importance of R-genes, the mechanisms regulating their expression are not well understood. Recently, small RNAs emerged as important gene expression regulators of various cellular processes. In plants, based on their biogenesis, we distinguish two main classes of endogeneous small RNAs: miRNAs and siRNAs. Despite differences in their biogenesis, one group of siRNAs also regulates protein-coding genes very similarly to miRNAs and these are called ta-siRNAs or secondary siRNAs. The formation of ta-siRNAs from a non-coding RNA (TAS) precursor is dependent on a miRNA trigger (Fig. 1). Recent studies have revealed the presence of a regulatory cascade that affects a large portion of NB-LRR resistance genes. The initiator of the cascade is a member of the miR-482 superfamily, and this targets the coding sequence for the P-loop motif in the mRNA sequences of NB-LRR genes. This targeting causes mRNA cleavage and initiates the production of RDR-6-dependent secondary siRNAs that also have the potential to regulate homologous NB-LRR resistance genes (Shivaprasad et al. 2012; Li et al. 2012; Zhai et al. 2011). The miR-482-mediated silencing of NB-LRR transcripts is lost in virus- and bacteria-infected leaves, very likely due to the action of pathogen-encoded proteins. Based on these data, it was proposed that miRNAs and secondary siRNAs regulate the pathogen-inducible expression of NB-LRR innate immune receptor proteins and the RNAi pathway could also function to achieve inducible expression of NB-LRR genes (Shivaprasad et al. 2012).

6 The Effect of Environmental Signals on RNAi-Based Defense Against Viral Pathogens

Plant–virus interactions are also strongly influenced by environmental factors, especially by temperature. Outbreaks of virus diseases are frequently associated with low temperature, while at high temperature, viral symptoms are often attenuated (heat masking) and the virus load of the infected plants is low (Johnson 1922; Hull 2002). Several studies showed that rapid spread of virus diseases and the development of severe symptoms is often associated with low temperatures (Hine et al. 1970; Gerik et al. 1990). Furthermore, thermotherapy is used to free vegetative material from infecting viruses (Manganaris et al. 2003). RNAi is the primary antiviral defense system in plants, and it was reported that RNAi-mediated plant defense is temperature dependent. At low ambient temperature (15 °C), the virus-triggered RNAi pathway is inhibited and the level of vsiRNAs is dramatically reduced, thus leading to enhanced virus susceptibility. In contrast, the antiviral RNAi pathway is activated and the amount of vsiRNAs gradually increases with rising temperature (Szittya et al. 2003). Moreover, it was reported that ssDNA geminivirus-induced RNAi is also altered by temperature, similar to RNA viruses. Cassava geminivirus-induced RNAi was increased by rising temperature, with the appearance of less symptomatic newly developed leaves, and vsiRNAs accumulated to higher levels. However, there was a difference in the vsiRNA accumulation level among the non-recovery-type and recovery-type geminiviruses at high ambient temperature (30 C). Recovery-type viruses accumulated moderately, while non-recovery-type viruses accumulated very high level of vsiRNAs during infection at high temperature (Chellappan et al. 2005). To determine the underlining mechanism of the enhanced RNAi-mediated antiviral defense at higher temperature, a TCV-*Arabidopsis* model system was used in a recent study. *Arabidopsis* Col-0 plants, together with plants with defects in RNAi pathway components, were infected with TCV and kept at 18–26 °C after inoculation. Contradicting previous reports (Qu et al. 2005; Scholthof et al. 2011), TCV actually produced higher levels of viral RNAs at 26° C than at 18 °C, suggesting that TCV RNA replication is stimulated by elevated growth temperature. However, as was expected, host RNAi activity was likewise stimulated at higher temperature. Plants containing loss-of-function mutations within DCL2, AGO2, and HEN1 died on TCV infection at higher temperature. This suggests the existence of an RNAi pathway comprised of at least DCL2, AGO2, and HEN1 that functions to partially attenuate the stress caused by TCV infections at higher temperature (Zhang et al. 2012).

7 Concluding Remarks and Future Direction

Plant viruses have provided a large contribution to our understanding of RNAi pathways. The high diversity of VSRs interfering with different steps of the plant RNAi pathway provides an invaluable molecular tool to dissect endogenous RNAi pathways, although in many cases, we are far from fully understanding the mechanisms behind this suppression. Indeed, the widely used transient expression assays or VSR-expressing transgenic plants may lead to an oversimplified view of their function. On the one hand, the use of these systems as molecular tools yields novel insights and broadens our understanding of the molecular details of the plant's RNAi pathways. As a result, during the last decade, we have learned a lot about RNAi-based antiviral immunity in plants. However, this is very likely just the beginning of a long journey. We still do not understand how viral RNAs and replication intermediates are recognized by the plant mRNA quality control and surveillance machinery. Since many viruses replicate in a well-defined compartment, it is also not clear when and where this recognition happens. We do not know whether VSRs have endogenous interactors that are essential for silencing suppression. Further understanding of the contribution of vsiRNAs to disease induction is needed, and we also need to investigate their role in possible epigenetic modifications. Since DCL4-produced tasiRNAs were shown to function non-cell autonomously (Chitwood et al. 2009), it also needs to be determined whether DCL4-produced vsiRNAs can move from cell to cell and their possible influence on gene expression in neighboring cells need to be understood.

Acknowledgments We apologize to those colleagues whose work we were unable to cite because of space restrictions. GS and JB are funded by grants from OTKA, K-101793 and NK-105850, respectively.

References

Ahlquist P (2002) RNA-dependent RNA polymerases, viruses, and RNA silencing. Science 296:1270–1273

Akbergenov R, Si-Ammour A, Blevins T, Amin I, Kutter C, Vanderschuren H, Zhang P, Gruissem W, Meins F Jr, Hohn T, Pooggin MM (2006) Molecular characterization of geminivirus-derived small RNAs in different plant species. Nucleic Acids Res 34:462–471

Al-Kaff NS, Covey SN, Kreike MM, Page AM, Pinder R, Dale PJ (1998) Transcriptional and posttranscriptional plant gene silencing in response to a pathogen. Science 279:2113–2115

Almeida R, Allshire RC (2005) RNA silencing and genome regulation. Trends Cell Biol 15:251–258

Ameres SL, Martinez J, Schroeder R (2007) Molecular basis for target RNA recognition and cleavage by human RISC. Cell 130:101–112

Anandalakshmi R, Pruss GJ, Ge X, Marathe R, Mallory AC, Smith TH, Vance VB (1998) A viral suppressor of gene silencing in plants. Proc Natl Acad Sci U S A 95:13079–13084

Aregger M, Borah BK, Seguin J, Rajeswaran R, Gubaeva EG, Zvereva AS, Windels D, Vazquez F, Blevins T, Farinelli L, Pooggin MM (2012) Primary and secondary siRNAs in geminivirus-induced gene silencing. PLoS Pathog 8:e1002941

Azevedo J, Garcia D, Pontier D, Ohnesorge S, Yu A, Garcia S, Braun L, Bergdoll M, Hakimi MA, Lagrange T, Voinnet O (2010) Argonaute quenching and global changes in Dicer homeostasis caused by a pathogen-encoded GW repeat protein. Genes Dev 24:904–915

Bao H, Guo H, Wang J, Zhou R, Lu X, Shi S (2009) MapView: visualization of short reads alignment on a desktop computer. Bioinformatics 25:1554–1555

Bartel DP (2004) MicroRNAs: genomics, biogenesis, mechanism, and function. Cell 116:281–297

Baulcombe D (2004) RNA silencing in plants. Nature 431:356–363

Baumberger N, Tsai CH, Lie M, Havecker E, Baulcombe DC (2007) The Polerovirus silencing suppressor P0 targets ARGONAUTE proteins for degradation. Curr Biol 17:1609–1614

Becker A, Lange M (2010) VIGS–genomics goes functional. Trends Plant Sci 15:1–4

Behm-Ansmant I, Rehwinkel J, Doerks T, Stark A, Bork P, Izaurralde E (2006) mRNA degradation by miRNAs and GW182 requires both CCR4:NOT deadenylase and DCP1:DCP2 decapping complexes. Genes Dev 20:1885–1898

Bernstein E, Caudy AA, Hammond SM, Hannon GJ (2001) Role for a bidentate ribonuclease in the initiation step of RNA interference. Nature 409:363–366
Bhattacharjee S, Zamora A, Azhar MT, Sacco MA, Lambert LH, Moffett P (2009) Virus resistance induced by NB-LRR proteins involves Argonaute4-dependent translational control. Plant J 58:940–951
Bies-Etheve N, Pontier D, Lahmy S, Picart C, Vega D, Cooke R, Lagrange T (2009) RNA-directed DNA methylation requires an AGO4-interacting member of the SPT5 elongation factor family. EMBO Rep 10:649–654
Blevins T, Rajeswaran R, Shivaprasad PV, Beknazariants D, Si-Ammour A, Park HS, Vazquez F, Robertson D, Meins F Jr, Hohn T, Pooggin MM (2006) Four plant Dicers mediate viral small RNA biogenesis and DNA virus induced silencing. Nucleic Acids Res 34:6233–6246
Blevins T, Rajeswaran R, Aregger M, Borah BK, Schepetilnikov M, Baerlocher L, Farinelli L, Meins F Jr, Hohn T, Pooggin MM (2011) Massive production of small RNAs from a non-coding region of Cauliflower mosaic virus in plant defense and viral counter-defense. Nucleic Acids Res 39:5003–5014
Bortolamiol D, Pazhouhandeh M, Marrocco K, Genschik P, Ziegler-Graff V (2007) The Polerovirus F box protein P0 targets ARGONAUTE1 to suppress RNA silencing. Curr Biol 17:1615–1621
Bouche N, Lauressergues D, Gasciolli V, Vaucheret H (2006) An antagonistic function for Arabidopsis DCL2 in development and a new function for DCL4 in generating viral siRNAs. EMBO J 25:3347–3356
Brigneti G, Voinnet O, Li WX, Ji LH, Ding SW, Baulcombe DC (1998) Viral pathogenicity determinants are suppressors of transgene silencing in Nicotiana benthamiana. EMBO J 17:6739–6746
Brodersen P, Voinnet O (2006) The diversity of RNA silencing pathways in plants. Trends Genet 22:268–280
Brodersen P, Sakvarelidze-Achard L, Bruun-Rasmussen M, Dunoyer P, Yamamoto YY, Sieburth L, Voinnet O (2008) Widespread translational inhibition by plant miRNAs and siRNAs. Science 320:1185–1190
Buchmann RC, Asad S, Wolf JN, Mohannath G, Bisaro DM (2009) Geminivirus AL2 and L2 proteins suppress transcriptional gene silencing and cause genome-wide reductions in cytosine methylation. J Virol 83:5005–5013
Buhler M, Verdel A, Moazed D (2006) Tethering RITS to a nascent transcript initiates RNAi- and heterochromatin-dependent gene silencing. Cell 125:873–886
Burgyan J, Havelda Z (2011) Viral suppressors of RNA silencing. Trends Plant Sci 16:265–272
Burgyan J, Rubino L, Russo M (1996) The 5′-terminal region of a tombusvirus genome determines the origin of multivesicular bodies. J Gen Virol 77:1967–1974
Carbonell A, Fahlgren N, Garcia-Ruiz H, Gilbert KB, Montgomery TA, Nguyen T, Cuperus JT, Carrington JC (2012) Functional analysis of three Arabidopsis ARGONAUTES using slicer-defective mutants. Plant Cell 24:3613–3629
Chao JA, Lee JH, Chapados BR, Debler EW, Schneemann A, Williamson JR (2005) Dual modes of RNA-silencing suppression by Flock House virus protein B2. Nat Struct Mol Biol 12:952–957
Chapman EJ, Carrington JC (2007) Specialization and evolution of endogenous small RNA pathways. Nat Rev Genet 8:884–896
Chapman EJ, Prokhnevsky AI, Gopinath K, Dolja VV, Carrington JC (2004) Viral RNA silencing suppressors inhibit the microRNA pathway at an intermediate step. Genes Dev 18:1179–1186
Chellappan P, Vanitharani R, Fauquet CM (2005) MicroRNA-binding viral protein interferes with Arabidopsis development. Proc Natl Acad Sci U S A 102:10381–10386
Chen HY, Yang J, Lin C, Yuan YA (2008) Structural basis for RNA-silencing suppression by Tomato aspermy virus protein 2b. EMBO Rep 9:754–760
Chisholm ST, Coaker G, Day B, Staskawicz BJ (2006) Host-microbe interactions: shaping the evolution of the plant immune response. Cell 124:803–814

Chitwood DH, Nogueira FT, Howell MD, Montgomery TA, Carrington JC, Timmermans MC (2009) Pattern formation via small RNA mobility. Genes Dev 23:549–554

Csorba T, Bovi A, Dalmay T, Burgyan J (2007) The p122 subunit of Tobacco Mosaic Virus replicase is a potent silencing suppressor and compromises both small interfering RNA- and microRNA-mediated pathways. J Virol 81:11768–11780

Csorba T, Pantaleo V, Burgyan J (2009) RNA silencing: an antiviral mechanism. Adv Virus Res 75:35–71

Csorba T, Lozsa R, Hutvagner G, Burgyan J (2010) Polerovirus protein P0 prevents the assembly of small RNA-containing RISC complexes and leads to degradation of ARGONAUTE1. Plant J 62:463–472

Cuellar WJ, Kreuze JF, Rajamaki ML, Cruzado KR, Untiveros M, Valkonen JP (2009) Elimination of antiviral defense by viral RNase III. Proc Natl Acad Sci U S A 106:10354–10358

Dalmay T, Hamilton A, Rudd S, Angell S, Baulcombe DC (2000) An RNA-dependent RNA polymerase gene in Arabidopsis is required for posttranscriptional gene silencing mediated by a transgene but not by a virus. Cell 101:543–553

Dalmay T, Horsefield R, Braunstein TH, Baulcombe DC (2001) SDE3 encodes an RNA helicase required for post-transcriptional gene silencing in Arabidopsis. EMBO J 20:2069–2078

Deleris A, Gallego-Bartolome J, Bao J, Kasschau KD, Carrington JC, Voinnet O (2006) Hierarchical action and inhibition of plant Dicer-like proteins in antiviral defense. Science 313:68–71

Derrien B, Baumberger N, Schepetilnikov M, Viotti C, De Cillia J, Ziegler-Graff V, Isono E, Schumacher K, Genschik P (2012) Degradation of the antiviral component ARGONAUTE1 by the autophagy pathway. Proc Natl Acad Sci U S A 109:15942–15946

Diaz-Pendon JA, Ding SW (2008) Direct and indirect roles of viral suppressors of RNA silencing in pathogenesis. Annu Rev Phytopathol 46:303–326

Diaz-Pendon JA, Li F, Li WX, Ding SW (2007) Suppression of antiviral silencing by cucumber mosaic virus 2b protein in Arabidopsis is associated with drastically reduced accumulation of three classes of viral small interfering RNAs. Plant Cell 19:2053–2063

Ding SW (2010) RNA-based antiviral immunity. Nat Rev Immunol 10:632–644

Ding SW, Voinnet O (2007) Antiviral immunity directed by small RNAs. Cell 130:413–426

Donaire L, Barajas D, Martinez-Garcia B, Martinez-Priego L, Pagan I, Llave C (2008) Structural and genetic requirements for the biogenesis of tobacco rattle virus-derived small interfering RNAs. J Virol 82:5167–5177

Donaire L, Wang Y, Gonzalez-Ibeas D, Mayer KF, Aranda MA, Llave C (2009) Deep-sequencing of plant viral small RNAs reveals effective and widespread targeting of viral genomes. Virology 392:203–214

Dunoyer P, Lecellier CH, Parizotto EA, Himber C, Voinnet O (2004) Probing the microRNA and small interfering RNA pathways with virus-encoded suppressors of RNA silencing. Plant Cell 16:1235–1250

Dunoyer P, Schott G, Himber C, Meyer D, Takeda A, Carrington JC, Voinnet O (2010) Small RNA duplexes function as mobile silencing signals between plant cells. Science 328:912–916

Eamens A, Vaistij FE, Jones L (2008) NRPD1a and NRPD1b are required to maintain post-transcriptional RNA silencing and RNA-directed DNA methylation in Arabidopsis. Plant J 55:596–606

Ebhardt HA, Thi EP, Wang MB, Unrau PJ (2005) Extensive 3′ modification of plant small RNAs is modulated by helper component-proteinase expression. Proc Natl Acad Sci U S A 102:13398–13403

El-Shami M, Pontier D, Lahmy S, Braun L, Picart C, Vega D, Hakimi MA, Jacobsen SE, Cooke R, Lagrange T (2007) Reiterated WG/GW motifs form functionally and evolutionarily conserved ARGONAUTE-binding platforms in RNAi-related components. Genes Dev 21:2539–2544

Eulalio A, Huntzinger E, Izaurralde E (2008) GW182 interaction with Argonaute is essential for miRNA-mediated translational repression and mRNA decay. Nat Struct Mol Biol 15:346–353

Fukunaga R, Doudna JA (2009) dsRNA with 5′ overhangs contributes to endogenous and antiviral RNA silencing pathways in plants. EMBO J 28:545–555
Fusaro AF, Matthew L, Smith NA, Curtin SJ, Dedic-Hagan J, Ellacott GA, Watson JM, Wang MB, Brosnan C, Carroll BJ, Waterhouse PM (2006) RNA interference-inducing hairpin RNAs in plants act through the viral defence pathway. EMBO Rep 7:1168–1175
Fusaro AF, Correa RL, Nakasugi K, Jackson C, Kawchuk L, Vaslin MF, Waterhouse PM (2012) The Enamovirus P0 protein is a silencing suppressor which inhibits local and systemic RNA silencing through AGO1 degradation. Virology 426:178–187
Garcia D, Garcia S, Pontier D, Marchais A, Renou JP, Lagrange T, Voinnet O (2012) Ago hook and RNA helicase motifs underpin dual roles for SDE3 in antiviral defense and silencing of nonconserved intergenic regions. Mol Cell 48:109–120
Garcia-Ruiz H, Takeda A, Chapman EJ, Sullivan CM, Fahlgren N, Brempelis KJ, Carrington JC (2010) Arabidopsis RNA-dependent RNA polymerases and dicer-like proteins in antiviral defense and small interfering RNA biogenesis during Turnip Mosaic Virus infection. Plant Cell 22:481–496
Gerik JS, Duffus JE, Perry R, Stenger DC, Van Maren AF (1990) Etiology of tomato plant decline in the California desert. Phytopathology 80:1352–1356
Giner A, Lakatos L, Garcia-Chapa M, Lopez-Moya JJ, Burgyan J (2010) Viral protein inhibits RISC activity by argonaute binding through conserved WG/GW motifs. PLoS Pathog 6:e1000996
Glick E, Zrachya A, Levy Y, Mett A, Gidoni D, Belausov E, Citovsky V, Gafni Y (2008) Interaction with host SGS3 is required for suppression of RNA silencing by tomato yellow leaf curl virus V2 protein. Proc Natl Acad Sci U S A 105:157–161
Goto K, Kobori T, Kosaka Y, Natsuaki T, Masuta C (2007) Characterization of silencing suppressor 2b of cucumber mosaic virus based on examination of its small RNA-binding abilities. Plant Cell Physiol 48:1050–1060
Guo HS, Ding SW (2002) A viral protein inhibits the long range signaling activity of the gene silencing signal. EMBO J 21:398–407
Haas G, Azevedo J, Moissiard G, Geldreich A, Himber C, Bureau M, Fukuhara T, Keller M, Voinnet O (2008) Nuclear import of CaMV P6 is required for infection and suppression of the RNA silencing factor DRB4. EMBO J 27:2102–2112
Hamilton AJ, Baulcombe DC (1999) A species of small antisense RNA in posttranscriptional gene silencing in plants. Science 286:950–952
Hammond SM, Bernstein E, Beach D, Hannon GJ (2000) An RNA-directed nuclease mediates post-transcriptional gene silencing in Drosophila cells. Nature 404:293–296
Harvey JJ, Lewsey MG, Patel K, Westwood J, Heimstadt S, Carr JP, Baulcombe DC (2011) An antiviral defense role of AGO2 in plants. PLoS ONE 6:e14639
He XJ, Hsu YF, Zhu S, Wierzbicki AT, Pontes O, Pikaard CS, Liu HL, Wang CS, Jin H, Zhu JK (2009) An effector of RNA-directed DNA methylation in arabidopsis is an ARGONAUTE 4- and RNA-binding protein. Cell 137:498–508
Hine RB, Osborne WE, Dennis RE (1970) Elevation and temperature effects on severity of maize dwarf mosaic virus in sorghum in Arizona. Plant Disease Reporter 54:1064–1068
Ho T, Pallett D, Rusholme R, Dalmay T, Wang H (2006) A simplified method for cloning of short interfering RNAs from Brassica juncea infected with Turnip mosaic potyvirus and Turnip crinkle carmovirus. J Virol Methods 136:217–223
Hull R (2002) Matthews' Plant Virology, 4th edn. Academic Press, San Diego
Hutvagner G, Simard MJ (2008) Argonaute proteins: key players in RNA silencing. Nat Rev Mol Cell Biol 9:22–32
Iki T, Yoshikawa M, Nishikiori M, Jaudal MC, Matsumoto-Yokoyama E, Mitsuhara I, Meshi T, Ishikawa M (2010) In vitro assembly of plant RNA-induced silencing complexes facilitated by molecular chaperone HSP90. Mol Cell 39:282–291
Iki T, Yoshikawa M, Meshi T, Ishikawa M (2011) Cyclophilin 40 facilitates HSP90-mediated RISC assembly in plants. EMBO J 31:267–278
Itaya A, Zhong X, Bundschuh R, Qi Y, Wang Y, Takeda R, Harris AR, Molina C, Nelson RS, Ding B (2007) A structured viroid RNA serves as a substrate for dicer-like cleavage to

produce biologically active small RNAs but is resistant to RNA-induced silencing complex-mediated degradation. J Virol 81:2980–2994
Jaubert M, Bhattacharjee S, Mello AF, Perry KL, Moffett P (2011) ARGONAUTE2 mediates RNA-silencing antiviral defenses against Potato virus X in Arabidopsis. Plant Physiol 156:1556–1564
Jay F, Wang Y, Yu A, Taconnat L, Pelletier S, Colot V, Renou JP, Voinnet O (2011) Misregulation of AUXIN RESPONSE FACTOR 8 underlies the developmental abnormalities caused by three distinct viral silencing suppressors in Arabidopsis. PLoS Pathog 7:e1002035
Johnson J (1922) The relation of air temperature to the mosaic disease of potatoes and other plants. Phytopathology 12:438–440
Jones JD, Dangl JL (2006) The plant immune system. Nature 444:323–329
Jones L, Hamilton AJ, Voinnet O, Thomas CL, Maule AJ, Baulcombe DC (1999) RNA-DNA interactions and DNA methylation in post-transcriptional gene silencing. Plant Cell 11:2291–2301
Jones L, Ratcliff F, Baulcombe DC (2001) RNA-directed transcriptional gene silencing in plants can be inherited independently of the RNA trigger and requires Met1 for maintenance. Curr Biol 11:747–757
Karlowski WM, Zielezinski A, Carrere J, Pontier D, Lagrange T, Cooke R (2010) Genome-wide computational identification of WG/GW Argonaute-binding proteins in Arabidopsis. Nucleic Acids Res 38:4231–4245
Kasschau KD, Carrington JC (1998) A counterdefensive strategy of plant viruses: suppression of posttranscriptional gene silencing. Cell 95:461–470
Kasschau KD, Xie Z, Allen E, Llave C, Chapman EJ, Krizan KA, Carrington JC (2003) P1/HC-Pro, a viral suppressor of RNA silencing, interferes with Arabidopsis development and miRNA unction. Dev Cell 4:205–217
Kumakura N, Takeda A, Fujioka Y, Motose H, Takano R, Watanabe Y (2009) SGS3 and RDR6 interact and colocalize in cytoplasmic SGS3/RDR6-bodies. FEBS Lett 583:1261–1266
Lakatos L, Szittya G, Silhavy D, Burgyan J (2004) Molecular mechanism of RNA silencing suppression mediated by p19 protein of tombusviruses. EMBO J 23:876–884 Epub 2004 Feb 2019
Lakatos L, Csorba T, Pantaleo V, Chapman EJ, Carrington JC, Liu YP, Dolja VV, Calvino LF, Lopez-Moya JJ, Burgyan J (2006) Small RNA binding is a common strategy to suppress RNA silencing by several viral suppressors. EMBO J 25:2768–2780
Lanet E, Delannoy E, Sormani R, Floris M, Brodersen P, Crete P, Voinnet O, Robaglia C (2009) Biochemical Evidence for Translational Repression by Arabidopsis MicroRNAs. Plant Cell 21:1762–1768
Li J, Yang Z, Yu B, Liu J, Chen X (2005) Methylation protects miRNAs and siRNAs from a 3′-end uridylation activity in Arabidopsis. Curr Biol 15:1501–1507
Li F, Pignatta D, Bendix C, Brunkard JO, Cohn MM, Tung J, Sun H, Kumar P, Baker B (2012) MicroRNA regulation of plant innate immune receptors. Proc Natl Acad Sci U S A 109:1790–1795
Liu J, Carmell MA, Rivas FV, Marsden CG, Thomson JM, Song JJ, Hammond SM, Joshua-Tor L, Hannon GJ (2004) Argonaute2 is the catalytic engine of mammalian RNAi. Science 305:1437–1441
Liu J, Rivas FV, Wohlschlegel J, Yates JR 3rd, Parker R, Hannon GJ (2005) A role for the P-body component GW182 in microRNA function. Nat Cell Biol 7:1261–1266
Llave C (2010) Virus-derived small interfering RNAs at the core of plant-virus interactions. Trends Plant Sci 15:701–707
Love AJ, Laird J, Holt J, Hamilton AJ, Sadanandom A, Milner JJ (2007) Cauliflower mosaic virus protein P6 is a suppressor of RNA silencing. J Gen Virol 88:3439–3444
Lozsa R, Csorba T, Lakatos L, Burgyan J (2008) Inhibition of 3′ modification of small RNAs in virus-infected plants require spatial and temporal co-expression of small RNAs and viral silencing-suppressor proteins. Nucleic Acids Res 36:4099–4107
Mallory AC, Reinhart BJ, Bartel D, Vance VB, Bowman LH (2002) A viral suppressor of RNA silencing differentially regulates the accumulation of short interfering RNAs and micro-RNAs in tobacco. Proc Natl Acad Sci U S A 99:15228–15233

Manganaris GA, Economou AS, Boubourakas IN, Katis NI (2003) Elimination of PPV and PNRSV through thermotherapy and meristem-tip culture in nectarine. Plant Cell Rep 22:195–200
Matzke M, Kanno T, Daxinger L, Huettel B, Matzke AJ (2009) RNA-mediated chromatin-based silencing in plants. Curr Opin Cell Biol 21:367–376
Mayers CN, Palukaitis P, Carr JP (2000) Subcellular distribution analysis of the cucumber mosaic virus 2b protein. J Gen Virol 81:219–226
Mayo MA, Ziegler-Graff V (1996) Molecular biology of luteoviruses. Adv Virus Res 46:413–460
Merai Z, Kerenyi Z, Molnar A, Barta E, Valoczi A, Bisztray G, Havelda Z, Burgyan J, Silhavy D (2005) Aureusvirus P14 is an efficient RNA silencing suppressor that binds double-stranded RNAs without size specificity. J Virol 79:7217–7226
Merai Z, Kerenyi Z, Kertesz S, Magna M, Lakatos L, Silhavy D (2006) Double-stranded RNA binding may be a general plant RNA viral strategy to suppress RNA silencing. J Virol 80:5747–5756
Meyers BC, Kaushik S, Nandety RS (2005) Evolving disease resistance genes. Curr Opin Plant Biol 8:129–134
Mi S, Cai T, Hu Y, Chen Y, Hodges E, Ni F, Wu L, Li S, Zhou H, Long C, Chen S, Hannon GJ, Qi Y (2008) Sorting of small RNAs into Arabidopsis argonaute complexes is directed by the 5′ terminal nucleotide. Cell 133:116–127
Miozzi L, Gambino G, Burgyan J, Pantaleo V (2013) Genome-wide identification of viral and host transcripts targeted by viral siRNAs in Vitis vinifera. Mol Plant Pathol 14:30–43
Moissiard G, Voinnet O (2006) RNA silencing of host transcripts by cauliflower mosaic virus requires coordinated action of the four Arabidopsis Dicer-like proteins. Proc Natl Acad Sci U S A 103:19593–19598
Molnar A, Csorba T, Lakatos L, Varallyay E, Lacomme C, Burgyan J (2005) Plant virus-derived small interfering RNAs originate predominantly from highly structured single-stranded viral RNAs. J Virol 79:7812–7818
Morel JB, Godon C, Mourrain P, Beclin C, Boutet S, Feuerbach F, Proux F, Vaucheret H (2002) Fertile hypomorphic ARGONAUTE (ago1) mutants impaired in post-transcriptional gene silencing and virus resistance. Plant Cell 14:629–639
Navarro B, Pantaleo V, Gisel A, Moxon S, Dalmay T, Bisztray G, Di Serio F, Burgyan J (2009) Deep sequencing of viroid-derived small RNAs from grapevine provides new insights on the role of RNA silencing in plant-viroid interaction. PLoS ONE 4:e7686
Navarro B, Gisel A, Rodio ME, Delgado S, Flores R, Di Serio F (2012) Small RNAs containing the pathogenic determinant of a chloroplast-replicating viroid guide the degradation of a host mRNA as predicted by RNA silencing. Plant J 70:991–1003
Omarov RT, Ciomperlik JJ, Scholthof HB (2007) RNAi-associated ssRNA-specific ribonucleases in Tombusvirus P19 mutant-infected plants and evidence for a discrete siRNA-containing effector complex. Proc Natl Acad Sci U S A 104:1714–1719
Pantaleo V, Szittya G, Burgyan J (2007) Molecular bases of viral RNA targeting by viral small interfering RNA-programmed RISC. J Virol 81:3797–3806
Pantaleo V, Saldarelli P, Miozzi L, Giampetruzzi A, Gisel A, Moxon S, Dalmay T, Bisztray G, Burgyan J (2010) Deep sequencing analysis of viral short RNAs from an infected Pinot Noir grapevine. Virology 408:49–56
Park MY, Wu G, Gonzalez-Sulser A, Vaucheret H, Poethig RS (2005) Nuclear processing and export of microRNAs in Arabidopsis. Proc Natl Acad Sci U S A 102:3691–3696
Pazhouhandeh M, Dieterle M, Marrocco K, Lechner E, Berry B, Brault V, Hemmer O, Kretsch T, Richards KE, Genschik P, Ziegler-Graff V (2006) F-box-like domain in the polerovirus protein P0 is required for silencing suppressor function. Proc Natl Acad Sci U S A 103:1994–1999
Pfeffer S, Dunoyer P, Heim F, Richards KE, Jonard G, Ziegler-Graff V (2002) P0 of beet Western yellows virus is a suppressor of posttranscriptional gene silencing. J Virol 76:6815–6824
Pham JW, Pellino JL, Lee YS, Carthew RW, Sontheimer EJ (2004) A Dicer-2-dependent 80s complex cleaves targeted mRNAs during RNAi in Drosophila. Cell 117:83–94

Qi X, Bao FS, Xie Z (2009) Small RNA deep sequencing reveals role for Arabidopsis thaliana RNA-dependent RNA polymerases in viral siRNA biogenesis. PLoS ONE 4:e4971
Qu F (2010) Antiviral role of plant-encoded RNA-dependent RNA polymerases revisited with deep sequencing of small interfering RNAs of virus origin. Mol Plant Microbe Interact 23:1248–1252
Qu F, Ye X, Hou G, Sato S, Clemente TE, Morris TJ (2005) RDR6 has a broad-spectrum but temperature-dependent antiviral defense role in Nicotiana benthamiana. J Virol 79:15209–15217
Qu F, Ye X, Morris TJ (2008) Arabidopsis DRB4, AGO1, AGO7, and RDR6 participate in a DCL4-initiated antiviral RNA silencing pathway negatively regulated by DCL1. Proc Natl Acad Sci U S A 105:14732–14737
Raja P, Sanville BC, Buchmann RC, Bisaro DM (2008) Viral genome methylation as an epigenetic defense against geminiviruses. J Virol 82:8997–9007
Ratcliff F, Harrison BD, Baulcombe DC (1997) A similarity between viral defense and gene silencing in plants. Science 276:1558–1560
Ratcliff FG, MacFarlane SA, Baulcombe DC (1999) Gene silencing without DNA. rna-mediated cross-protection between viruses. Plant Cell 11:1207–1216
Ruiz-Ferrer V, Voinnet O (2009) Roles of plant small RNAs in biotic stress responses. Annu Rev Plant Biol 60:485–510
Scholthof HB, Alvarado VY, Vega-Arreguin JC, Ciomperlik J, Odokonyero D, Brosseau C, Jaubert M, Zamora A, Moffett P (2011) Identification of an ARGONAUTE for antiviral RNA silencing in Nicotiana benthamiana. Plant Physiol 156:1548–1555
Schott G, Mari-Ordonez A, Himber C, Alioua A, Voinnet O, Dunoyer P (2012) Differential effects of viral silencing suppressors on siRNA and miRNA loading support the existence of two distinct cellular pools of ARGONAUTE1. EMBO J 31:2553–2565
Schuck J, Gursinsky T, Pantaleo V, Burgyán J, Behrens SE (2013) AGO/RISC-mediated antiviral RNA silencing in a plant in vitro system. Nucleic Acids Res doi:10.1093/nar/gkt193
Shen B, Goodman HM (2004) Uridine addition after microRNA-directed cleavage. Science 306:997
Shimura H, Pantaleo V, Ishihara T, Myojo N, Inaba J, Sueda K, Burgyan J, Masuta C (2011) A viral satellite RNA induces yellow symptoms on tobacco by targeting a gene involved in chlorophyll biosynthesis using the RNA silencing machinery. PLoS Pathog 7:e1002021
Shivaprasad PV, Chen HM, Patel K, Bond DM, Santos BA, Baulcombe DC (2012) A microRNA superfamily regulates nucleotide binding site-leucine-rich repeats and other mRNAs. Plant Cell 24:859–874
Silhavy D, Burgyan J (2004) Effects and side-effects of viral RNA silencing suppressors on short RNAs. Trends Plant Sci 9:76–83
Silhavy D, Molnar A, Lucioli A, Szittya G, Hornyik C, Tavazza M, Burgyan J (2002) A viral protein suppresses RNA silencing and binds silencing-generated, 21- to 25-nucleotide double-stranded RNAs. EMBO J 21:3070–3080
Smith NA, Eamens AL, Wang MB (2011) Viral small interfering RNAs target host genes to mediate disease symptoms in plants. PLoS Pathog 7:e1002022
Song JJ, Smith SK, Hannon GJ, Joshua-Tor L (2004) Crystal structure of Argonaute and its implications for RISC slicer activity. Science 305:1434–1437
Szittya G, Molnar A, Silhavy D, Hornyik C, Burgyan J (2002) Short defective interfering RNAs of tombusviruses are not targeted but trigger post-transcriptional gene silencing against their helper virus. Plant Cell 14:359–372
Szittya G, Silhavy D, Molnar A, Havelda Z, Lovas A, Lakatos L, Banfalvi Z, Burgyan J (2003) Low temperature inhibits RNA silencing-mediated defence by the control of siRNA generation. EMBO J 22:633–640
Szittya G, Moxon S, Pantaleo V, Toth G, Rusholme Pilcher RL, Moulton V, Burgyan J, Dalmay T (2010) Structural and functional analysis of viral siRNAs. PLoS Pathog 6:e1000838
Takeda A, Tsukuda M, Mizumoto H, Okamoto K, Kaido M, Mise K, Okuno T (2005) A plant RNA virus suppresses RNA silencing through viral RNA replication. EMBO J 24:3147–3157
Takeda A, Iwasaki S, Watanabe T, Utsumi M, Watanabe Y (2008) The mechanism selecting the guide strand from small RNA duplexes is different among argonaute proteins. Plant Cell Physiol 49:493–500

Tian D, Traw MB, Chen JQ, Kreitman M, Bergelson J (2003) Fitness costs of R-gene-mediated resistance in Arabidopsis thaliana. Nature 423:74–77
Till S, Ladurner AG (2007) RNA Pol IV plays catch with Argonaute 4. Cell 131:643–645
Tomari Y, Zamore PD (2005) Perspective: machines for RNAi. Genes Dev 19:517–529
Vaistij FE, Jones L (2009) Compromised virus-induced gene silencing in RDR6-deficient plants. Plant Physiol 149:1399–1407
van Wezel WR, Dong X, Liu H, Tien P, Stanley J, Hong Y (2002) Mutation of three cysteine residues in Tomato yellow leaf curl virus-China C2 protein causes dysfunction in pathogenesis and posttranscriptional gene-silencing suppression. Mol Plant Microbe Interact 15:203–208
Varallyay E, Valoczi A, Agyi A, Burgyan J, Havelda Z (2010) Plant virus-mediated induction of miR168 is associated with repression of ARGONAUTE1 accumulation. EMBO J 29:3507–3519
Vargason J, Szittya G, Burgyan J, Hall TM (2003) Size selective recognition of siRNA by an RNA silencing suppressor. Cell 115:799–811
Vaucheret H (2006) Post-transcriptional small RNA pathways in plants: mechanisms and regulations. Genes Dev 20:759–771
Vaucheret H (2008) Plant Argonautes. Trends Plant Sci 13:350–358
Vaucheret H, Mallory AC, Bartel DP (2006) AGO1 homeostasis entails coexpression of MIR168 and AGO1 and preferential stabilization of miR168 by AGO1. Mol Cell 22:129–136
Vazquez F (2006) Arabidopsis endogenous small RNAs: highways and byways. Trends Plant Sci 11:460–468
Verdel A, Jia S, Gerber S, Sugiyama T, Gygi S, Grewal SI, Moazed D (2004) RNAi-mediated targeting of heterochromatin by the RITS complex. Science 303:672–676
Voinnet O (2005) Induction and suppression of RNA silencing: insights from viral infections. Nat Rev Genet 6:206–220
Voinnet O (2009) Origin, biogenesis, and activity of plant microRNAs. Cell 136:669–687
Voinnet O, Pinto YM, Baulcombe DC (1999) Suppression of gene silencing: a general strategy used by diverse DNA and RNA viruses of plants. Proc Natl Acad Sci U S A 96:14147–14152
Wang H, Buckley KJ, Yang X, Buchmann RC, Bisaro DM (2005) Adenosine kinase inhibition and suppression of RNA silencing by geminivirus AL2 and L2 proteins. J Virol 79:7410–7418
Wang XB, Wu Q, Ito T, Cillo F, Li WX, Chen X, Yu JL, Ding SW (2010) RNAi-mediated viral immunity requires amplification of virus-derived siRNAs in Arabidopsis thaliana. Proc Natl Acad Sci U S A 107:484–489
Wang XB, Jovel J, Udomporn P, Wang Y, Wu Q, Li WX, Gasciolli V, Vaucheret H, Ding SW (2011) The 21-Nucleotide, but Not 22-Nucleotide, Viral Secondary Small Interfering RNAs Direct Potent Antiviral Defense by Two Cooperative Argonautes in Arabidopsis thaliana. Plant Cell 23:1625–1638
Wassenegger M, Krczal G (2006) Nomenclature and functions of RNA-directed RNA polymerases. Trends Plant Sci 11:142–151
Wu Q, Wang X, Ding SW (2010) Viral suppressors of RNA-based viral immunity: host targets. Cell Host Microbe 8:12–15
Xiong R, Wu J, Zhou Y, Zhou X (2009) Characterization and subcellular localization of an RNA silencing suppressor encoded by Rice stripe tenuivirus. Virology 387:29–40
Yang LP, Fang YY, An CP, Dong L, Zhang ZH, Chen H, Xie Q, Guo HS (2013) C2-mediated decrease in DNA methylation, accumulation of siRNAs, and increase in expression for genes involved in defense pathways in plants infected with beet severe curly top virus. Plant J 73:910–917
Ye K, Malinina L, Patel DJ (2003) Recognition of small interfering RNA by a viral suppressor of RNA silencing. Nature 426:874–878
Yu B, Yang Z, Li J, Minakhina S, Yang M, Padgett RW, Steward R, Chen X (2005) Methylation as a crucial step in plant microRNA biogenesis. Science 307:932–935
Yu B, Chapman EJ, Yang Z, Carrington JC, Chen X (2006) Transgenically expressed viral RNA silencing suppressors interfere with microRNA methylation in Arabidopsis. FEBS Lett 580:3117–3120

Zhai J, Jeong DH, De Paoli E, Park S, Rosen BD, Li Y, Gonzalez AJ, Yan Z, Kitto SL, Grusak MA, Jackson SA, Stacey G, Cook DR, Green PJ, Sherrier DJ, Meyers BC (2011) MicroRNAs as master regulators of the plant NB-LRR defense gene family via the production of phased, trans-acting siRNAs. Genes Dev 25:2540–2553
Zhang X, Yuan YR, Pei Y, Lin SS, Tuschl T, Patel DJ, Chua NH (2006) Cucumber mosaic virus-encoded 2b suppressor inhibits Arabidopsis Argonaute1 cleavage activity to counter plant defense. Genes Dev 20:3255–3268
Zhang X, Zhao H, Gao S, Wang WC, Katiyar-Agarwal S, Huang HD, Raikhel N, Jin H (2011) Arabidopsis Argonaute 2 regulates innate immunity via miRNA393(*)-mediated silencing of a Golgi-localized SNARE gene, MEMB12. Mol Cell 42:356–366
Zhang X, Zhang X, Singh J, Li D, Qu F (2012) Temperature-dependent survival of Turnip crinkle virus-infected arabidopsis plants relies on an RNA silencing-based defense that requires dcl2, AGO2, and HEN1. J Virol 86:6847–6854
Zhou Z, Dell'Orco M, Saldarelli P, Turturo C, Minafra A, Martelli GP (2006) Identification of an RNA-silencing suppressor in the genome of Grapevine virus A. J Gen Virol 87:2387–2395
Zrachya A, Glick E, Levy Y, Arazi T, Citovsky V, Gafni Y (2007) Suppressor of RNA silencing encoded by Tomato yellow leaf curl virus-Israel. Virology 358:159–165
Zvereva AS, Pooggin MM (2012) Silencing and innate immunity in plant defense against viral and non-viral pathogens. Viruses 4:2578–2597

RNA Interference-Mediated Intrinsic Antiviral Immunity in Invertebrates

Arabinda Nayak, Michel Tassetto, Mark Kunitomi and Raul Andino

Abstract In invertebrates such as insects and nematodes, RNA interference (RNAi) provides RNA-based protection against viruses. This form of immunity restricts viral replication and dissemination from infected cells and viruses, in turn, have evolved evasion mechanisms or RNAi suppressors to counteract host defenses. Recent advances indicate that, in addition to RNAi, other related small RNA pathways contribute to antiviral functions in invertebrates. This has led to a deeper understanding of fundamental aspects of small RNA-based antiviral immunity in invertebrates and its contribution to viral spread and pathogenesis.

Contents

1 Introduction

RNA interference (RNAi) is a gene expression regulatory mechanism in eukaryotes that mediates RNA degradation in a sequence-specific manner. In both plants and invertebrates (*Drosophila*, mosquito, and *Caenorhabditis elegans*), RNAi

A. Nayak · M. Tassetto · M. Kunitomi · R. Andino (✉)
Department of Microbiology and Immunology, University of California,
San Francisco 94122-2280, USA
e-mail: raul.andino@ucsf.edu

B. R. Cullen (ed.), *Intrinsic Immunity*, Current Topics in Microbiology and Immunology 371, DOI: 10.1007/978-3-642-37765-5_7,

is also a major antiviral mechanism (Ding 2010). In invertebrate cells, antiviral RNAi is initiated in response to viral replication, which generates long double-stranded RNAs (dsRNAs). These long dsRNAs are recognized as pathogen-associated molecular patterns (PAMPs) by the dsRNA sensor protein Dicer (Dcr). Dcr processes these long viral dsRNAs into 21-nucleotide small duplex RNAs, called small interfering RNA or siRNA. Throughout the text, we refer to viral dsRNA-derived siRNAs as viRNAs. Argonaute (Ago), the RNAi effector protein, binds to these viRNAs and uses their sequence to identify single-stranded viral genomes or mRNAs for degradation (Zamore et al. 2000; Bernstein et al. 2001; Hammond et al. 2001; Aliyari et al. 2008; Aliyari and Ding 2009). The entire repertoire of RNAi proteins including Dcr and Ago is constitutively expressed in cells, thereby conferring intrinsic immunity against viruses. To survive and replicate efficiently, many *Drosophila* viruses actively inhibit the RNAi machinery by encoding RNAi suppressor proteins in their genome (Wu et al. 2010b). In contrast, infectious arthropod-borne viruses (arboviruses) that are transmitted to mammals by mosquito vectors do not seem to encode RNAi suppressor proteins and thus must use alternate mechanisms to evade mosquito defenses (Blair 2011). Recent studies show that in addition to RNAi, related small RNA pathways mediated by micro-RNA (miRNA) and Piwi-interacting RNA (piRNA) contribute to antiviral function in invertebrates (Skalsky et al. 2010; Hussain et al. 2012; Morazzani et al. 2012; Vodovar et al. 2012). Here, we review our current understanding of the role of RNAi and related small RNA pathways in antiviral immunity in invertebrates. Finally, we discuss the salient questions and future directions in RNAi research.

2 siRNA-Mediated Antiviral Immunity

2.1 Mechanism of RNAi

The mechanism of RNAi is best characterized in *Drosophila*, in which the Dcr2 enzyme cleaves long dsRNAs into siRNAs (Bernstein et al. 2001). A dsRNA-binding protein (dsRBP), Loquacious isoform PD (Loqs-PD), and Arsenite resistant protein 2 (Ars2) (Fig. 1) aid this process by increasing the affinity of Dcr2 for its substrates (Marques et al. 2009; Sabin et al. 2009; Cenik et al. 2011). R2D2, another dsRBP, facilitates siRNA loading into Ago2 to assemble the RNA inducing silencing complex (RISC) (Tomari et al. 2004b; Liu et al. 2006). Armitage, an RNA helicase, also participates in siRNA loading into the Ago2 RISC (Tomari et al. 2004a). Ago2, with the help of the endoribonuclease C3PO, cleaves one strand (passenger strand) of the siRNA duplex (Matranga et al. 2005; Rand et al. 2005; Liu et al. 2009). The remaining strand (guide strand) is subsequently 2′-O-methylated at its 3′ terminus by the enzyme Hen-1 (an S-adenosylmethionine-dependent methyltransferase) (Figs. 1 and 2). Methylation of the guide strand increases its stability and facilitates the maturation of siRISC (Horwich et al. 2007). siRISC identifies its target by sequence homology and, in

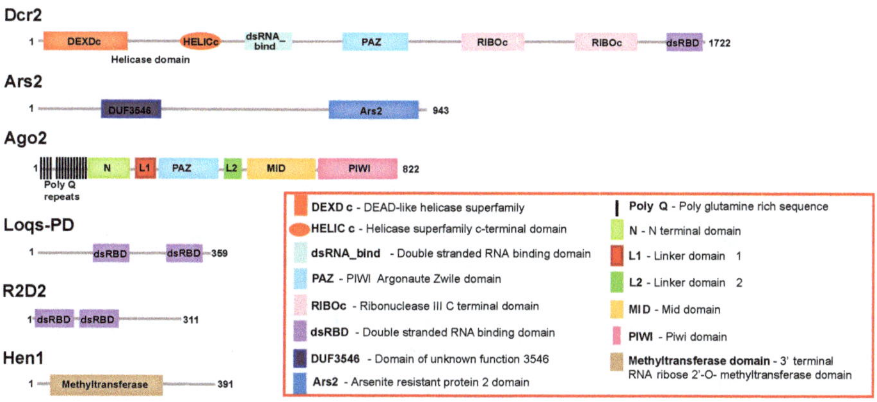

Fig. 1 Domain architecture of *Drosophila* RNAi proteins and their roles in the RISC assembly process. Dcr2 contains an amino-terminal DExDc domain, a HELICc domain, a dsRNA-binding domain, a PAZ domain, two-tandem ribonuclease III domains, and a carboxy-terminal dsRNA-binding domain (*dsRBD*). The dsRBD binds to dsRNA substrates in a sequence independent manner. During dsRNA processing, two RNAse III domains in Dcr2 create a catalytic center at which each domain cleaves one strand of the dsRNA. Cleavage by Dcr2 leaves signature 3′-OH with dinucleotide overhangs and 5′ monophosphates. The PAZ domain of Dcr2 binds to 3′ dinucleotide overhangs and acts as a molecular ruler for successive siRNA molecule generation at 21-nucleotide intervals, equaling the distance between the PAZ domain and RNAse III active site. The dsRNA-binding domain is a divergent class of dsRBD, which binds to dsRNA with low affinity, and its precise role in Dcr2 function is not fully understood (Qin et al. 2010). The DExDC helicase domain mediates ATP hydrolysis and causes translocation of Dcr2 over the length of the dsRNA substrate to generate many siRNAs from single dsRNA substrate (Cenik et al. 2011). The HELICc domain forms the carboxy-terminal component of the helicase domain. Ars2 contains an amino-terminal DUF3546 domain and a carboxy-terminal Arsenite resistant protein 2 (Ars2) domain and acts as a cofactor for Dcr2-mediated dsRNA cleavage. Both Loqs-PD and R2D2 contain two dsRBDs each and interact with Dcr2 during RISC assembly process. Loqs-PD promotes siRNA production by Dcr2, whereas R2D2 promotes siRNA loading from Dcr2 to Ago2. Ago2 comprises an amino-terminal poly Q repeat domain, an N domain, a PAZ domain, a mid-domain (*MID*), and a carboxy-terminal Piwi domain. Although Ago2 without most of the polyQ repeats is competent in RISC-mediated target cleavage, the polyQ repeats domain is believed to play a role in antiviral defense (Liu et al. 2009; Hain et al. 2010). The N, PAZ, mid, and Piwi domains are connected through two-structured linker domains L1 and L2. The Ago2-PAZ domain binds to siRNA 3′-OH 2-nucleotide overhangs, and the Ago2-MID binds to the 5′ monophosphate group of the guide siRNA. Hen1 contains a single-strand RNA-specific methyltransferase domain, and methylate the guide siRNAs at their 3′ termini. The Ago2-Piwi domain possesses RNAse H activity and catalyzes the guide siRNA-mediated target RNA cleavage

the presence of fully complementary base-pairing between the guide strand and its target, induces Ago2-mediated RNA nucleolysis (Fig. 2). In *Drosophila*, additional proteins including Tudor staphylococcal nuclease (Tudor-SN), Vasa intronic gene (VIG), and Rm62 (a DEAD box RNA helicase) associate with siRISC (Caudy et al. 2002; Ishizuka et al. 2002; Caudy et al. 2003). The mechanistic roles of these proteins in siRISC function are not fully understood, although they have been shown to contribute to antiviral function (Zambon et al. 2006).

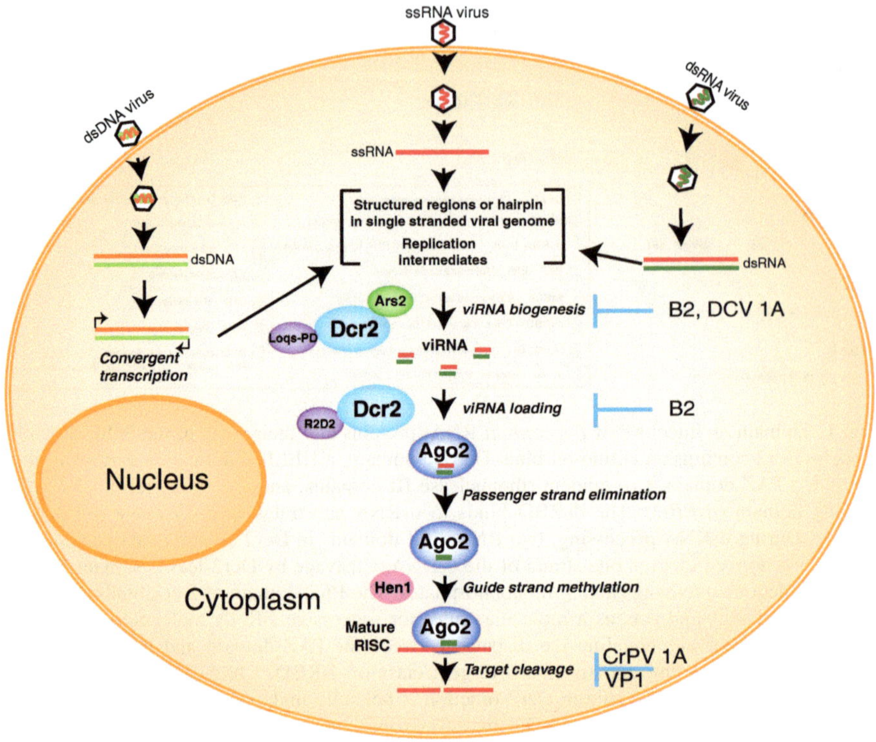

Fig. 2 Activation of the *Drosophila* RNAi machinery and viral countermeasures. Replication of single-stranded RNA (*ssRNA*), dsRNA, and dsDNA viruses generates dsRNA replication intermediates. Double-stranded replication intermediates, and structured regions or hairpins in single-stranded viral genomes are cleaved by Dcr2 to produce viRNAs, a process that is aided by both Ars2 and Loqs-PD. Each viRNA contains a passenger strand (colored in *red*) and a guide strand (colored in *green*). In the next step, viRNAs bound to Dcr2 in the presence of R2D2 are loaded into Ago2-containing complex. The passenger strand is eliminated, and the Ago2 bound guide strand is methylated by Hen1, at which point the Ago2 complex is considered mature RISC. B2 proteins of FHV and WuNV target both viRNA biogenesis and viRNA loading into Ago2-RISC. DCV 1A inhibits the production of viRNA by protecting the long dsRNA. In contrast, CrPV 1A and Noravirus VP1 prevent viral genome degradation by interfering with Ago2 cleavage

2.2 Antiviral Function of RNAi

The critical role of RNAi in invertebrate antiviral immunity was first reported in *Drosophila* infected with flock house virus (FHV), a positive-strand RNA virus (Li et al. 2002; Galiana-Arnoux et al. 2006; Wang et al. 2006). The FHV genome consists of RNA1 and RNA2, which encode the viral RNA-dependent RNA polymerase (RdRP) and the viral capsid protein, respectively. FHV replication triggers a Dcr2-initiated RNAi response with the production of 21-nucleotide viRNAs. To efficiently replicate, FHV encodes a potent RNAi suppressor protein, B2. B2 inhibits the production of viRNAs by Dcr2, thus effectively blocking the host RNAi response (see Sect. 2.5.1). Moreover, mutant flies defective for RNAi proteins Dcr2 or R2D2 are hypersensitive to FHV infection. The antiviral role of RNAi in *Drosophila* was further confirmed when Ago2 was shown to be essential to combat two other insect positive–strand RNA viruses, Drosophila C virus (DCV) and cricket paralysis virus (CrPV) (van Rij et al. 2006; Wang et al. 2006). Although divergent from the nodavirus FHV, the two dicistroviruses DCV and CrPV each encode an RNAi suppressor protein called 1A to antagonize the RNAi response, providing evidence that the host RNAi response can exert significant selective pressure on viruses to encode RNAi suppressors in their genomes (Li et al. 2002; van Rij et al. 2006; Bonning and Miller 2010; Nayak et al. 2010). Since these seminal studies, the antiviral role of RNAi in *Drosophila* has been extended to various types of RNA and DNA viruses. For example, vesicular stomatitis virus (VSV), a negative-sense single-stranded RNA virus, replicates more efficiently in Dcr2, Ago2, or R2D2 mutant flies (Mueller et al. 2010). In addition, Dcr2 has been shown to protect flies against challenges from Sindbis virus (SINV), an enveloped positive-sense single-stranded RNA virus (Galiana-Arnoux et al. 2006). Furthermore, increased susceptibility to *Drosophila* X virus (DXV), a double-stranded RNA virus, is reported in flies deficient for Ago2, R2D2, VIG, Armitage, or Rem 62 (Zambon et al. 2006). Flies depleted of Ars2 also show increased susceptibility to a diverse set of RNA viruses including FHV, DCV, SINV, and VSV (Sabin et al. 2009). Most recently, RNAi has been shown to protect flies from Invertebrate iridescent virus 6 (IIV-6), a dsDNA virus. RNAi- defective (Dcr2 or Ago2 mutant) flies show increased susceptibility to IIV-6 infection compared to wild type flies (Bronkhorst et al. 2012; Kemp et al. 2013).

Although the majority of work showing RNAi as a primary antiviral immune response is from *Drosophila*, there are several lines of evidence for a similar response in other Diptera, specifically mosquitoes. Mosquitoes encode homologues of all of the canonical RNAi components described in *Drosophila* (Campbell et al. 2008). viRNAs, the signature of active Dcr2 cleavage, are readily detectable both in the adult mosquito and mosquito-derived cell lines, indicating that RNAi in mosquitoes targets invading viruses (Blair 2011). Accordingly, the replication of mosquito viruses can be inhibited by the introduction of exogenous cognate dsRNAs targeting the viral genome (Keene et al. 2004). In addition, RNAi of host RNAi components, through injection of dsRNA targeting Ago2, Dcr2, or R2D2 in the adult mosquito, increases dissemination of the virus and viral load (Keene et al. 2004; Sánchez-Vargas et al. 2009). Inhibition of RNAi during infection has also been achieved through ectopic expression of the RNAi suppressor protein B2 in positive-strand alphaviruses SINV and O' nyong–nyong virus (ONNV) (Myles et al. 2008). Viruses that encode this RNAi suppressor replicate to a greater extent and disseminate better throughout the host. These observations—that viRNAs can be detected during the course of infection, that RNAi can target viruses and reduce viral replication, and that the inhibition of RNAi exacerbates viral infection—strongly suggest that RNAi is a critical arm of the mosquito antiviral immune response.

RNAi operates similarly in non-insect invertebrates, as demonstrated in the nematode *C. elegans*. In *C. elegans,* RNAi provides protection against exogenous viral infection. *C. elegans* cells defective for rde-1 (Ago2 homolog), rde-4 (R2D2 homolog), or rrf-1 (RdRP required for RNAi amplification) are more sensitive to VSV infection (Schott et al. 2005; Wilkins et al. 2005). In addition, *C. elegans* lacking rde-1 supports efficient replication of B2-deficient FHV (Lu et al. 2005). Furthermore, infection of *C. elegans* lacking rde-1 results in higher Orsay virus (a positive-strand RNA virus related to nodavirus) replication than infection of wild type, strongly suggesting a role for nematode RNAi against natural virus infection (Félix et al. 2011).

2.3 Initiation of RNAi: Recognition of Viral PAMPs

Multiple lines of evidence implicate Dcr as the sensor molecule in the recognition of viral dsRNAs as PAMPs. The PAZ domain of Dcr recognizes the termini of the dsRNA and orients the RNAse III domain for dsRNA cleavage (Fig. 1) (Liu et al. 2003; Zhang et al. 2004; MacRae et al. 2007; Lau et al. 2012). Additionally, involvement of the helicase motif of *Drosophila* Dcr2 (Fig. 1) and *C. elegans* Dcr1 in dsRNA recognition has been recently reported (Cenik et al. 2011; Welker et al. 2011). The helicase motif of *Drosophila* Dcr2 (Fig. 1) efficiently senses dsRNA substrates bearing either blunt or 5′ overhanging termini, but is less efficient on substrates with 3′ overhanging termini. During positive-strand RNA virus replication, a 50–100-fold greater quantity of sense-strand (positive-strand) RNAs are generated relative to the antisense strand (negative strand). The antisense strands are exclusively associated with the sense strands as double-stranded replication intermediates (Baltimore and Girard 1966; Cleaves et al. 1981; Chu and Westaway 1985). Replication intermediates are thought to be major substrates of Dcr2. Biochemical study suggests that Dcr2 is highly processive in vitro, cleaving long dsRNAs into siRNAs without dissociating from its substrate (Cenik et al. 2011; Welker et al. 2011). This property of Dcr2 is well suited for processing long viral replication intermediates substrates and presumably should lead to a contiguous viRNA profile covering the entire length. In contrast, the viRNA profiles from *Drosophila* cells infected with FHV show hot spots at the 5′ and 3′ terminal regions of RNA1. These viRNAs match to sense and antisense strands, suggesting that Dcr2 predominantly targets specific regions of the replication intermediates (Aliyari et al. 2008; Czech et al. 2008; Flynt et al. 2009). It also suggests that structured regions associated with positive-/sense-strand RNA, which accumulates to high levels in FHV-infected cells (Kopek et al. 2007), are not the preferred substrate for Dcr2. Similar viRNA profiles matching both the sense and antisense strand of viral replication intermediates have been reported from S2 cells persistently infected with positive-strand RNA viruses [*Drosophila* A virus (DAV), DCV, Noravirus, and American nodavirus (ANV)] and double-stranded RNA viruses [DXV, *Drosophila* birnavirus (DBV), *Drosophila* totivirus (DTV)]

(Wu et al. 2010a). Importantly, viRNA-profiling studies using a negative-strand RNA virus, VSV, also show a distribution pattern derived from viral replication intermediates (Mueller et al. 2010; Sabin et al. 2013). Recently, it has been shown that infection of *Drosophila* cells with a DNA virus, IIV-6, generates a viRNA profile matching both sense and antisense strands in equal proportions, similar to RNA viruses (Bronkhorst et al. 2012; Kemp et al. 2013). This suggests that, like RNA viruses, DNA viruses produce dsRNAs as a consequence of convergent transcription (Fig. 2), which are recognized as PAMPs by Dcr2. In contrast to the above studies, a viRNA-profiling study using DCV-infected *Drosophila* cells showed a strong bias for sense strand over the antisense strand, suggesting that structured regions and hairpins within the single-stranded genomic-strand RNA are the target of the Dcr2 enzyme (Sabin et al. 2013).

A small RNA-sequencing profile of a dengue virus type 2 (DENV2)-infected *Aedes aegypti* mosquito cell line (Aag2) showed a viRNA profile matching both sense- and antisense-strand RNA molecules with roughly equal proportion, indicating that viral replication intermediates are substrates for mosquito Dcr2 enzyme (Scott et al. 2010). In contrast, a viRNA profile from SINV-infected mosquito and West Nile virus (WNV)-infected mosquito mid-gut showed a significant bias for sense over antisense strand, indicating that both replication intermediates and structured regions present in the single-stranded sense viral genomes are substrates for Dcr2 (Myles et al. 2008; Brackney et al. 2009). Similarly, a viRNA profile from a Semliki forest virus (SFV)-infected *Aedes albopictus*-derived mosquito cell line (U4.4) showed a slight bias for sense over antisense polarity, suggesting that most viRNA are derived from viral replication intermediates and a minority are derived from structured regions of the viral genome (Siu et al. 2011). Of note, the viRNA accumulation in infected cells varies between and within invertebrate species. For example, the viRNA population in FHV-infected *C. elegans* constitutes a smaller proportion of the total small RNA population than the viRNA population in FHV-infected *Drosophila* (Lu et al. 2009). Similarly, in mosquito, positive-strand alphavirus (SINV, SFV) replication triggers more robust viRNA production than positive-strand flavivirus (WNV and dengue virus) replication (Blair 2011).

2.4 Amplification and Systemic Spread of RNAi

In some cases (e.g., nematodes and plants), the initiation of RNAi, which leads to production of siRNA, is further amplified to provide a robust RNAi response. In nematodes, the primary pool of siRNAs generated by initial Dcr-1 cleavage is further amplified to generate a secondary pool of siRNAs. Following the recognition of the cognate target RNA by primary siRISC, the cellular RdRP enzyme rrf-1 synthesizes the secondary pool of siRNA to sustain the initial RNAi response, as well as facilitate intercellular or systemic spread of the RNAi signal throughout the animal (Carthew and Sontheimer 2009). This systemic spread is mediated by a set of systemic RNAi-defective or *Sid* proteins. *Sid-1, Sid-2, and Sid-3* are trans-membrane channel proteins and mediate the uptake or import of dsRNA into cells, whereas *Sid-5*, an endosome-associated protein, exclusively regulates the export of dsRNA from cells (Winston et al. 2002, 2007; Hinas et al. 2012). In contrast to nematodes, neither a cellular RdRP nor any sid orthologs have been reported in *Drosophila*. Nevertheless, systemic RNAi in *Drosophila* has been shown to play a critical role in antiviral defense (Saleh et al. 2009). *Drosophila* strains with mutation in genes (*egghead*, *ninaC*, and *CG4572*) involved in the endosome-associated dsRNA uptake pathway exhibit defective systemic RNAi and are highly susceptible to DCV and SINV infection. However, little is known about the intercellular mechanism of transport of RNAi signals between cells. A recent study suggests that the *Drosophila* RNAi machinery localizes to multivesicular bodies (MVBs) associated with late endosomes (Lee et al. 2009). Early endosomes in progression to late endosomes in the endosomal sorting complex required for transport (ESCRT) pathway inwardly bud as intraluminal vesicles (ILVs) to form MVBs. MVBs either fuse with the lysosome for cargo recycling or fuse with the plasma membrane to releases the ILVs into the extracellular space as exosomes. Recent studies suggest that exosomes are vehicles of intercellular communication and contain a variety of RNA molecules including mRNAs and miRNAs (Simons and Raposo 2009; Théry 2011). Since the RNAi machinery colocalizes with MVBs, it is plausible that, in *Drosophila*, exosomes might serve as a vehicle for the spread of dsRNA signals between cells. Intercellular spread of antiviral signals has also been reported in *A. albopictus* mosquito cells infected with SFV, although the underlying mechanism is not understood (Attarzadeh-Yazdi et al. 2009).

In mammals, one of the features of antiviral mechanisms is immunological memory, which is acquired after initial challenge with a viral pathogen. This leads to a more robust and effective immune response to subsequent exposures to the same pathogen. In *C. elegans*, expression of an FHV transgene leads to production and inheritance of FHV viRNA down to several generations, which provides protection from subsequent FHV challenge (Rechavi et al. 2011). More importantly, these viRNAs are maintained in individuals with genetic backgrounds that prevent the synthesis of new viRNAs. Interestingly, long-term immunity against FHV infection in these animals depends upon rrf-1. This form of trans-generational inheritance of an extra-chromosomally encoded immunity trait has not been reported in other invertebrates. It would be important to determine whether trans-generational inheritance also occurs in the context of natural FHV infection, in contrast to FHV transgene expression.

2.5 Viral Countermeasures to RNAi

To counteract RNAi-based immunity, many insect viruses express RNAi suppressor proteins. The existence of a competitive arms race between RNAi immunity and viruses is supported by strong evidence of adaptive evolution in the

Drosophila RNAi genes Dcr2, Ago2, and R2D2 as well as in viral RNAi suppressor proteins (Obbard et al. 2006, 2009). Based on their mechanism of RNAi inhibition, RNAi suppressors are broadly classified into two different groups: Dcr2-targeting RNAi suppressors and Ago2-targeting RNAi suppressors.

2.5.1 Dcr2-Targeting RNAi Suppressors

One of the most well-studied Dcr2-targeting RNAi suppressors is the FHV B2 protein. B2 binds both long dsRNAs and siRNAs with high affinity (Chao et al. 2005). During FHV replication in *Drosophila* cells, B2 interacts with the viral RdRP and forms an integral component of the viral replication initiation complex. By doing so, B2 protects FHV dsRNA replication intermediates from Dcr2 processing (Aliyari et al. 2008). Besides interacting with dsRNAs, FHV B2 interacts with the PAZ domain of Dcr2 (Singh et al. 2009). Similarly, the B2 RNAi suppressor protein encoded by Wuhan nodavirus (WhNV) directly interacts with the PAZ domain of Dcr2 and suppresses interaction of siRNA with Dcr2 (Qi et al. 2012). Because of their ability to bind long dsRNA, siRNA, and Dcr2, B2 proteins inhibit multiple steps in the RISC assembly process including siRNA biogenesis and subsequent incorporation of siRNA into Ago2-containing RISC (Fig. 2). A separate RNAi suppressor protein encoded by DCV (DCV 1A) has also been shown to interfere with Dcr2 processing. In vitro mechanistic studies suggest that DCV 1A inhibits the Dcr2-mediated processing of long dsRNA into siRNA and also interferes with the siRNA-initiated RISC assembly process (van Rij et al. 2006; Nayak et al. 2010). Unlike B2, however, DCV 1A is not known to interact with Dcr2 or any other component of the RNAi machinery (Fig. 2).

2.5.2 Ago2-Targeting RNAi Suppressors

In contrast to B2 and DCV 1A, the RNAi suppressor CrPV 1A does not bind to dsRNA. As a result, CrPV 1A does not affect Dcr2-mediated siRNA biogenesis in *Drososphila* cells and subsequent incorporation of siRNA into RISC. Instead, CrPV 1A interacts with Ago2 and blocks the target RNA cleavage function of RISC (Fig. 2) (Nayak et al. 2010; Nayak and Andino 2011; van Mierlo et al. 2012). CrPV 1A is believed to interfere with the Piwi cleavage function of Ago2. Similarly, the RNAi suppressor protein encoded by Noravirus (VP1) has been shown to interfere with Ago2-mediated RNA cleavage (Fig. 2) (van Mierlo et al. 2012).

2.5.3 Role of RNAi Suppressors in Viral Pathogenesis

Most of the *Drosophila* viruses that encode RNAi suppressors (e.g., FHV, DCV and Noravirus) cause an initial acute infection followed by the establishment of a chronic persistent infection in both S2 cells and flies. In contrast, CrPV induces high mortality in S2 cells, adult flies, and a broad range of other insects (Bonning and Miller 2010). This raises an important question: what factors lead to persistent *versus* lethal infection? A recent study suggests that during infection of S2 cells with FHV and DCV, the viral RNAs are reverse transcribed by endogenous retrotransposon-encoded reverse transcriptase activity (Goic et al. 2013). The resulting viral DNA molecules are integrated into the host genome and transcribed to generate dsRNA. These dsRNAs are processed by the RNAi machinery to generate viRNAs that attenuate lethal viral infection, causing establishment of persistent infection. What, then, differentiates CrPV from other viruses? All the RNAi suppressors described above, including B2, DCV 1A, Noravirus VP1, and CrPV 1A, cause efficient RNAi suppression in *Drosophila* S2 cells. In vivo, however, CrPV 1A is more potent than the others (Nayak et al. 2010; van Mierlo et al. 2012) (Christophe Antoniewski, personal communication). It has been postulated that the potency of the RNAi suppressor in vivo determines overall viral virulence. This is strongly supported by the finding that SINV expressing CrPV 1A causes significant virus spread, pathogenesis, and mortality in flies. Conversely, SINV expressing DCV 1A, a weaker RNAi suppressor, only induces mild pathogenesis and low mortality (Nayak et al. 2010). The differential pathogenicity caused by CrPV 1A and DCV 1A cannot be attributed to the RNAi protein with which they interact (Ago2 or Dcr2, respectively). Like CrPV 1A, Noravirus VP1 interferes with Ago2 function; however, Noravirus establishes a persistent infection in flies. It is possible that besides interfering with Ago2 function, CrPV 1A affects additional cellular processes that explain its increased virulence.

To date, no arboviruses have been shown to encode RNAi suppressor proteins, yet these viruses are found to persistently infect both mosquito adults and cell lines. This raises the question of how arboviruses manage to replicate in the host, despite active RNAi defense. The prototypical flaviviruses, WNV and DENV, do appear to encode a sub-genomic RNA derived from the 3′ untranslated region (UTR) that suppresses the siRNA pathway in mosquito cells (Schnettler et al. 2012). An additional evasion mechanism involving positive selection of point mutations in genomic regions actively targeted by the RNAi machinery allows WNV to escape the host response (Brackney et al. 2009). Further investigation is needed to fully understand the molecular basis of persistent infection by flaviviruses.

3 Related Small RNA Pathways in Antiviral Immunity

Despite the predominant role of the siRNA pathway in invertebrate antiviral immunity, recent evidence suggests a possible interdependence between siRNA and related small RNA pathways (e.g., miRNA and piRNA), as well as redundant roles for these small RNAs in mediating antiviral function in invertebrates (Sabin et al. 2009; Wu et al. 2010a). In mammals, both host and virally encoded miRNAs are known to regulate viral replication in cells (Gottwein and Cullen

2008; Skalsky and Cullen 2010). Current studies indicate that virally encoded miRNA and viral misregulation of host miRNAs modulate host gene expression in a way that may facilitate viral replication in mosquito cells (Skalsky et al. 2010; Hussain et al. 2012). Similarly in *Drosophila* and mosquito, the piRNA pathway has recently emerged as an antiviral mechanism alongside the siRNA pathway (Morazzani et al. 2012; Vodovar et al. 2012; Léger et al. 2013).

3.1 miRNA-Mediated Antiviral Immunity

miRNAs are approximately 22 nucleotides long and are synthesized in two steps. In *Drosophila,* the nuclear RNAse III enzyme, Drosha, and the dsRNA-binding protein, Pasha, cleave primary miRNA transcripts into pre-miRNAs, a process promoted by Ars2 (Denli et al. 2004; Sabin et al. 2009). In the next step, the pre-miRNAs are recognized by a nuclear export protein, Exportin 5, for transport to the cytoplasm (Okada et al. 2009). Dcr1, with the help of a cytoplasmic dsRBD protein Loquacious PB (Loqs-PB), cleaves the pre-miRNAs into a miRNA duplex containing a miRNA and a miRNA* (miRNA star) strand (Jiang et al. 2005; Saito et al. 2005). In *Drosophila*, most of the miRNA strands in complex with Ago1 bind to partially complementary sites in the 3′ UTR of target mRNAs (Brennecke et al. 2005; Stark et al. 2007). GW182 (glycine tryptophan-rich protein 182) protein interacts with Ago1 and localizes the Ago1-miRNA-mRNA complex to P bodies (processing bodies/PBs) where mRNA translation repression or mRNA decay takes place (Rehwinkel et al. 2005; Behm-Ansmant et al. 2006). The Kunjin strain of $\mathrm{WNV}_{\mathrm{KUN}}$ was recently reported to express a miRNA, KUN-miR-1, from the terminal 3′ stem-loop structure of its genome. KUN-miR-1 upregulates the expression of the host gene GATA-4, a Zn-finger transcription factor (Hussain et al. 2012). Depletion of Ago1 or inhibition of KUN-miR-1 with a miRNA-specific RNA inhibitor in mosquito cells downregulates viral replication. In *Aedes aegypti*, GATA-4 is involved in lipid trafficking, and it is hypothesized that KUN-miR-1 facilitates the formation of the virus-induced membrane structures required for RNA replication (Hussain et al. 2012). In addition, infection of *Culex quinquefasciatus* mosquitoes with WNV causes notable upregulation of miR-92 and downregulation of miR-989, indicating possible modulation of host gene expression by flavivirus infection (Skalsky et al. 2010).

3.2 piRNA-Mediated Antiviral Immunity

In addition to miRNAs, piRNAs have recently been shown to play important roles in antiviral function. The unexpected role of piRNAs in antiviral defense was first reported in a *Drosophila* ovary somatic sheath (OSS) cell line persistently infected with six different viruses including DCV (Lau et al. 2009). piRNAs targeting all of these viruses were identified in addition to canonical viRNAs by small RNA sequencing. This led to the assumption that, besides protecting the germ line from transposable elements, piRNAs in *Drosophila* also protect the germ line-associated somatic tissue from virus infection. From a biogenesis point of view, piRNAs are larger in size than siRNAs and miRNAs (24–32 nucleotides) and are derived, in a Dcr-independent manner, from single-stranded RNA precursors transcribed from genomic regions containing defective transposons (Brennecke et al. 2007). Based on their association with germ line-specific Piwi sub-clade proteins [Piwi or Aubergine (Aub) or Ago3], they are classified into two distinct categories. Primary piRNAs are found associated with Piwi or Aub and show a strong bias for uridine at position 1 (1U) from the 5′ end. The secondary piRNAs, amplified from the primary piRNA through a ping-pong amplification loop (Siomi et al. 2011), are associated with Ago3 and show a strong bias for adenine (A) at position 10 (10A) from the 5′ end. *Drosophila* OSS cells express Piwi but not Aub or Ago3, and accordingly viral piRNAs originating from OSS cells show a characteristic 5′ 1U bias with no secondary amplification. Several subsequent studies of an arbovirus-infected mosquito cell line deficient for Dcr2 (C6/36 derived from *A. albopictus*) show the importance the piRNA pathway in antiviral immunity in the mosquito (Brackney et al. 2010; Scott et al. 2010; Hess et al. 2011). Small RNA profiles from WNV-, SINV-, DENV-2-, and La Crosse virus (a negative-strand arbovirus)-infected C6/36 cells show the production of abundant 23–30 nucleotide piRNAs. Moreover, infection of RNAi-competent adult *Aedes aegypti* and *A. albopictus* mosquitoes with chikungunya virus (CHIKV), an alphavirus, triggers the production of viral piRNAs in addition to viRNAs, indicating that both the siRNA and piRNA pathway targets CHIKV in somatic cells (Morazzani et al. 2012). Unlike viral piRNAs derived from *Drosophila* OSS cells, the viral piRNAs derived from CHIKV-infected mosquitos show signature 5′ 1U and 10A biases, suggesting both primary and secondary piRNA production. Expression study analysis shows that all the Piwi-subfamily proteins (Ago3 and 7 distinct Piwi proteins) are expressed in mosquito somatic cells, in contrast to germ line-specific expression in *Drosophila* (Morazzani et al. 2012). Similar observations for the role of siRNA and piRNA in antiviral defense have also been reported in an RNAi-competent mosquito cell line infected with a variety of other RNA viruses, including DENV, SINV, VSV, La Crosse virus, and Rift Valley fever (Vodovar et al. 2012; Léger et al. 2013). These studies support a role for the piRNA pathway, alongside the siRNA pathway in antiviral defense.

4 Salient Questions and Future Directions

Ample evidence demonstrates that RNAi is the major intrinsic antiviral mechanism in invertebrates. Nevertheless, there remain a number of questions that need further investigation. (1) Do RNAi and other small RNA pathways in invertebrates interact or cross talk to fight viral infection? (2) Does the RNAi machinery in

Drosophila re-localize to cellular compartments other than endosomes during viral infection? (3) Recent studies indicate that turnover of Dcr and Ago by selective autophagy regulates miRNA function in mammalian cells (Gibbings et al. 2012). Do invertebrate antiviral RNAi components (e.g., Dcr2 and Ago2) undergo similar homeostatic regulation in cells? (4) In *Drosophila*, chromatin-associated Dcr2 and Ago2 interact with the core transcription machinery to regulate the cellular transcriptome (Cernilogar et al. 2011). Do Dcr2- and Ago2-antagonizing RNAi suppressor proteins alter the cellular transcriptome to favor viral replication? (5) Several studies have shown that stress granules (SGs) and P bodies (PBs) are important cytoplasmic foci that regulate viral replication in cells (Beckham and Parker 2008). Does the RNAi machinery cross talk with SGs and PBs to regulate viral replication? (6) Studies in *Drosophila* and mosquito show Dcr2 involvement in JAK-STAT pathway activation, which leads to secretion of the antiviral peptide Vago in infected cells (Deddouche et al. 2008; Paradkar et al. 2012). Does recognition of dsRNA by Dcr2 initiate a downstream signaling process that leads to cross talk with the JAK-STAT pathway?

5 Conclusions

RNAi forms an evolutionarily ancient front line of defense against viral invasion in most, if not all, invertebrate phyla. Emerging evidence suggests that other branches of the RNA silencing machinery, including those guided by miRNA and piRNA, also contribute to antiviral functions in *Drosophila* and mosquito. Viral replication intermediates and structured regions associated with single-stranded viral genomes are sensed as PAMPs by the Dcr enzyme, and systemic RNAi evidently plays an essential role in providing robust immunity in all invertebrates. The differential potency of virally encoded RNAi suppressors seems to be a consequence of evolutionary adaptation to establish either an acute (and frequently lethal) or persistent (less virulent) infection in hosts. Association of the RNAi machinery with the endosomal compartment could function to establish a secure, two-way defensive checkpoint. The initial checkpoint eliminates invading viruses at the point of cell entry while the second checkpoint prevents endosomal-associated exit mechanisms of newly budding viruses. More importantly, RNAi-mediated immunity in invertebrates is evolutionarily ancient, yet exhibits many features common to mammalian immunity including target specificity (similar to adaptive immunity in mammals), systemic spread, and even immunological memory as evidenced in nematodes. As RNAi research progresses further, it is surely bound to bring many exciting and unexpected findings.

Acknowledgments We apologize to colleagues whose relevant publications were not cited because of space constraints. We thank Cecily Burrill, member of the RA laboratory for critical reading of the manuscript. We thank Kirk Ehmsen and Ashesh Saraiya for their helpful comments and discussion. Research in RA laboratory is financially supported by National Institute Health grants R01 AI40085 and AI064738.

References

Aliyari R, Ding S-W (2009) RNA-based viral immunity initiated by the Dicer family of host immune receptors. Immunol Rev 227:176–188

Aliyari R, Wu Q, Li HW, Wang XH, Li F, Green LD, Han CS, Li WX, Ding SW (2008) Mechanism of induction and suppression of antiviral immunity directed by virus-derived small RNAs in Drosophila. Cell Host Microbe 4:387–397

Attarzadeh-Yazdi G, Fragkoudis R, Chi Y, Siu RWC, Ülper L, Barry G, Rodriguez-Andres J, Nash AA, Bouloy M, Merits A, Fazakerley JK, Kohl A (2009) Cell-to-cell spread of the RNA interference response suppresses Semliki Forest virus (SFV) infection of mosquito cell cultures and cannot be antagonized by SFV. J Virol 83:5735–5748

Baltimore D, Girard M (1966) An intermediate in the synthesis of poliovirus RNA. Proc Natl Acad Sci USA 56:741

Beckham CJ, Parker R (2008) P bodies, stress granules, and viral life cycles. Cell Host Microbe 3:206–212

Behm-Ansmant I, Rehwinkel J, Doerks T, Stark A, Bork P, Izaurralde E (2006) mRNA degradation by miRNAs and GW182 requires both CCR4:NOT deadenylase and DCP1:DCP2 decapping complexes. Genes Dev 20:1885–1898

Bernstein E, Caudy AA, Hammond SM, Hannon GJ (2001) Role for a bidentate ribonuclease in the initiation step of RNA interference. Nature 409:363–365

Blair CD (2011) Mosquito RNAi is the major innate immune pathway controlling arbovirus infection and transmission. Future Microbiol 6:265–277

Bonning BC, Miller WA (2010) Dicistroviruses. Annu Rev Entomol 55:129–150

Brackney DE, Beane JE, Ebel GD (2009) RNAi targeting of West Nile virus in mosquito midguts promotes virus diversification. PLoS Pathog 5:e1000502

Brackney DE, Scott JC, Sagawa F, Woodward JE, Miller NA, Schilkey FD, Mudge J, Wilusz J, Olson KE, Blair CD (2010) C6/36 Aedes albopictus cells have a dysfunctional antiviral RNA interference response. PLOS Neglected Trop Dis 4:e856

Brennecke J, Stark A, Russell RB, Cohen SM (2005) Principles of MicroRNA–target recognition. PLoS Biol 3:e85

Brennecke J, Aravin AA, Stark A, Dus M, Kellis M, Sachidanandam R, Hannon GJ (2007) Discrete small RNA-generating loci as master regulators of transposon activity in Drosophila. Cell 128:1089–1103

Bronkhorst AW, van Cleef KWR, Vodovar N, İnce İA, Blanc H, Vlak JM, Saleh M-C, van Rij RP (2012) The DNA virus Invertebrate iridescent virus 6 is a target of the Drosophila RNAi machinery. Proc Natl Acad Sci 109:E3604–E3613

Campbell CL, Black WC, Hess AM, Foy BD (2008) Comparative genomics of small RNA regulatory pathway components in vector mosquitoes. BMC Genomics 9:425

Carthew RW, Sontheimer EJ (2009) Origins and mechanisms of miRNAs and siRNAs. Cell 136:642–655

Caudy AA, Myers M, Hannon GJ, Hammond SM (2002) Fragile X-related protein and VIG associate with the RNA interference machinery. Genes Dev 16:2491–2496

Caudy AA, Ketting RF, Hammond SM, Denli AM, Bathoorn AM, Tops BB, Silva JM, Myers MM, Hannon GJ, Plasterk RH (2003) A micrococcal nuclease homologue in RNAi effector complexes. Nature 425:411–414

Cenik ES, Fukunaga R, Lu G, Dutcher R, Wang Y, Tanaka Hall TM, Zamore PD (2011) Phosphate and R2D2 restrict the substrate specificity of Dicer-2, an ATP-driven ribonuclease. Mol Cell 42:172–184

Cernilogar FM, Onorati MC, Kothe GO, Burroughs AM, Parsi KM, Breiling A, Sardo FL, Saxena A, Miyoshi K, Siomi H (2011) Chromatin-associated RNA interference components contribute to transcriptional regulation in Drosophila. Nature 480:391–395

Chao JA, Lee JH, Chapados BR, Debler EW, Schneemann A, Williamson JR (2005) Dual modes of RNA-silencing suppression by flock house virus protein B2. Nat Struct Mol Biol 12:952–957

Chu PW, Westaway EG (1985) Replication strategy of Kunjin virus: evidence for recycling role of replicative form RNA as template in semiconservative and asymmetric replication. Virology 140:68–79

Cleaves GR, Ryan TE, Walter Schlesinger R (1981) Identification and characterization of type 2 dengue virus replicative intermediate and replicative form RNAs. Virology 111:73–83

Czech B, Malone CD, Zhou R, Stark A, Schlingeheyde C, Dus M, Perrimon N, Kellis M, Wohlschlegel JA, Sachidanandam R, Hannon GJ, Brennecke J (2008) An endogenous small interfering RNA pathway in Drosophila. Nature 453:798–802

Deddouche S, Matt N, Budd A, Mueller S, Kemp C, Galiana-Arnoux D, Dostert C, Antoniewski C, Hoffmann JA, Imler J-L (2008) The DExD/H-box helicase Dicer-2 mediates the induction of antiviral activity in drosophila. Nat Immunol 9:1425–1432

Denli AM, Tops BBJ, Plasterk RHA, Ketting RF, Hannon GJ (2004) Processing of primary microRNAs by the microprocessor complex. Nat Cell Biol 432:231–235

Ding S-W (2010) RNA-based antiviral immunity. Nat Rev Immunol 10:632–644

Félix M-A, Ashe A, Piffaretti J, Wu G, Nuez I, Bélicard T, Jiang Y, Zhao G, Franz CJ, Goldstein LD (2011) Natural and experimental infection of Caenorhabditis nematodes by novel viruses related to nodaviruses. PLoS Biol 9:e1000586

Flynt A, Liu N, Martin R, Lai EC (2009) Dicing of viral replication intermediates during silencing of latent Drosophila viruses. Proc Natl Acad Sci 106:5270–5275

Galiana-Arnoux D, Dostert C, Schneemann A, Hoffmann JA, Imler J-L (2006) Essential function in vivo for Dicer-2 in host defense against RNA viruses in drosophila. Nat Immunol 7:590–597

Gibbings D, Mostowy S, Jay F, Schwab Y, Cossart P, Voinnet O (2012) Selective autophagy degrades DICER and AGO2 and regulates miRNA activity. Nat Cell Biol 14:1314–1321

Goic B, Vodovar N, Mondotte JA, Monot C, Frangeui L, Blanc H, Gausson V, Vera-Otarols J, Cristofari G Saleh MC (2013) RNA virus persistence in Drosophila is controlled by small interfering RNAs produced from *de novo* synthesized virus cDNA. Nature Immunol 14:396–403

Gottwein E, Cullen BR (2008) Viral and cellular microRNAs as determinants of viral pathogenesis and immunity. Cell Host Microbe 3:375–387

Hain D, Bettencourt BR, Okamura K, Csorba T, Meyer W, Jin Z, Biggerstaff J, Siomi H, Hutvagner G, Lai EC, Welte M, Muller HA (2010) Natural variation of the amino-terminal glutamine-rich domain in Drosophila argonaute2 is not associated with developmental defects. PLoS ONE 5:e15264

Hammond SM, Boettcher S, Caudy AA, Kobayashi R, Hannon GJ (2001) Argonaute2, a link between genetic and biochemical analyses of RNAi. Sci Signal 293:1146

Hess AM, Prasad AN, Ptitsyn A, Ebel GD, Olson KE, Barbacioru C, Monighetti C, Campbell CL (2011) Small RNA profiling of Dengue virus-mosquito interactions implicates the Piwi RNA pathway in anti-viral defense. BMC Microbiol 11:45

Hinas A, Wright AJ, Hunter CP (2012) SID-5 is an endosome-associated protein required for efficient systemic RNAi in C. elegans. Curr Biol 22:1938–1943

Horwich MD, Li C, Matranga C, Vagin V, Farley G, Wang P, Zamore PD (2007) The Drosophila RNA methyltransferase, DmHen1, modifies germline piRNAs and single-stranded siRNAs in RISC. Curr Biol 17:1265–1272

Hussain M, Torres S, Schnettler E, Funk A, Grundhoff A, Pijlman GP, Khromykh AA, Asgari S (2012) West Nile virus encodes a microRNA-like small RNA in the 3′ untranslated region which up-regulates GATA4 mRNA and facilitates virus replication in mosquito cells. Nucleic Acids Res 40:2210–2223

Ishizuka A, Siomi MC, Siomi H (2002) A Drosophila fragile X protein interacts with components of RNAi and ribosomal proteins. Genes Dev 16:2497–2508

Jiang F, Ye X, Liu X, Fincher L, McKearin D, Liu Q (2005) Dicer-1 and R3D1-L catalyze microRNA maturation in Drosophila. Genes Dev 19:1674–1679

Keene KM, Foy BD, Sanchez-Vargas I, Beaty BJ, Blair CD, Olson KE (2004) RNA interference acts as a natural antiviral response to O'nyong-nyong virus (Alphavirus; Togaviridae) infection of Anopheles gambiae. Proc Natl Acad Sci USA 101:17240–17245

Kemp C, Mueller S, Goto A, Barbier V, Paro S, Bonnay F, Dostert C, Troxler L, Hetru C, Meignin C, Pfeffer S, Hoffmann JA, Imler JL (2013) Broad RNA interference-mediated antiviral immunity and virus-specific inducible responses in Drosophila. J Immunol 190:650–658

Kopek BG, Perkins G, Miller DJ, Ellisman MH, Ahlquist P (2007) Three-dimensional analysis of a viral RNA replication complex reveals a virus-induced mini-organelle. PLoS Biol 5:e220

Lau NC, Robine N, Martin R, Chung WJ, Niki Y, Berezikov E, Lai EC (2009) Abundant primary piRNAs, endo-siRNAs, and microRNAs in a Drosophila ovary cell line. Genome Res 19:1776–1785

Lau PW, Guiley KZ, De N, Potter CS, Carragher B, MacRae IJ (2012) The molecular architecture of human Dicer. Nat Struct Mol Biol 19:436–440

Lee YS, Pressman S, Andress AP, Kim K, White JL, Cassidy JJ, Li X, Lubell K, Lim DH, Cho IS, Nakahara K, Preall JB, Bellare P, Sontheimer EJ, Carthew RW (2009) Silencing by small RNAs is linked to endosomal trafficking. Nat Cell Biol 11:1150–1156

Léger P, Lara E, Jagla B, Sismeiro O, Mansuroglu Z, Coppée JY, Bonnefoy E, Bouloy M (2013) Dicer-2- and piwi-mediated RNA interference in rift valley fever virus-infected mosquito cells. J Virol 87:1631–1648

Li H, Li W-X, Ding S-W (2002) Induction and suppression of RNA silencing by an animal virus. Sci Signal 296:1319

Liu Q, Rand TA, Kalidas S, Du F, Kim H-E, Smith DP, Wang X (2003) R2D2, a bridge between the initiation and effector steps of the Drosophila RNAi pathway. Science 301:1921–1925

Liu X, Jiang F, Kalidas S, Smith D, Liu Q (2006) Dicer-2 and R2D2 coordinately bind siRNA to promote assembly of the siRISC complexes. RNA (New York) 12:1514–1520

Liu Y, Ye X, Jiang F, Liang C, Chen D, Peng J, Kinch LN, Grishin NV, Liu Q (2009) C3PO, an endoribonuclease that promotes RNAi by facilitating RISC activation. Science 325:750–753

Lu R, Maduro M, Li F, Li HW, Broitman-Maduro G, Li WX, Ding SW (2005) Animal virus replication and RNAi-mediated antiviral silencing in Caenorhabditis elegans. Nat Cell Biol 436:1040–1043

Lu R, Yigit E, Li W-X, Ding S-W (2009) An RIG-I-Like RNA helicase mediates antiviral RNAi downstream of Viral siRNA biogenesis in Caenorhabditis elegans. PLoS Pathog 5:e1000286

MacRae IJ, Zhou K, Doudna JA (2007) Structural determinants of RNA recognition and cleavage by Dicer. Nat Struct Mol Biol 14:934–940

Marques JT, Kim K, Wu PH, Alleyne TM, Jafari N, Carthew RW (2009) Loqs and R2D2 act sequentially in the siRNA pathway in Drosophila. Nat Struct Mol Biol 17:24–30

Matranga C, Tomari Y, Shin C, Bartel DP, Zamore PD (2005) Passenger-strand cleavage facilitates assembly of siRNA into Ago2-containing RNAi enzyme complexes. Cell 123:607–620

Morazzani EM, Wiley MR, Murreddu MG, Adelman ZN, Myles KM (2012) Production of virus-derived ping-pong-dependent pirna-like small RNAs in the mosquito soma. PLoS Pathog 8:e1002470

Mueller S, Gausson V, Vodovar N, Deddouche S, Troxler L, Perot J, Pfeffer S, Hoffmann JA, Saleh M-C, Imler J-L (2010) RNAi-mediated immunity provides strong protection against the negative-strand RNA vesicular stomatitis virus in Drosophila. Proc Natl Acad Sci USA 107:19390–19395

Myles KM, Wiley MR, Morazzani EM, Adelman ZN (2008) Alphavirus-derived small RNAs modulate pathogenesis in disease vector mosquitoes. Proc Natl Acad Sci 105:19938–19943

Nayak A, Andino R (2011) Slicer activity in Drosophila melanogaster S2 extract. Methods in molecular biology (Clifton, NJ) 721: 231–244

Nayak A, Berry B, Tassetto M, Kunitomi M, Acevedo A, Deng C, Krutchinsky A, Gross J, Antoniewski C, Andino R (2010) Cricket paralysis virus antagonizes Argonaute 2 to modulate antiviral defense in Drosophila. Nat Struct Mol Biol 17: 547–554

Obbard DJ, Jiggins FM, Halligan DL, Little TJ (2006) Natural selection drives extremely rapid evolution in antiviral RNAi genes. Curr Biol 16:580–585

Obbard DJ, Gordon KHJ, Buck AH, Jiggins FM (2009) The evolution of RNAi as a defence against viruses and transposable elements. Philos Trans Roy Soc B: Biol Sci 364:99–115

Okada C, Yamashita E, Lee SJ, Shibata S, Katahira J, Nakagawa A, Yoneda Y, Tsukihara T (2009) A high-resolution structure of the pre-microRNA nuclear export machinery. Science 326:1275–1279

Paradkar PN, Trinidad L, Voysey R, Duchemin J-B, Walker PJ (2012) Secreted Vago restricts West Nile virus infection in Culex mosquito cells by activating the Jak-STAT pathway. Proc Natl Acad Sci 109:18915–18920

Qi N, Zhang L, Qiu Y, Wang Z, Si J, Liu Y, Xiang X, Xie J, Qin CF, Zhou X, Hu Y (2012) Targeting of Dicer-2 and RNA by a Viral RNA silencing suppressor in Drosophila cells. J Virol 86:5763–5773

Qin H, Chen F, Huan X, Machida S, Song J, Yuan YA (2010) Structure of the Arabidopsis thaliana DCL4 DUF283 domain reveals a noncanonical double-stranded RNA-binding fold for protein–protein interaction. RNA (New York) 16: 474–481

Rand TA, Petersen S, Du F, Wang X (2005) Argonaute2 cleaves the anti-guide strand of siRNA during RISC activation. Cell 123:621–630

Rechavi O, Minevich G, Hobert O (2011) Transgenerational inheritance of an acquired small RNA-based antiviral response in C. elegans. Cell 147:1248–1256

Rehwinkel J, Behm-Ansmant I, Gatfield D, Izaurralde E (2005) A crucial role for GW182 and the DCP1:DCP2 decapping complex in miRNA-mediated gene silencing. RNA (New York) 11: 1640–1647

Sabin LR, Zhou R, Gruber JJ, Lukinova N, Bambina S, Berman A, Lau CK, Thompson CB, Cherry S (2009) Ars2 regulates both miRNA- and siRNA- dependent silencing and suppresses RNA virus infection in Drosophila. Cell 138:340–351

Sabin LR, Zheng Q, Thekkat P, Yang J, Hannon GJ, Gregory BD, Tudor M, Cherry S (2013) Dicer-2 processes diverse viral RNA species. PLoS ONE 8:e55458

Saito K, Ishizuka A, Siomi H, Siomi MC (2005) Processing of Pre-microRNAs by the Dicer-1–Loquacious complex in Drosophila cells. PLoS Biol 3:e235

Saleh M-C, Tassetto M, van Rij RP, Goic B, Gausson V, Berry B, Jacquier C, Antoniewski C, Andino R (2009) Antiviral immunity in Drosophila requires systemic RNA interference spread. Nature 458:346–350

Sánchez-Vargas I, Scott JC, Poole-Smith BK, Franz AW, Barbosa-Solomieu V, Wilusz J, Olson KE, Blair CD (2009) Dengue virus type 2 infections of Aedes aegypti are modulated by the mosquito's RNA interference pathway. PLoS Pathog 5:e1000299

Schnettler E, Sterken MG, Leung JY, Metz SW, Geertsema C, Goldbach RW, Vlak JM, Kohl A, Khromykh AA, Pijlman GP (2012) Noncoding flavivirus RNA displays RNA interference suppressor activity in insect and Mammalian cells. J Virol 86:13486–13500

Schott DH, Cureton DK, Whelan SP, Hunter CP (2005) An antiviral role for the RNA interference machinery in Caenorhabditis elegans. Proc Natl Acad Sci USA 102:18420–18424

Scott JC, Brackney DE, Campbell CL, Bondu-Hawkins V, Hjelle B, Ebel GD, Olson KE, Blair CD (2010) Comparison of Dengue Virus type 2-specific small RNAs from RNA interference-competent and—incompetent mosquito cells. PLOS Neglected Trop Dis 4:e848

Simons M, Raposo G (2009) Exosomes–vesicular carriers for intercellular communication. Curr Opin Cell Biol 21:575–581

Singh G, Popli S, Hari Y, Malhotra P, Mukherjee S, Bhatnagar RK (2009) Suppression of RNA silencing by Flock house virus B2 protein is mediated through its interaction with the PAZ domain of Dicer. FASEB J: Off Publ Federation Am Soc Exp Biol 23:1845–1857

Siomi MC, Sato K, Pezic D, Aravin AA (2011) Piwi-interacting small RNAs: the vanguard of genome defence. Nat Rev Mol Cell Biol 12:246–258

Siu RW, Fragkoudis R, Simmonds P, Donald CL, Chase-Topping ME, Barry G, Attarzadeh-Yazdi G, Rodriguez-Andres J, Nash AA, Merits A (2011) Antiviral RNA interference responses induced by Semliki Forest virus infection of mosquito cells: characterization, origin, and frequency-dependent functions of virus-derived small interfering RNAs. J Virol 85:2907–2917

Skalsky RL, Cullen BR (2010) Viruses, microRNAs, and host interactions. Annual Rev Microbiol 64(1):123

Skalsky RL, Vanlandingham DL, Scholle F, Higgs S, Cullen BR (2010) Identification of microRNAs expressed in two mosquito vectors, Aedes albopictus and Culex quinquefasciatus. BMC Genomics 11:119

Stark A, Lin MF, Kheradpour P, Pedersen JS, Parts L, Carlson JW, Crosby MA, Rasmussen MD, Roy S, Deoras AN, Ruby JG, Brennecke J, Matthews BB, Schroeder AJ, Gramates LS, Pierre SES, Roark M, Wiley KL Jr, Kulathinal RJ, Zhang P, Myrick KV, Antone JV, Gelbart WM, Yu C, Park S, Wan KH, Celniker SE, Hodges E, Hinrichs AS, Caspi A, Paten B, Park S-W, Han MV, Maeder ML, Polansky BJ, Robson BE, Aerts S, van Helden J, Hassan B, Gilbert DG, Eastman DA, Rice M, Weir M, Hahn MW, Park Y, Dewey CN, Pachter L, Kent WJ, Haussler D, Lai EC, Bartel DP, Hannon GJ, Kaufman TC, Eisen MB, Clark AG, Smith D, Kellis M (2007) Discovery of functional elements in 12 Drosophila genomes using evolutionary signatures. Nature 450:219–232

Théry C (2011) Exosomes: secreted vesicles and intercellular communications. F1000 biology reports 3

Tomari Y, Du T, Haley B, Schwarz DS, Bennett R, Cook HA, Koppetsch BS, Theurkauf WE, Zamore PD (2004a) RISC assembly defects in the Drosophila RNAi mutant armitage. Cell 116:831–841

Tomari Y, Matranga C, Haley B, Martinez N, Zamore PD (2004b) A protein sensor for siRNA asymmetry. Science 306:1377–1380

van Mierlo JT, Bronkhorst AW, Overheul GJ, Sadanandan SA, Ekström J-O, Heestermans M, Hultmark D, Antoniewski C, van Rij RP (2012) Convergent evolution of argonaute-2 slicer antagonism in two distinct insect RNA viruses. PLoS Pathog 8:e1002872

van Rij RP, Saleh MC, Berry B, Foo C, Houk A, Antoniewski C, Andino R (2006) The RNA silencing endonuclease Argonaute 2 mediates specific antiviral immunity in Drosophila melanogaster. Genes Dev 20:2985–2995

Vodovar N, Bronkhorst AW, van Cleef KWR, Miesen P, Blanc H, van Rij RP, Saleh M-C (2012) Arbovirus-derived piRNAs exhibit a ping-pong signature in mosquito cells. PLoS ONE 7:e30861

Wang X-H, Aliyari R, Li W-X, Li H-W, Kim K, Carthew R, Atkinson P, Ding S-W (2006) RNA interference directs innate immunity against viruses in adult Drosophila. Sci Signal 312:452

Welker NC, Maity TS, Ye X, Aruscavage PJ, Krauchuk AA, Liu Q, Bass BL (2011) Dicer's helicase domain discriminates dsRNA termini to promote an altered reaction mode. Mol Cell 41:589–599

Wilkins C, Dishongh R, Moore SC, Whitt MA, Chow M, Machaca K (2005) RNA interference is an antiviral defence mechanism in Caenorhabditis elegans. Nat Cell Biol 436:1044–1047

WinstonWM, Molodowitch C, Hunter CP (2002) Systemic RNAi in C. elegans requires the putative transmembrane protein SID-1. Science 295:2456-2459

Winston WM, Sutherlin M, Wright AJ, Feinberg EH, Hunter CP (2007) Caenorhabditis elegans SID-2 is required for environmental RNA interference. Proc Natl Acad Sci 104:10565–10570

Wu Q, Luo Y, Lu R, Lau N, Lai EC, Li WX, Ding SW (2010a) Virus discovery by deep sequencing and assembly of virus-derived small silencing RNAs. Proc Natl Acad Sci 107:1606–1611

Wu Q, Wang X, Ding SW (2010b) Viral suppressors of RNA-based viral immunity: host targets. Cell Host Microbe 8:12–15

Zambon RA, Vakharia VN, Wu LP (2006) RNAi is an antiviral immune response against a dsRNA virus in Drosophila melanogaster. Cell Microbiol 8:880–889

Zamore PD, Tuschl T, Sharp PA, Bartel DP (2000) RNAi: double-stranded RNA directs the ATP-dependent cleavage of mRNA at 21 to 23 nucleotide intervals. Cell 101:25–34

Zhang H, Kolb FA, Jaskiewicz L, Westhof E, Filipowicz W (2004) Single processing center models for human Dicer and bacterial RNase III. Cell 118:57–68

Roles of MicroRNAs in the Life Cycles of Mammalian Viruses

Eva Gottwein

Abstract MicroRNAs (miRNAs) are a class of small noncoding RNAs expressed by plants, animals, and some viruses. miRNAs generally function as part of miRNA-induced silencing complexes to modestly repress mRNAs with imperfect sequence complementarity. Over the last years, many different roles of miRNA mediated regulation in the life cycles of mammalian viruses have been uncovered. In this chapter, I will mainly explore four different examples of how cellular miRNAs interact with viruses: the role of miR-155 in viral oncogenesis, viral strategies to eliminate cellular miR-27, the contribution of miR-122 to the replication of hepatitis C virus, and miRNAs as an experimental tool to control virus replication and vector transgene expression. In the final part of this chapter, I will give a brief overview of virally encoded microRNAs.

Contents

1 Introduction

RNA silencing is mediated by small RNAs of ~20–31 nucleotides (nts) in complex with a member of the Argonaute family of proteins. RNA silencing occurs in plants, fungi, and animals, but its biological functions and the small RNA

E. Gottwein (✉)
Department of Microbiology-Immunology, Feinberg School of Medicine,
Northwestern University, 303 E. Chicago Ave, Chicago, IL 60611, USA
e-mail: e-gottwein@northwestern.edu

B. R. Cullen (ed.), *Intrinsic Immunity*, Current Topics in Microbiology and Immunology 371, DOI: 10.1007/978-3-642-37765-5_8,

and protein components of the RNA-induced silencing complexes (RISCs) have diverged between organisms. Approximately 21-nt small interfering RNAs (siRNAs) processed from viral double-stranded RNAs (dsRNAs) form the basis of antiviral immunity in plants, nematodes, and insects (Ding 2010), where they guide the endonucleolytic destruction of complementary viral RNAs in a process called RNA interference (RNAi). In contrast, RNAi is not thought to operate as an antiviral defense mechanism in mammals (Cullen 2006a). Small silencing RNAs expressed by mammals include the large family of microRNAs (miRNAs) and the less understood endogenous siRNAs (endo-siRNAs), which both associate with the Ago subfamily of Argonaute proteins. Piwi-interacting RNAs (piRNAs), which are slightly longer (24–31 nts) than miRNAs, are expressed in the germ line and associate with the Piwi subfamily of Argonaute proteins. This chapter is focused on the roles of miRNAs in the life cycles of mammalian viruses. The biogenesis and functions of other classes of small RNAs have been reviewed in detail elsewhere (Ghildiyal and Zamore 2009; Kim et al. 2009; Saito and Siomi 2010).

miRNAs are ~22-nt small RNAs expressed by all animals and plants, and by some viruses. I will introduce the biogenesis and effectors of vertebrate miRNAs, which differ in some aspects from those in other organisms (for a detailed review see Kim et al. 2009). miRNAs are derived from longer, primary transcripts (pri-miRNAs) through a stepwise maturation process. In the canonical miRNAs biogenesis pathway observed by the large majority of miRNAs, pri-miRNAs are transcribed by RNA polymerase II (Pol II) and cleaved in the nucleus by the microprocessor, a complex of the RNase III enzyme Drosha and its cofactor, DiGeorge syndrome critical region gene 8 (DGCR8), to liberate an imperfect hairpin structure called the pre-miRNA. Following its translocation to the cytoplasm, the pre-miRNA is further processed by the RNase III enzyme Dicer, resulting in an imperfect duplex of two ~22-nt-long RNAs with 2-nt 3′-overhangs. The strand with the weaker 5′-base pairing is preferentially loaded as the mature miRNA into one of four mammalian Ago proteins (Ago1-4), while the unincorporated strand (called the passenger or miRNA* strand) is degraded. A small number of mammalian miRNAs follow non-canonical routes of biogenesis, including the production of the pri-miRNA by RNA polymerase III (Pol III), and Drosha- or Dicer-independent pathways (Okamura et al. 2007; Ruby et al. 2007; Babiarz et al. 2008; Cheloufi et al. 2010; Chong et al. 2010; Cifuentes et al. 2010). With few exceptions (Yekta et al. 2004; Davis et al. 2005), miRNA programmed RISC complexes (miRISC) recognize imperfect binding sites resulting in a modest mRNA destabilization and/or repression of translation (Bartel 2009). Effective miRNA binding sites are most commonly located in 3′ untranslated regions and exhibit perfect base pairing of the miRNA seed sequence, that is, nts 2–8 from the miRNA 5′-end or nts 2–7 if there is an A across from nt 1 of the miRNA. Suboptimal seed base pairing can also lead to target repression if other supplementary interactions, for example with the 3′ portion of the miRNA, are present (Shin et al. 2010; Chi et al. 2012). The four Ago proteins are thought to function interchangeably in miRNA-mediated repression, but only Ago2 retains the endonuclease activity that mediates RNAi in invertebrates. In mammals, this endonucleolytic activity of Ago2 is essential for the Dicer-independent maturation of miR-451 and also facilitates experimentally induced RNAi (Liu et al. 2004; Meister et al. 2004; Cheloufi et al. 2010; Cifuentes et al. 2010). Several hundred conserved mammalian miRNAs have been identified (Griffiths-Jones et al. 2006; Griffiths-Jones et al. 2008; Chiang et al. 2010), and most mammalian transcripts are likely subject to miRNA-mediated regulation in some biological context (Friedman et al. 2009). It is therefore not surprising that miRNAs are also important determinants of mammalian virus replication.

Cellular miRNAs can negatively or positively affect viral replication as a consequence of their physiological functions. Viruses may benefit from inducing miRNAs that have advantageous effects and from repressing miRNAs that negatively affect the virus life cycle. Accordingly, the alteration of the cellular miRNA repertoire by many different virus families has been reported. I will discuss this paradigm using the induction of cellular miR-155 by a number of oncogenic viruses and the degradation of cellular miR-27 during herpesviral infection as examples. Cellular miRNAs may also have direct effects on viral replication, through positive or negative interactions with viral RNA genomes or other viral RNAs. To illustrate such direct interactions, I will give a detailed summary of what is known about the role of cellular miR-122 in HCV replication. In addition, I will discuss how differences in miRNA expression between tissues can be exploited to manipulate viral tropism. Finally, some viruses have been found to express their own miRNAs. In the last part of this review, I will therefore give a brief summary of virally encoded miRNAs.

2 Induction of miR-155 by Oncogenic Viruses

The human γ-herpesvirus Epstein-Barr virus (EBV) is known to extensively regulate cellular miRNA expression in latently infected B cells (Mrazek et al. 2007; Cameron et al. 2008; Forte et al. 2012) and also encodes more than 40 viral miRNAs (Pfeffer et al. 2004; Cai et al. 2006; Zhu et al. 2009). The dramatically altered miRNA repertoire in EBV-infected cells likely reflects both the exquisite adaptation of this virus to its host and the importance of fine tuning cellular gene expression for successful EBV infection and persistence. EBV latency in B cells can lead to the development of B cell lymphomas, including Hodgkin's disease, Burkitt's lymphoma, and diffuse large B cell lymphomas (DLBCLs) in immunocompromised patients (Kieff and Rickinson 2007). EBV also immortalizes primary B cells in vitro, resulting in so-called lymphoblastoid cell lines (LCLs). LCLs express the six EBV nuclear antigen proteins (EBNAs), the latent membrane proteins (LMP1, LMP2A/B), and all EBV miRNAs, a gene expression program referred to as latency III (Cai et al. 2006; Kieff and Rickinson 2007). The latency III program, which is also seen in EBV-associated DLBCLs, is thought to induce cellular proliferation through signaling events that recapitulate those induced by antigen, IL4 and CD40 ligand (Kieff and Rickinson 2007). In newly infected individuals, this EBV growth program may help expand the population of infected cells and drive their differentiation into memory-type B cells (Thorley-Lawson, 2005), which then serve as the latent reservoir of the virus and exhibit more restricted viral gene expression.

The most dramatic EBV-induced change in cellular miRNA expression is the induction of cellular miR-155 as part of the latency III program, resulting in high levels of miR-155 expression in LCLs and in DLBCLs (Yin et al. 2008; Linnstaedt et al. 2010). miR-155 is encoded by the noncoding B cell integration cluster (*bic*) gene, which was first identified as a common integration site of avian leukosis virus (ALV) in ALV-induced B cell lymphomas in chickens (Tam et al. 1997). miR-155 expression is transiently induced following the activation of macrophages, dendritic cells, B cells, and T cells (Haasch et al. 2002; O'Connell et al. 2007; Thai et al. 2007), through the AP-1 and NFκB transcription factors (O'Connell et al. 2007; Thai et al. 2007). In B cells, miR-155 is important for the germinal center (GC) reaction and the generation of T cell-dependent humoral immunity (Thai et al. 2007; Vigorito et al. 2007). miR-155 has well-recognized oncogenic properties. High expression of miR-155 is observed in several different types of B cell lymphomas in humans and in some solid tumors (Metzler et al. 2004; Eis et al. 2005; Kluiver et al. 2005; Volinia et al. 2006). The constitutive transgenic expression of miR-155 in B cells (Eμ-miR155) caused pre-B cell proliferation in the spleen and in bone marrow and resulted in high-grade B cell malignancies at later times (Costinean et al. 2006). Similarly, the forced expression of miR-155 in transplanted murine or human hematopoietic stem cells resulted in proliferative disease (O'Connell et al. 2008; Boss et al. 2011; Dahlke et al. 2012). The induction of high levels of cellular miR-155, through transcriptional activation of *bic* (Motsch et al. 2007; Yin et al. 2008), is likely also important for the oncogenic potential of EBV. Both LCLs and a DLBCL-derived B cell line depend on miR-155 activity for their proliferation and survival (Linnstaedt et al. 2010). The inhibition of miR-155 in LCLs was accompanied by the accumulation of cells in the G1 phase of the cell cycle and induction of apoptosis. Interestingly, the replication-defective B cell oncogenic avian retrovirus reticuloendotheliosis virus strain T (REV-T) probably also induces miR-155, through a viral homolog of the NFκB family member Rel (Bolisetty et al. 2009).

While EBV induces the expression of cellular miR-155, the oncogenic herpesviruses Kaposi's Sarcoma–associated herpesvirus (KSHV) and Marek's disease virus type 1 (MDV-1) appear to have evolved an alternative strategy to supply miR-155-like activities and encode viral mimics of miR-155. KSHV is the second human oncogenic γ-herpesvirus and causes the B cell proliferative disorders primary effusion lymphoma (PEL) and multicentric Castleman's disease (MCD). Marek's disease virus type 1 is an α-herpesvirus which causes T cell lymphomas in chickens. KSHV miR-K11 and MDV-1 miR-M4 both share the first eight nucleotides with miR-155, which is conserved over the entire sequence between human and chicken and only differs in nt 13 in mouse miR-155. Because the first eight nucleotides encompass the entire seed sequence critical for target regulation, both miR-K11 and miR-M4 regulate seed targets of miR-155. This has been directly demonstrated for a subset of targets and is also confirmed by transcriptome-wide analyses of miR-K11 binding sites (Gottwein et al. 2007; Skalsky et al. 2007; Zhao et al. 2009; Gottwein et al. 2011; Haecker et al. 2012). While a role of KSHV miR-K11 in the transformation of human B cells by KSHV remains to be demonstrated, miR-K11 was shown to phenocopy miR-155-induced B cell hyperproliferation following the forced expression in human or murine HSCs in vivo (Boss et al. 2011; Dahlke et al. 2012). Deletion of MDV-1 miR-4, on the other hand, abolished the induction of T cell lymphomas by MDV-1 in chickens (Zhao et al. 2011). This study provided the first in vivo evidence that a viral miRNA can play an essential role in virally induced tumorigenesis.

The mechanism underlying the oncogenic potential of miR-155 and its viral analogs miR-K11 and miR-M4 is still under investigation. Compared to other miRNAs, miR-155 targets are relatively well studied and extensive lists of high confidence candidate targets have been derived from mRNA and protein expression data as well as from CLIP-Seq (cross-linking and immunoprecipitation of RISC complexes in conjunction with deep sequencing) experiments (Gottwein et al. 2007; Rodriguez et al. 2007; Skalsky et al. 2007; O'Connell et al. 2008; Selbach et al. 2008; Xu et al. 2010; Gottwein et al. 2011; Loeb et al. 2012; Skalsky et al. 2012). Many of these candidate targets have undergone at least some validation, and there are several good candidates for mediators of the oncogenic properties of miR-155. Targets that have been associated with miR-155-induced proliferation through functional experiments in vitro and/or in vivo include histone deacetylase 4 (HDAC4) in B cells (Sandhu et al. 2012), Src homology-2 domain–containing inositol 5-phosphatase 1 (SHIP1) in myeloid cells (Costinean et al. 2009; O'Connell et al. 2009), and TP53INP1 in prostate cancer (Gironella et al. 2007). Not all of these targets, however, have miR-155 binding sites that are conserved in humans, mice, and chickens; that is, the organisms in which miR-155 has known oncogenic activity. It is therefore an open question whether details of the mechanism of miR-155 action in these three organisms and in different cell types are the same.

miR-155 is not the only example of a virally induced cellular miRNA. EBV alone induces several miRNAs (Mrazek et al. 2007; Cameron et al. 2008; Lin and Flemington 2011; Forte et al. 2012), and at least the induction of miR-34a is also important for the survival and proliferation of LCLs (Forte et al. 2012). Among several other examples of virally induced miRNAs are the upregulation of cellular miR-132 early upon infection by a number of herpesviruses, (Lagos et al. 2010), which dampens the induction of the innate immune response, and the upregulation of cellular miR-141 by the picornavirus enterovirus 71 to shut off the cap-dependent translation initiation of host mRNAs (Ho et al. 2011).

3 Viral Inhibitors of Cellular miR-27

The paradigm of viral inhibition of cellular miRNAs is exemplified by the destabilization of mature miR-27a and miR-27b by two evolutionary distant herpesviruses. miR-27a and miR-27b (collectively referred to as miR-27 from hereon) differ only by one nt near their 3′-end and are expressed from two distinct miRNA clusters that also encode miR-23a and miR-24-2 or miR-23b and miR-24-1, respectively. Herpesvirus saimiri (HVS) is a T cell tropic γ-herpesvirus that naturally infects squirrel monkeys, causes aggressive T cell malignancies in New World monkeys, and transforms T cells in vitro (Damania 2004). HVS expresses seven Sm-class small nuclear (sn) RNAs (HSURs 1-7, for HVS U-rich RNAs), which are the most abundantly expressed viral transcripts in HVS-transformed T cells (Murthy et al. 1986; Lee et al. 1988). Like cellular snRNAs, the HSURs bear 5′-trimethyl caps, 3′-terminal stem-loops and assemble with host Sm proteins in the cytoplasm followed by the nuclear localization of the resulting complex (Golembe et al. 2005). Of the 7 HSURs, only HSUR1 and 2 are also present in the genome of the related γ-herpesvirus Herpesvirus ateles (Albrecht 2000), suggesting a conserved and important function in viral infection. Cazalla et al. reported that HSUR1 and HSUR2, but not the other HSURs, carry seed complementary sites to cellular miR-27, miR-16, and miR-142-3p (Fig. 1a) and assessed whether HSURs could bind these miRNAs (Cazalla et al. 2010). HSUR1 and HSUR2, but not the other HSURs or cellular snRNPs, co-precipitated with Ago2. This complex likely also contains Sm proteins, because anti-Sm immunoprecipitates also contained miR-16, miR-27, and miR-142-3p in a manner dependent on the expression and co-precipitation of HSUR1/2 (Cazalla et al. 2010). HSUR1/2 are dispensable for HVS replication and T cell transformation in vitro (Murthy et al. 1989; Ensser et al. 1999), but resulting cell lines were reported to grow slower (Murthy et al. 1989). ΔHSUR1/2-immortalized T cell lines overexpressed miR-27 compared to T cells immortalized by wt HVS, but miR-16 or miR-142-3p levels were unchanged. This effect was specific to mature miR-27 miRNAs, because the expression of the miRNAs derived from the same primary transcript or the

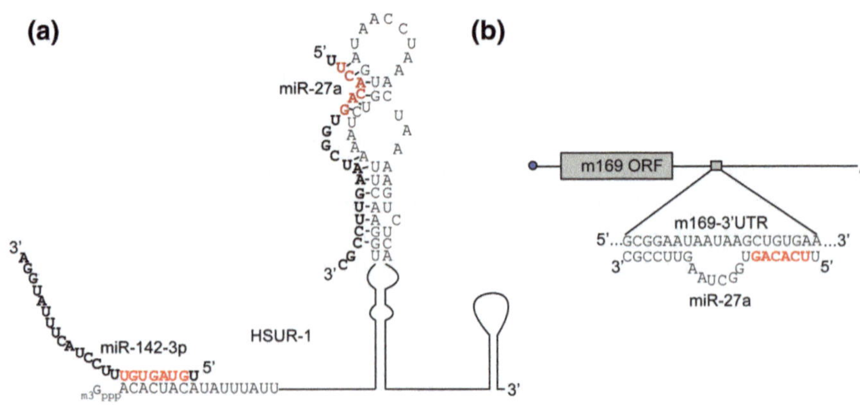

Fig. 1 **a** Schematic of HVS HSUR1 in complex with miR-27a and miR-142-3p. The seed sequences of miR-27a and miR-142-3p are highlighted in red. **b** Schematic of mCMV m169. Also shown are the location of the miR-27 binding site in the 3′UTR of m169 and its base pairing with miR-27a

passenger strand was unaffected. Pulse–chase experiments confirmed that the turnover of mature miR-27 was accelerated in the presence of HSUR1/2. This function is mediated by a direct interaction of HSUR1 with miR-27, because miR-27 was recovered with Sm complexes in HVS-negative T cells that expressed wt HSUR1, but not in T cells expressing a miR-27 binding site mutant. HSUR1 can be retargeted to the cellular miRNA miR-20, which further supported that the binding of miR-27 to HSUR1 is sufficient to induce its degradation. The extent of miR-27 downregulation by HSV is sufficient to result in the derepression of the validated miR-27 target FOXO1, indicating that HSUR1 is likely to result in a global derepression of miR-27 targets. The functional consequences of miR-27 antagonism by HVS remain to be established, however. Gene expression profiling data suggest that HSURs1/2 cause the upregulation of several mRNAs coding for proteins with roles in T cell activation and proliferation (Cook et al. 2005), but conflicting data were presented elsewhere (Ensser et al. 1999). The consequences of the association of HSUR1 and 2 with miR-16 and miR-142-3p are still unknown. Because miR-16 and miR-142-3p are both abundant miRNAs in T cells, HSUR expression may not be high enough to allow sponge-like sequestration and no derepression of miR-16 targets was observed in the absence of HSUR1/2 (Cazalla et al. 2010).

Intriguingly, HVS is not the only herpesvirus to antagonize miR-27. Buck et al. uncovered from miRNA profiling data that miR-27 expression is rapidly reduced after infection of murine fibroblasts with murine cytomegalovirus (mCMV) (Buck et al. 2010). As for HVS, this effect was specific to the mature miR-27 miRNAs because other miRNAs from the same pri-miRNA transcript and the miR-27 passenger strands were unaffected. Two independent reports have mapped this effect to a single miR-27 binding site in the 3′UTR of the previously uncharacterized viral m169 transcript (Fig. 1b), which is both necessary and sufficient for miR-27 degradation (Libri et al. 2012; Marcinowski et al. 2012). Like HSUR1, m169 can be redirected to other cellular miRNAs. Base pairing to m169 induced miR-27 3′-tailing, that is, the addition of 3′ nucleotides not present in the miR-27 precursors, and the subsequent degradation of miR-27 (Marcinowski et al. 2012). This interaction also resulted in reduced m169 mRNA expression and miR-27 repressed luciferase reporters bearing the m169 3′UTR, suggesting that m169 could be repressed by miR-27 before miR-27 is fully degraded. The m169-encoded protein remains uncharacterized, however. Mutation of the miR-27 binding sites in the context of the viral genome resulted in significant attenuation of mCMV replication in vivo, suggesting that the degradation of miR-27 by m169 RNA and/or the regulation of m169 expression by miR-27 is indeed important for mCMV infection.

These two viral inhibitors of cellular miR-27 share several properties that are likely important for their function. They are both expressed at very high levels and engage in both seed base pairing and extensive base pairing to the 3′-end of the miRNA (Fig. 1). These properties are not dissimilar to artificial sites shown to induce miRNA degradation in Drosophila and mammalian cells (Ameres et al. 2010; Ebert and Sharp 2010), suggesting that seed base pairing in combination with extensive 3′-end pairing can trigger miRNA degradation. It will be interesting to see whether HSUR1 and m169 are only the first examples of viral inhibitors of mature cellular miRNAs. In addition to the antagonism of mature miR-27 described in detail here, other notable examples of viral antagonism of cellular miRNAs include the global downregulation of mature miRNAs by poxviruses (Backes et al. 2012) and the competition of adenovirus VA1 noncoding RNAs with miRNA precursors for the nuclear export factor Exportin 5 and Dicer processing (Lu and Cullen 2004; Andersson et al. 2005).

4 Role of miR-122 in HCV Replication

The most striking example yet of a virus that directly interacts with a host miRNA to promote its replication is the dependence of hepatitis C virus (HCV) on the liver-specific miRNA miR-122 (Jopling et al. 2005). HCV is an important human pathogen with more than 170 million people chronically infected. There are seven distinct HCV genotypes, with sequence variation of up to 30 % within individual genotypes. During the last decade, the standard of care for chronic HCV infection consisted of polyethylene glycol–conjugated interferon α and ribavirin (Manns et al. 2006). This treatment is frequently ineffective, however, and also carries serious side effects. Chronic HCV infection can lead to cirrhosis and liver cancer. Novel compounds that directly target the viral proteins are currently being developed and/or are in clinical trials (Sarrazin et al. 2012). In 2011, two such antivirals, which directly target the viral protease, were licensed for the treatment for chronic HCV genotype 1 infection. However, the high mutation rate of HCV is known to lead to rapid viral escape from such compounds. Therefore, combination therapy will be required and also the development of additional treatment strategies, including therapies that are targeted against cellular factors necessary for HCV replication. The clinical inhibition of miR-122 is one such highly promising strategy (see below), and an antisense inhibitor of miR-122 is currently in clinical trials (Jopling 2012; Machlin et al. 2012).

HCV is a hepatotropic, positive-sense single-stranded RNA virus with a ~9.6 kb genome (Fig. 2a) (Lemon et al. 2007). The HCV genome contains a single open reading frame encoding the polyprotein precursor, flanked by 5′ and 3′ untranslated regions (5′UTR/3′UTR). Upon viral entry, the incoming genome serves as messenger RNA, whose translation depends on an internal ribosomal entry site (IRES) comprised by stem-loops 2–4 of the 5′UTR. The resulting polyprotein is processed to ten mature proteins, including the core and envelope proteins, and the non-structural proteins required for RNA replication. HCV RNA replication proceeds through a negative-sense RNA intermediate and takes place in close association with virally induced membrane structures. Because HCV isolates from patients do not replicate readily in cell culture, studies on HCV employ a number of cell culture models that have been developed over time (reviewed in Murray and Rice 2011). In subgenomic replicon systems (Lohmann et al. 1999), the 5′UTR of HCV allows the translation of a selectable marker, while a non-HCV

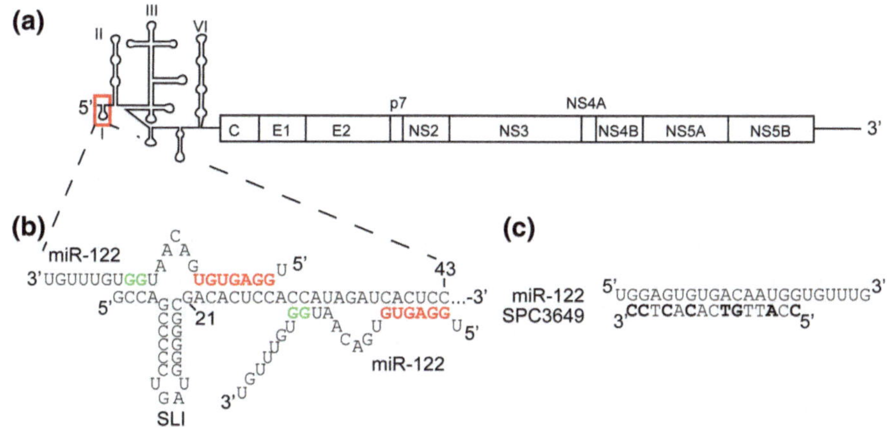

Fig. 2 **a** Schematic of the HCV genome, not drawn to scale. The size of the 5′UTR is exaggerated for a more detailed representation; 5′UTR stem-loops and mature HCV proteins are indicated. The red box marks the first 43 nts of the HCV genome, which are shown in detail in panel b. **b** Structure of the trimolecular complex formed between nts 1–43 of the HCV genome and two copies of miR-122. Base pairing shown in this panel is supported by structural and mutational analyses of this region (see text). The seed region of miR-122 is highlighted in red. miR-122 nts 15/16 are in green. **c** Complex of miR-122 with SPC3649. LNA-modified bases are in bold

IRES initiates the translation of the HCV proteins required for RNA replication. The identification and introduction of cell culture adaptive mutations (Murray and Rice 2011), which enhance the efficiency of RNA replication in culture, at the expense of infectious virus production, further facilitated the study of HCV replication. Finally, the generation of the molecular HCV genotype 2 clone JFH-1, which allows both RNA replication and infectious virus production in culture, has enabled studies of the full viral life cycle (Kato et al. 2001; Wakita et al. 2005; Zhong et al. 2005). As a consequence, however, mechanistic studies of the HCV life cycle have come to largely depend on this clone and intergenotypic chimeras thereof. Similarly, these studies have relied largely on a single human hepatoma cell line (Huh7) and its subclones, which are permissive for the full life cycle of HCV. While these culture models are indispensable for HCV research, it is important to keep in mind that the genetic features that allow efficient replication and propagation of HCV in culture can differ from those in chimpanzees or humans (Murray and Rice 2011).

miR-122 is a liver-specific miRNA (Lagos-Quintana et al. 2002), which may at least partially account for the liver tropism of HCV (Chang et al. 2008; Lin et al. 2010; Fukuhara et al. 2012). miR-122 is the most abundant miRNA in the human liver, where it is expressed at more than 10^5 copies per cell (Chang et al. 2004; Randall et al. 2007). The physiological function of miR-122 is primarily in the regulation of lipid and cholesterol metabolism, and consequently, the inhibition of miR-122 in mice and primates lowered total plasma cholesterol levels (Krutzfeldt et al. 2005; Esau et al. 2006; Elmen et al. 2008; Lanford et al. 2010). In 2005,

Jopling, Sarnow, and colleagues reported that the inhibition of miR-122 in Huh7 cells impaired HCV RNA accumulation and protein expression from a bi-cistronic replicon and a full-length viral genomic RNA with cell culture–adaptive mutations (Jopling et al. 2005). A requirement for miR-122 for HCV replication has since been demonstrated for all seven genotypes (Li et al. 2011), in the context of HCV infection (Randall et al. 2007; Jangra et al. 2010b), and in vivo in chimpanzees (Lanford et al. 2010).

The function of miR-122 in the HCV life cycle depends on the direct interaction of miR-122 with two sites located between stem-loop 1 and the HCV IRES (Fig. 2b) (Jopling et al. 2005; Jopling et al. 2008). The two miR-122 seed matches are only 9 nucleotides apart, similar to the 8–40 nts typically observed between cooperative 3′UTR miRNA binding sites (Grimson et al. 2007; Saetrom et al. 2007). Both sites are occupied by miR-122 in the same HCV RNA copy both in vitro and in cell culture (Jopling et al. 2008; Mortimer and Doudna 2013), because distinct mutations within the two sites could only be rescued by co-transfection of two distinct compensatory miR-122 mutants (Jopling et al. 2008). Combined mutation of both sites abolished infectious virus production, while individual mutation of either site 1 or site 2 severely attenuated virus production (Jangra et al. 2010b). These experiments also revealed that integrity of the seed interaction at site 1 is considerably more important than that at site 2, resulting in a more pronounced reduction in HCV infectivity titers. While seed base pairing of miR-122 to both sites is clearly necessary for the stimulatory effect of miR-122, it is not sufficient and additional 3′-interactions with at least cytosines 15 and 16 of miR-122 are important (Fig. 2b) (Machlin et al. 2011; Roberts et al. 2011; Shimakami et al. 2012b). Finally, the 3′-end of miR-122 is critical as well, because the deletion of nts 20–23 of miR-122 abolished miR-122 function at either site and the substitution of nts 22/23 with deoxynucleotides abolished its function at site 2 (Machlin et al. 2011). Since these nts are not known to pair with the HCV genome, it was proposed that they serve to stabilize miR-122 (Jopling 2012).

Compared to canonical miRNA binding sites, the HCV:miR-122 interaction is clearly unusual not just because it enhances RNA accumulation rather than resulting in mRNA repression, but also because the interaction takes place in the HCV 5'UTR, requires the integrity of both sites for regulation, and critically depends on non-seed regions of miR-122. These unusual features are not due to novel properties of miR-122 itself, because miR-122 engages in canonical regulatory interactions with its cellular targets (e.g., Krutzfeldt et al. 2005; Esau et al. 2006; Elmen et al. 2008; Lanford et al. 2010). They are also not due to the arrangement and/or sequence of the miR-122 binding sites, because the first 45 nts of the HCV genome repress reporter expression like other miRNA binding sites when placed into the 3′UTR context of a luciferase reporter. Thus, the regulatory outcome of the HCV:miR-122 interaction appears to depend on the context of the HCV genome. Additional features of the HCV genome that are important for regulation by miR-122 include the absence of a 5′ cap (Roberts et al. 2011) and 5′UTR stem-loop 3 (Roberts et al. 2011). Indirect effects of miR-122 on HCV replication in culture can be largely excluded, because deleterious mutations of the miR-122 binding sites can be rescued by compensatory mutations in miR-122 (Jopling et al. 2005, 2008), which are expected to disrupt physiological functions of miR-122. When directly tested, the regulation of HCV RNA abundance was furthermore found to be independent of the effects of miR-122 on lipid metabolism (Norman and Sarnow 2010).

The mechanism of miR-122 action in the HCV life cycle has proven hard to identify and remains unsolved. Potential functions of miR-122 in HCV translation, RNA stability and replication have been investigated. Early experiments by Jopling et al. (2005) showed that core protein expression from non-replicating, cell culture–adapted HCV genomic RNA is comparable for wt and miR-122 site mutant HCV, suggesting that the stability and translation of non-replicating HCV RNAs are not affected by miR-122 binding. Others have observed a modest enhancement of translation from the HCV IRES by miR-122, in the absence of measurable effects on RNA abundance (Henke et al. 2008; Jangra et al. 2010b; Roberts et al. 2011). Specifically, miR-122 was able to stimulate luciferase expression from minireporters that contain the HCV 5′UTR and the first codons of the core protein fused to luciferase by ~3–4 fold (Henke et al. 2008; Roberts et al. 2011). However, the stimulatory effect of miR-122 on translation in the context of full-length HCV genome was only up to twofold (Henke et al. 2008; Jangra et al. 2010b). Contrasting with these reports is the finding that miR-122 acts to protect the HCV RNA from degradation by 5′–3′ exoribonuclease 1 (Xrn1) (Li et al. 2013), resulting in a modest stabilization of the HCV RNA (Shimakami et al. 2012a). However, even if these effects are functional during HCV infection, they are unlikely to account for the dramatic phenotype observed following the mutation of the miR-122 sites or the sequestration of miR-122, because inhibition of Xrn1 did not rescue miR-122 binding-defective HCV mutants (Li et al. 2013) and HCV IRES mutants with translational defects similar to those induced by loss of miR-122 interaction were not nearly as impaired for replication (Jangra et al. 2010b). Given the absence of pronounced effects on either RNA stability or translation, a role for miR-122 in HCV replication seems likely (Jopling et al. 2005). RNA elongation was unaffected by miR-122 in in vitro replication assays (Villanueva et al. 2010) and in cell culture when the amount of newly synthesized HCV RNA was measured (Norman and Sarnow 2010). It was therefore proposed that miR-122 could be important for the assembly of replication complexes or may play yet unexplored roles in the initiation of RNA replication (Norman and Sarnow 2010; Li et al. 2013). It is also possible that miR-122 has an additional role to shield the 5’-end of the HCV genome from recognition by the pattern recognition receptor retinoic acid-inducible gene I (RIG-I). Although such an effect remains to be investigated, it cannot account for the strong requirement for miR-122 in Huh7.5 cells, which are defective for RIG-I signaling (Blight et al. 2002; Sumpter et al. 2005; Binder et al. 2007).

Due to the unusual regulatory outcome of this interaction, it is an interesting question whether the action of miR-122 is mediated by the canonical RISC. Existing data show that wt but not miR-122 site mutant HCV RNAs associate with Ago2 (Shimakami et al. 2012a). While these and other data suggest that miR-122 binds to HCV within a RISC-like complex, the exact composition of this complex remains to be identified. RNAi of each of the four individual Ago proteins followed by subsequent infection with HCV reduced HCV RNA levels and the production of infectious HCV (Randall et al. 2007). Similarly, the simultaneous depletion of Ago1/2 reduced HCV RNA levels in cells transfected with replication competent HCV RNA, but HCV RNA levels could be restored by overexpression of miR-122, suggesting that reduced miR-122 expression resulting from the knockdown of RISCs may have mediated the observed effect and that Ago3 and/or Ago4 can potentially compensate for the loss of Ago1/2 (Machlin et al. 2011). Consistent with this interpretation, the decreased luciferase expression from a minimal, non-replicating reporter following the simultaneous repression of all four Ago proteins could not be rescued by overexpression of miR-122 (Roberts et al. 2011). In contrast to these reports, others have reported that only the knockdown of Ago2, but not of the other Ago proteins, reduced HCV RNA abundance in the context of a preexisting infection (Shimakami et al. 2012a). A specific requirement for Ago2 over the other Ago proteins was also suggested by the inability of miR-122 to stabilize transfected HCV RNA in Ago2-/- murine embryonic fibroblasts (Shimakami et al. 2012a), the preferential association of the HCV RNA with Ago2 over Ago1 (Shimakami et al. 2012a), and the importance of Ago2 for HCV genome replication (Wilson et al. 2011). Additional studies point to a role for P-body and stress granule components in HCV replication that is independent of the role of miR-122, and therefore, such mechanistically unrelated effects will have to be separated from those that are related to the direct interaction of miR-122 with HCV (Jangra et al. 2010a).

As an essential cellular cofactor of HCV replication, miR-122 has become an attractive therapeutic target for chronic HCV infection (Machlin et al. 2012). The knockdown of miR-122 in vivo has been achieved in mice and non-human primates using antisense inhibitors, including antagomirs and locked nucleic acid (LNA)–modified antisense oligonucleotides (Krutzfeldt et al. 2005; Esau et al. 2006; Elmen et al. 2008; Lanford et al. 2010). The LNA-based miR-122-inhibitor SPC3649 developed by Santaris Pharma A/S (Fig. 2c) (Esau et al. 2006; Elmen et al. 2008; Lanford et al. 2010), which is now known as miravirsen, is currently in clinical trials, with good success to date (Jopling 2012; Machlin et al. 2012). In chimpanzees, SPC3649 achieved an effective suppression of HCV viral loads, while side effects were undetectable or mild (Lanford et al. 2010). The emergence of SPC3649-resistant viruses was not observed either during the 12-week treatment period or during the 16-week follow-up, suggesting that there is a high threshold toward developing SPC3649 resistance. That HCV mutants capable of miR-122-independent replication could potentially emerge, however, is suggested by the recent finding that the acquisition of cellular U3 nucleolar RNA sequences in place of SL1 can render HCV independent of miR-122 (Li et al. 2011). As miR-122 inhibitors move through clinical trials in humans, it will therefore be important to understand the mechanism of miR-122 action in HCV infection and how resistance mutations could be acquired. The reported miR-122-independent virus could provide an interesting starting point to mechanistically characterize the role of miR-122 in HCV replication (Li et al. 2011). Finally, it will also be important to better understand the physiological roles of miR-122, for example as a tumor suppressor (Jopling 2012).

5 Exploiting Differences in miRNA Expression to Control Viral Tropism

Viral mRNAs and the RNA genomes of many viruses are accessible to the RISC and experimentally induced RNAi (reviewed in Umbach and Cullen 2009). It is therefore anticipated that unavoidable encounters with cellular miRNAs affect the evolution of viral sequences, such that advantageous interactions are preferentially conserved, while disadvantageous interactions are avoided. It is likely that at least some viruses use cellular miRNAs to achieve robust or cell-type-specific viral gene expression and CLIP-Seq experiments have revealed candidates for such interactions between cellular miRNAs and herpesviral mRNAs (Lo et al. 2007; Gottwein et al. 2011; Riley et al. 2012; Skalsky et al. 2012). The functional significance of these interactions remains to be established, however.

The exploration of the interactions between mammalian viruses and the RNAi machinery has also opened up strategies to devise miRNA-regulated viral vectors and replicating viruses. These viruses are engineered to carry perfect or imperfect miRNA binding sites in order to restrict transgene expression or viral replication in cells expressing the targeting miRNA. miRNAs are frequently differentially expressed between tissues (Lagos-Quintana et al. 2002), and some miRNAs are specific to certain tissues, such as the hematopoietic-specific miR-142-3p (Chen et al. 2004) and liver-specific miR-122. The paradigm of a miRNA-regulated gene therapy vector was first introduced by Brown et al., who placed perfect miR-142-3p binding sites into a non-replicating lentiviral vector expressing enhanced green fluorescent protein (eGFP) (Brown et al. 2006). This miR-142-3p-targeted vector showed selective silencing of eGFP expression in the hematopoietic cell lineage. Restricting transgene expression in the hematopoietic lineage also avoids expression in professional antigen-presenting cells (APCs), which may lead to a transgene-specific adaptive immune response (Brown et al. 2006, 2007). Accordingly, this design also facilitated stable eGFP expression in immunocompetent mice (Brown et al. 2006). Brown et al. further extended these studies to the therapeutic transgene coagulation factor IX (F.IX) in combination with a liver-specific promoter. The miR-142-3p-silenced F.IX-encoding lentiviral vector, but not the non-targeted vector, enabled sustained transgene expression, and rescued defective blood clotting in hemophilia B mice (Brown et al. 2007). This vector furthermore induced immunological tolerance to F.IX.

Since these initial studies, similar design strategies have been applied to control the tropism and pathogenicity of oncolytic viruses and to vaccine design. Cancer therapy with oncolytic viruses (virotherapy) relies on the ability of these viruses to destroy cancer cells without harming healthy cells. A number of viral systems have been developed for this purpose, several of which are in clinical trials (Russell et al. 2012). The selectivity of the oncolytic virus to the targeted cancer cells is essential for the success of each of these strategies. miRNA-induced silencing of virus replication in non-target tissues therefore represents a novel and broadly applicable approach to achieve the repression of virus replication in unintended tissues. In one approach, the targeting miRNA is selectively expressed in tissues where toxicity is observed, such as the liver or the muscle. This approach was first explored by Kelly et al. (2008) using the picornavirus coxsackievirus A21 (CVA21) as an example. CVA21 effectively clears human myeloma xenograft tumors in immunodeficient mice, but the virus also spreads to muscle tissue where its replication leads to lethal myositis. CVA21 engineered to carry two perfect sites each for two muscle-specific miRNAs effectively treated the xenograft tumors but did not result in myelitis, due to its newly acquired inability to replicate in muscle cells. A second approach can be applied to cancers that critically dependent on the decreased expression or loss of an otherwise ubiquitous miRNA. To test this approach, Edge et al. introduced let-7 binding sites into the 3′UTR of the vesicular stomatitis virus mRNA encoding the viral matrix protein, which is essential for virus replication and pathogenesis (Edge et al. 2008). This design resulted in attenuation of VSV in let-7-expressing normal cells, but toxicity in cancer cells with reduced let-7 expression (Edge et al. 2008). Similarly, an oncolytic measles virus with a miR-7-repressed mRNA conferred specificity of oncolytic replication to glioblastoma xenografts with downregulation of miR-7 (Leber et al. 2011).

The introduction of miRNA binding sites into viruses could be particularly useful for the design of attenuated vaccines. This was first demonstrated in a proof-of-concept study using poliovirus (Barnes et al. 2008). Introduction of binding sites to broadly expressed let-7 or neuron-specific miR-124 resulted in effective viral attenuation and the acquisition of protective immunity. The presence of binding sites for miR-93, which is broadly expressed in humans and mice, similarly attenuated Influenza A virus in mice, which were protected from subsequent challenge with wt influenza (Perez et al. 2009). Because miR-93 is not conserved in chickens, using miR-93 for attenuation furthermore allowed the production of the experimental vaccine strain at high titers in embryonated chicken eggs. One potential concern with any therapeutic virus strategy is the escape from attenuation. The combination of miRNA-mediated regulation with other attenuation strategies may therefore offer increased safeguard from viral escape. A second concern is that the expression of abundant perfect or extensive imperfect targets of specific miRNAs could lead to either the degradation of the targeting miRNA or its sequestration from its endogenous targets (Ameres et al. 2010; Ebert and Sharp 2010). These effects will have to be taken into account as these strategies are developed further.

Similar strategies can also be exploited for experimental design in basic research. For example, the presence of miRNA binding sites in plasmid or lentiviral reporter vectors has been used extensively to assess and control for miRNA activity (Zeng et al. 2002; Doench et al. 2003; Zeng et al. 2003; Gottwein et al. 2006). In an elegant extension of the viral systems outlined above, the introduction of binding sites for tissue-specific miRNAs can be used to assess whether a specific tissue is important for the course of viral infection and the host response to the virus in vivo. For example, miR-142-targeted influenza virus is restricted for replication in hematopoietic cells, including APCs, which resulted in a decreased production of type I interferon, thereby showing that virus replication in APCs is important for the antiviral response to influenza A in vivo (Langlois et al. 2012).

6 Viral miRNAs

In 2004, Pfeffer and colleagues reported that EBV expresses at least five viral miRNAs during latency, suggesting that this and other viruses may encode their own miRNAs to regulate cellular or viral gene expression (Pfeffer et al. 2004). Subsequent studies have shown that most herpesviruses encode several viral miRNAs, while some small DNA viruses encode individual miRNAs and RNA viruses typically do not express miRNAs (Pfeffer et al. 2005; Cullen 2006b). Examples of viruses that encode miRNAs are shown in Table 1. This list is not comprehensive, but includes all human viruses known to express miRNAs and other mammalian viruses that express miRNAs using unusual strategies outlined below. What is known about the expression and function of viral miRNAs has been reviewed in great detail elsewhere (Skalsky and Cullen 2010; Grundhoff and Sullivan 2011; Lin and Flemington 2011; Gottwein 2012; Kincaid and Sullivan 2012), and I will therefore limit my discussion of viral miRNAs to a few key concepts that have been uncovered.

Most viral miRNAs are produced like cellular miRNAs, that is, from Pol II transcripts through the stepwise processing by Drosha and Dicer. However, there are examples of unconventional expression strategies. The miRNAs of the murine γ-herpesvirus MHV-68 are derived from RNA Pol III transcripts that contain a non-aminoacylated tRNA sequence at the 5′-end and pre-miRNA stem-loops at the 3′-end (Bowden et al. 1997; Pfeffer et al. 2005). These pri-miRNAs are not recognized by the microprocessor complex, but are instead processed to pre-miRNAs by the tRNA processing enzyme tRNaseZ (Bogerd et al. 2010). Similarly, HVS expresses several viral miRNAs from chimeric transcripts with HSURs 2, 4, and 5 and uses the cellular integrator complex, which processes cellular and viral Sm-class RNAs, to liberate the pre-miRNAs (Cazalla et al. 2011). Specific reasons for why MHV68 and HVS would bypass Drosha processing remain unknown. It has been argued that nuclear DNA viruses are most likely to encode miRNAs because the canonical production of miRNAs from RNA viruses that access the nucleus, such as retroviruses or influenza virus, would presumably result in a loss of fitness, due to the destruction of viral RNAs by Drosha (Cullen 2006b). That retroviruses can overcome this caveat, however, is illustrated by the recently identified miRNAs of the complex retrovirus bovine leukemia virus (BLV) (Kincaid et al. 2012). Retroviruses have a +strand RNA genome, which is reverse transcribed into an integrated, nuclear DNA copy, called the provirus. The proviral 5′ long terminal repeat (5′ LTR) serves as a Pol II promoter that drives the production of viral mRNAs and genomic RNAs.

Table 1 List of human viruses known to encode miRNAs

Family and *subfamily*	Species	No. of pre-miRs	Key observations and references
Herpesviruses (dsDNA, nuclear)			
α-Herpesviruses	Herpes simplex virus 1 (HSV1)	16	miRNA expression mostly from latency-associated transcripts (LATs) (Umbach et al. 2008; Jurak et al. 2011)
	Herpes simplex virus 2 (HSV2)	18	miRNA expression mostly from latency-associated transcripts (LATs) (Jurak et al. 2011)
	Varicella zoster virus (VZV)	0	VZV probably does not encode viral miRNAs (Pfeffer et al. 2005; Umbach et al. 2009)
β-Herpesviruses	Human cytomegalovirus (HCMV)	11	miRNAs expressed from various locations in the genome (Pfeffer et al. 2005; Tuddenham and Pfeffer 2011)
	Human herpesvirus 6 (HHV6, Roseolovirus)	4	miRNAs expressed from repeat regions (Tuddenham et al. 2012)
γ-Herpesviruses	Epstein–Barr virus (EBV)	25	miRNA expression from two clusters: 3 BHRF1 miRNA hairpins, which are expressed only during latency III, and 22 BART miRNA hairpins (Pfeffer et al. 2004; Cai et al. 2006; Zhu et al. 2009). The BHRF1 miRNAs promote B cell immortalization by EBV (Seto et al. 2010; Feederle et al. 2011a; Feederle et al. 2011b)
	Kaposi's sarcoma–associated herpesvirus (KSHV)	12	miRNA expression from one cluster during latency (Cai et al. 2005; Pfeffer et al. 2005; Samols et al. 2005; Grundhoff et al. 2006). miR-K11 is a viral mimic of cellular miR-155 (Gottwein et al. 2007; Skalsky et al. 2007)
	[a]Herpesvirus saimiri (HVS)	3	pre-miRNAs are excised by the integrator complex (Cazalla et al. 2011)
	[a]Murine γ-herpesvirus 68 (MHV68)	15	pri-miRNAs driven by Pol III, pre-miRNAs excised by tRNaseZ (Pfeffer et al. 2005; Bogerd et al. 2010)

Family and *subfamily*	Species	No. of pre-miRs	Key observations and references
Polyomaviruses (dsDNA, nuclear)			
	JC polyomavirus, BK polyomavirus, Merkel cell polyomavirus	1	miRNAs have perfect complementarity to early viral transcripts, resulting in their endonucleolytic cleavage at later time points during infection (Sullivan et al. 2005)
Retroviruses (+strand RNA, nuclear DNA intermediate)			
	[a]Bovine leukemia virus (BLV)	5	pre-miRs are directly produced from Pol III promoters (Kincaid et al. 2012)

[a]Listed are animal viruses that express miRNAs using non-conventional strategies. This table lists virus families and subfamilies, virus species, the number of pre-miRNA hairpins, as well as key observations and references. Note that the number of mature miRNAs generally exceeds the number of encoded pre-miRNA hairpins

If retroviruses were to express Drosha-dependent miRNAs from the viral LTR promoter, this would result in at least partial degradation of the viral genome. BLV instead uses internal Pol III promoters to express several short hairpin RNAs, which can directly act as pre-miRNAs (Kincaid et al. 2012). These RNA stem-loops lack features necessary for recognition by the microprocessor, therefore leaving the LTR-driven RNAs intact. Non-canonical, Drosha-independent expression strategies have also been reported for a small number of cellular miRNAs. These include mirtrons, that is, small introns that serve as pre-miRNAs after splicing, endogenous shRNAs, and Dicer processing products from longer dsRNA hairpins (Berezikov et al. 2007; Okamura et al. 2007; Ruby et al. 2007; Babiarz et al. 2008). There are also examples of miRNAs that bypass the requirement for Dicer (Cheloufi et al. 2010; Cifuentes et al. 2010). Proof-of-concept studies have suggested that miRNA processing in the cytoplasm might occur when pri-miRNA sequences are artificially inserted into RNA viruses, likely due to the cytoplasmic re-localization of the microprocessor complex (Shapiro et al. 2010, 2012). It therefore remains possible that other non-conventional miRNAs from RNA viruses, including those with cytoplasmic lifestyles, will be identified by future studies.

While the functional investigation of viral miRNAs is still in its infancy, several paradigms of viral miRNA function are beginning to emerge from existing studies. A number of viruses encode miRNAs that mimic the activity of specific cellular miRNAs through their identical seed regions. Consequences of the resulting functional mimicry have been discussed above for KSHV miR-K11 and MDV-1 miR-M4 (Gottwein et al. 2007; Skalsky et al. 2007; Zhao et al. 2009). Similarly, one of the BLV miRNAs (miR-B4) shares its first 8 nts with cellular miR-29 (Kincaid et al. 2012), a miRNA associated with indolent B cell chronic lymphocytic leukemia. The transgenic expression of miR-29 in B cells and HSCs leads to CLL and acute myeloid leukemia (AML), respectively (Han et al. 2010; Santanam et al. 2010), which suggests that miR-B4 could be important for BLV-induced B cell tumorigenesis. Interestingly, the EBV miRNA miR-BART1-3p shares its minimal seed sequence (nts 2–7) with miR-29, suggesting that at least some targets are likely to be common for miR-BART1-3p and miR-29 (Skalsky et al. 2012). Candidates for several additional mimics of cellular miRNAs exist (Gottwein et al. 2011; Grundhoff and Sullivan 2011; Kincaid and Sullivan 2012; Skalsky et al. 2012). Viruses presumably encode mimics of cellular miRNAs to access evolutionarily conserved regulatory networks. Most viral miRNAs, on the other hand, do not share seed sequences with cellular miRNAs and are therefore likely to have evolved novel regulatory interactions and functions. Some of these viral miRNAs are expressed from locations antisense to viral mRNAs and consequently autoregulate viral gene expression (Pfeffer et al. 2004; Sullivan et al. 2005; Grey et al. 2007; Umbach et al. 2008). These autoregulatory interactions may confer robustness to the transitions between different stages of viral infection. This may be the case for the switch between latent and lytic replication modes of α-herpesviruses (Umbach et al. 2008; Jurak et al. 2011) and between the early and late stages of polyomavirus gene expression (Sullivan et al. 2005).

To understand potential functions of viral miRNAs, target-first and phenotype-first approaches have been pursued. The rationale underlying target-first approaches is that a detailed understanding of the viral miRNA targets will lead to insights into their functions. Recently, the transcriptome-wide identification of binding sites for viral miRNAs using CLIP-Seq approaches has been reported for EBV and KSHV (Gottwein et al. 2011; Haecker et al. 2012; Riley et al. 2012; Skalsky et al. 2012). These studies have uncovered hundreds of binding sites of the more than 40 EBV and ~20 KSHV miRNAs in cellular mRNAs, suggesting that these miRNAs engage in highly complex regulatory interactions with the cellular transcriptome. Future studies will have to begin to uncover how many and which of these identified interactions matter to the outcomes of viral infection. Currently, functional studies of only few individual targets are available, and these have uncovered roles of the viral miRNAs in immune evasion, the control of proliferation, cellular survival, and signaling (Grundhoff and Sullivan 2011; Kincaid and Sullivan 2012). Because the targets uncovered by targetomics approaches are quite diverse, it would be premature to assume that these already characterized functions are the main functions of viral miRNAs. In phenotype-first approaches, viral miRNA mutants are analyzed for consequences of miRNA inactivation. The functional analysis of miRNAs encoded by human herpesviruses in vivo is largely impossible, because the targets of these miRNAs are unlikely to be conserved in animal models, where these exist. This caveat makes non-human mammalian viruses valuable models for analysis of miRNA mutants in vivo. Such an analysis has uncovered that the miRNAs encoded by murine cytomegalovirus, a model system for HCMV, are important for viral replication in the salivary glands of mice with a certain genetic background, which is important for the persistence and transmission of this virus (Dolken et al. 2010). This phenotype was rescued by the combined depletion of natural killer cells and CD4$^+$ T cells, suggesting that the mCMV miRNAs are important for immune evasion in vivo.

Acknowledgments EG would like to thank Judith Gottwein, Mark Manzano, and Eleonora Forte for comments on the manuscript. This work was supported by the Zell Family Foundation and the Robert H. Lurie Comprehensive Cancer Center. The content is solely the responsibility of the author and does not necessarily represent the official views of the Zell Family Foundation and the Robert H. Lurie Comprehensive Cancer Center.

References

Albrecht JC (2000) Primary structure of the Herpesvirus ateles genome. J Virol 74:1033–1037

Ameres SL, Horwich MD, Hung JH, Xu J, Ghildiyal M, Weng Z, Zamore PD (2010) Target RNA-directed trimming and tailing of small silencing RNAs. Science 328:1534–1539. doi:10.1126/science.1187058 328/5985/1534 [pii]

Andersson MG, Haasnoot PC, Xu N, Berenjian S, Berkhout B, Akusjarvi G (2005) Suppression of RNA interference by adenovirus virus-associated RNA. J Virol 79:9556–9565. doi:10.1128/JVI.79.15.9556-9565.2005 79/15/9556 [pii]

Babiarz JE, Ruby JG, Wang Y, Bartel DP, Blelloch R (2008) Mouse ES cells express endogenous shRNAs, siRNAs, and other microprocessor-independent, Dicer-dependent small RNAs. Genes Dev 22:2773–2785. doi:10.1101/gad.1705308 22/20/2773 [pii]

Backes S, Shapiro JS, Sabin LR et al (2012) Degradation of host microRNAs by poxvirus poly (A) polymerase reveals terminal RNA methylation as a protective antiviral mechanism. Cell Host Microbe 12:200–210. doi:10.1016/j.chom.2012.05.019 S1931-3128(12)00233-8 [pii]

Barnes D, Kunitomi M, Vignuzzi M, Saksela K, Andino R (2008) Harnessing endogenous miRNAs to control virus tissue tropism as a strategy for developing attenuated virus vaccines. Cell Host Microbe 4:239–248. doi:10.1016/j.chom.2008.08.003 S1931-3128(08)00259-X [pii]

Bartel DP (2009) MicroRNAs: target recognition and regulatory functions. Cell 136:215–233. doi:10.1016/j.cell.2009.01.002 S0092-8674(09)00008-7 [pii]

Berezikov E, Chung WJ, Willis J, Cuppen E, Lai EC (2007) Mammalian mirtron genes. Mol Cell 28:328–336. doi:10.1016/j.molcel.2007.09.028 S1097-2765(07)00669-7 [pii]

Binder M, Kochs G, Bartenschlager R, Lohmann V (2007) Hepatitis C virus escape from the interferon regulatory factor 3 pathway by a passive and active evasion strategy. Hepatology 46:1365–1374. doi:10.1002/hep.21829

Blight KJ, McKeating JA, Rice CM (2002) Highly permissive cell lines for subgenomic and genomic hepatitis C virus RNA replication. J Virol 76:13001–13014

Bogerd HP, Karnowski HW, Cai X, Shin J, Pohlers M, Cullen BR (2010) A mammalian herpesvirus uses noncanonical expression and processing mechanisms to generate viral microRNAs. Mol Cell 37:135–142. doi:10.1016/j.molcel.2009.12.016 S1097-2765(09)00920-4 [pii]

Bolisetty MT, Dy G, Tam W, Beemon KL (2009) Reticuloendotheliosis virus strain T induces miR-155, which targets JARID2 and promotes cell survival. J Virol 83:12009–12017. doi:10.1128/JVI.01182-09 JVI.01182-09 [pii]

Boss IW, Nadeau PE, Abbott JR, Yang Y, Mergia A, Renne R (2011) A Kaposi's sarcoma-associated herpesvirus-encoded ortholog of microRNA miR-155 induces human splenic B-cell expansion in NOD/LtSz-scid IL2Rgammanull mice. J Virol 85:9877–9886. doi:10.1128/JVI.05558-11 JVI.05558-11 [pii]

Bowden RJ, Simas JP, Davis AJ, Efstathiou S (1997) Murine gammaherpesvirus 68 encodes tRNA-like sequences which are expressed during latency. J Gen Virol 78(Pt 7):1675–1687

Brown BD, Venneri MA, Zingale A, Sergi Sergi L, Naldini L (2006) Endogenous microRNA regulation suppresses transgene expression in hematopoietic lineages and enables stable gene transfer. Nat Med 12:585–591. doi:10.1038/nm1398 nm1398 [pii]

Brown BD, Cantore A, Annoni A et al (2007) A microRNA-regulated lentiviral vector mediates stable correction of hemophilia B mice. Blood 110:4144–4152. doi:10.1182/blood-2007-03-078493 blood-2007-03-078493 [pii]

Buck AH, Perot J, Chisholm MA et al (2010) Post-transcriptional regulation of miR-27 in murine cytomegalovirus infection. RNA 16:307–315. doi:10.1261/rna.1819210 rna.1819210 [pii]

Cai X, Lu S, Zhang Z, Gonzalez CM, Damania B, Cullen BR (2005) Kaposi's sarcoma-associated herpesvirus expresses an array of viral microRNAs in latently infected cells. Proc Natl Acad Sci USA 102:5570–5575

Cai X, Schafer A, Lu S et al (2006) Epstein-Barr virus microRNAs are evolutionarily conserved and differentially expressed. PLoS Pathog 2:e23

Cameron JE, Fewell C, Yin Q, McBride J, Wang X, Lin Z, Flemington EK (2008) Epstein-Barr virus growth/latency III program alters cellular microRNA expression. Virology 382:257–266. doi:10.1016/j.virol.2008.09.018 S0042-6822(08)00602-8 [pii]

Cazalla D, Yario T, Steitz JA (2010) Down-regulation of a host microRNA by a Herpesvirus saimiri noncoding RNA. Science 328:1563–1566. doi:10.1126/science.1187197 328/5985/1563 [pii]

Cazalla D, Xie M, Steitz JA (2011) A primate herpesvirus uses the integrator complex to generate viral microRNAs. Mol Cell 43:982–992. doi:10.1016/j.molcel.2011.07.025 S1097-2765(11)00590-9 [pii]

Chang J, Nicolas E, Marks D et al (2004) miR-122, a mammalian liver-specific microRNA, is processed from hcr mRNA and may downregulate the high affinity cationic amino acid transporter CAT-1. RNA Biol 1:106–113. doi:1066 [pii]

Chang J, Guo JT, Jiang D, Guo H, Taylor JM, Block TM (2008) Liver-specific microRNA miR-122 enhances the replication of hepatitis C virus in nonhepatic cells. J Virol 82:8215–8223. doi:10.1128/JVI.02575-07 JVI.02575-07 [pii]

Cheloufi S, Dos Santos CO, Chong MM, Hannon GJ (2010) A dicer-independent miRNA biogenesis pathway that requires Ago catalysis. Nature 465:584–589. doi:10.1038/nature09092 nature09092 [pii]

Chen CZ, Li L, Lodish HF, Bartel DP (2004) MicroRNAs modulate hematopoietic lineage differentiation. Science 303:83–86. doi:10.1126/science.1091903 1091903 [pii]

Chi SW, Hannon GJ, Darnell RB (2012) An alternative mode of microRNA target recognition. Nat Struct Mol Biol 19:321–327. doi:10.1038/nsmb.2230 nsmb.2230 [pii]

Chiang HR, Schoenfeld LW, Ruby JG et al (2010) Mammalian microRNAs: experimental evaluation of novel and previously annotated genes. Genes Dev 24:992–1009. doi:10.1101/gad.1884710 gad.1884710 [pii]

Chong MM, Zhang G, Cheloufi S, Neubert TA, Hannon GJ, Littman DR (2010) Canonical and alternate functions of the microRNA biogenesis machinery. Genes Dev 24:1951–1960. doi:10.1101/gad.1953310 gad.1953310 [pii]

Cifuentes D, Xue H, Taylor DW et al (2010) A novel miRNA processing pathway independent of dicer requires Argonaute2 catalytic activity. Science 328:1694–1698. doi:10.1126/science.1190809 science.1190809 [pii]

Cook HL, Lytle JR, Mischo HE et al (2005) Small nuclear RNAs encoded by Herpesvirus saimiri upregulate the expression of genes linked to T cell activation in virally transformed T cells. Curr Biol 15:974–979. doi:10.1016/j.cub.2005.04.034 S0960-9822(05)00434-3 [pii]

Costinean S, Zanesi N, Pekarsky Y, Tili E, Volinia S, Heerema N, Croce CM (2006) Pre-B cell proliferation and lymphoblastic leukemia/high-grade lymphoma in Eμ-miR155 transgenic mice. Proc Natl Acad Sci USA 103:7024–7029. doi:10.1073/pnas.0602266103 0602266103 [pii]

Costinean S, Sandhu SK, Pedersen IM et al (2009) Src homology 2 domain-containing inositol-5-phosphatase and CCAAT enhancer-binding protein beta are targeted by miR-155 in B cells of Emicro-MiR-155 transgenic mice. Blood 114:1374–1382. doi:10.1182/blood-2009-05-220814 blood-2009-05-220814 [pii]

Cullen BR (2006a) Is RNA interference involved in intrinsic antiviral immunity in mammals? Nat Immunol 7:563–567. doi:10.1038/ni1352 ni1352 [pii]
Cullen BR (2006b) Viruses and microRNAs. Nat Genet 38(Suppl):S25–S30
Dahlke C, Maul K, Christalla T, Walz N, Schult P, Stocking C, Grundhoff A (2012) A microRNA Encoded by Kaposi Sarcoma-Associated Herpesvirus Promotes B-Cell Expansion In Vivo. PLoS One 7:e49435. doi:10.1371/journal.pone.0049435 PONE-D-12-20948 [pii]
Damania B (2004) Oncogenic gamma-herpesviruses: comparison of viral proteins involved in tumorigenesis. Nat Rev Microbiol 2:656–668. doi:10.1038/nrmicro958 nrmicro958 [pii]
Davis E, Caiment F, Tordoir X et al (2005) RNAi-mediated allelic trans-interaction at the imprinted Rtl1/Peg11 locus. Curr Biol 15:743–749. doi:10.1016/j.cub.2005.02.060 S0960-9822(05)00233-2 [pii]
Ding SW (2010) RNA-based antiviral immunity. Nat Rev Immunol 10:632–644. doi:10.1038/nri2824 nri2824 [pii]
Doench JG, Petersen CP, Sharp PA (2003) siRNAs can function as miRNAs. Genes Dev 17:438–442
Dolken L, Krmpotic A, Kothe S et al (2010) Cytomegalovirus microRNAs facilitate persistent virus infection in salivary glands. PLoS Pathog 6:e1001150. doi:10.1371/journal.ppat.1001150
Ebert MS, Sharp PA (2010) MicroRNA sponges: progress and possibilities. RNA 16:2043–2050. doi:10.1261/rna.2414110 rna.2414110 [pii]
Edge RE, Falls TJ, Brown CW, Lichty BD, Atkins H, Bell JC (2008) A let-7 MicroRNA-sensitive vesicular stomatitis virus demonstrates tumor-specific replication. Mol Ther 16:1437–1443. doi:10.1038/mt.2008.130 mt2008130 [pii]
Eis PS, Tam W, Sun L et al (2005) Accumulation of miR-155 and BIC RNA in human B cell lymphomas. Proc Natl Acad Sci USA 102:3627–3632. doi:10.1073/pnas.0500613102 0500613102 [pii]
Elmen J, Lindow M, Schutz S et al (2008) LNA-mediated microRNA silencing in non-human primates. Nature 452:896–899. doi:10.1038/nature06783 nature06783 [pii]
Ensser A, Pfinder A, Muller-Fleckenstein I, Fleckenstein B (1999) The URNA genes of herpesvirus saimiri (strain C488) are dispensable for transformation of human T cells in vitro. J Virol 73:10551–10555
Esau C, Davis S, Murray SF et al (2006) miR-122 regulation of lipid metabolism revealed by in vivo antisense targeting. Cell Metab 3:87–98. doi:10.1016/j.cmet.2006.01.005 S1550-4131(06)00029-5 [pii]
Feederle R, Haar J, Bernhardt K et al (2011a) The members of an Epstein-Barr virus microRNA cluster cooperate to transform B lymphocytes. J Virol 85:9801–9810. doi:10.1128/JVI.05100-11 JVI.05100-11 [pii]
Feederle R, Linnstaedt SD, Bannert H, Lips H, Bencun M, Cullen BR, Delecluse HJ (2011b) A viral microRNA cluster strongly potentiates the transforming properties of a human herpesvirus. PLoS Pathog 7:e1001294. doi:10.1371/journal.ppat.1001294
Forte E, Salinas R, Chang C et al (2012) The Epstein-Barr virus induced tumor suppressor miR-34a is growth promoting in EBV-infected B cells. J Virol. doi:10.1128/JVI.07056-11 JVI.07056-11 [pii]
Friedman RC, Farh KK, Burge CB, Bartel DP (2009) Most mammalian mRNAs are conserved targets of microRNAs. Genome Res 19:92–105. doi:10.1101/gr.082701.108 gr.082701.108 [pii]
Fukuhara T, Kambara H, Shiokawa M et al (2012) Expression of microRNA miR-122 facilitates an efficient replication in nonhepatic cells upon infection with hepatitis C virus. J Virol 86:7918–7933. doi:10.1128/JVI.00567-12 JVI.00567-12 [pii]
Ghildiyal M, Zamore PD (2009) Small silencing RNAs: an expanding universe. Nat Rev Genet 10:94–108. doi:10.1038/nrg2504 nrg2504 [pii]
Gironella M, Seux M, Xie MJ et al (2007) Tumor protein 53-induced nuclear protein 1 expression is repressed by miR-155, and its restoration inhibits pancreatic tumor development. Proc Natl Acad Sci USA 104:16170–16175. doi:10.1073/pnas.0703942104 0703942104 [pii]

Golembe TJ, Yong J, Battle DJ, Feng W, Wan L, Dreyfuss G (2005) Lymphotropic Herpesvirus saimiri uses the SMN complex to assemble Sm cores on its small RNAs. Mol Cell Biol 25:602–611. doi:10.1128/MCB.25.2.602-611.2005 25/2/602 [pii]

Gottwein E (2012) Kaposi's Sarcoma-Associated Herpesvirus microRNAs. Front Microbiol 3:165. doi:10.3389/fmicb.2012.00165

Gottwein E, Cai X, Cullen BR (2006) A novel assay for viral microRNA function identifies a single nucleotide polymorphism that affects Drosha processing. J Virol 80:5321–5326

Gottwein E, Mukherjee N, Sachse C et al (2007) A viral microRNA functions as an orthologue of cellular miR-155. Nature 450:1096–1099. doi:10.1038/nature05992 nature05992 [pii]

Gottwein E, Corcoran DL, Mukherjee N et al (2011) Viral microRNA targetome of KSHV-infected primary effusion lymphoma cell lines. Cell Host Microbe 10:515–526. doi:10.1016/j.chom.2011.09.012 S1931-3128(11)00331-3 [pii]

Grey F, Meyers H, White EA, Spector DH, Nelson J (2007) A human cytomegalovirus-encoded microRNA regulates expression of multiple viral genes involved in replication. PLoS Pathog 3:e163. doi:10.1371/journal.ppat.0030163 07-PLPA-RA-0411 [pii]

Griffiths-Jones S, Grocock RJ, van Dongen S, Bateman A, Enright AJ (2006) miRBase: microRNA sequences, targets and gene nomenclature. Nucleic Acids Res 34:D140–D144. doi:10.1093/nar/gkj112 34/suppl_1/D140 [pii]

Griffiths-Jones S, Saini HK, van Dongen S, Enright AJ (2008) miRBase: tools for microRNA genomics. Nucleic Acids Res 36:D154–D158. doi:10.1093/nar/gkm952 gkm952 [pii]

Grimson A, Farh KK, Johnston WK, Garrett-Engele P, Lim LP, Bartel DP (2007) MicroRNA targeting specificity in mammals: determinants beyond seed pairing. Mol Cell 27:91–105. doi:10.1016/j.molcel.2007.06.017 S1097-2765(07)00407-8 [pii]

Grundhoff A, Sullivan CS (2011) Virus-encoded microRNAs. Virology 411:325–343. doi:10.1016/j.virol.2011.01.002 S0042-6822(11)00007-9 [pii]

Grundhoff A, Sullivan CS, Ganem D (2006) A combined computational and microarray-based approach identifies novel microRNAs encoded by human gamma-herpesviruses. RNA 12:733–750. doi:10.1261/rna.2326106 rna.2326106 [pii]

Haasch D, Chen YW, Reilly RM et al (2002) T cell activation induces a noncoding RNA transcript sensitive to inhibition by immunosuppressant drugs and encoded by the proto-oncogene, BIC. Cell Immunol 217:78–86 S0008874902005063 [pii]

Haecker I, Gay LA, Yang Y, Hu J, Morse AM, McIntyre LM, Renne R (2012) Ago HITS-CLIP expands understanding of Kaposi's sarcoma-associated herpesvirus miRNA function in primary effusion lymphomas. PLoS Pathog 8:e1002884. doi:10.1371/journal.ppat.1002884 PPATHOGENS-D-12-00522 [pii]

Han YC, Park CY, Bhagat G et al (2010) microRNA-29a induces aberrant self-renewal capacity in hematopoietic progenitors, biased myeloid development, and acute myeloid leukemia. J Exp Med 207:475–489. doi:10.1084/jem.20090831 jem.20090831 [pii]

Henke JI, Goergen D, Zheng J et al (2008) microRNA-122 stimulates translation of hepatitis C virus RNA. EMBO J 27:3300–3310. doi:10.1038/emboj.2008.244 emboj2008244 [pii]

Ho BC, Yu SL, Chen JJ et al (2011) Enterovirus-induced miR-141 contributes to shutoff of host protein translation by targeting the translation initiation factor eIF4E. Cell Host Microbe 9:58–69. doi:10.1016/j.chom.2010.12.001 S1931-3128(10)00411-7 [pii]

Jangra RK, Yi M, Lemon SM (2010a) DDX6 (Rck/p54) is required for efficient hepatitis C virus replication but not for internal ribosome entry site-directed translation. J Virol 84:6810–6824. doi:10.1128/JVI.00397-10 JVI.00397-10 [pii]

Jangra RK, Yi M, Lemon SM (2010b) Regulation of hepatitis C virus translation and infectious virus production by the microRNA miR-122. J Virol 84:6615–6625. doi:10.1128/JVI.00417-10 JVI.00417-10 [pii]

Jopling C (2012) Liver-specific microRNA-122: Biogenesis and function. RNA Biol 9:137–142. doi:10.4161/rna.18827 18827 [pii]

Jopling CL, Yi M, Lancaster AM, Lemon SM, Sarnow P (2005) Modulation of hepatitis C virus RNA abundance by a liver-specific MicroRNA. Science 309:1577–1581

Jopling CL, Schutz S, Sarnow P (2008) Position-dependent function for a tandem microRNA miR-122-binding site located in the hepatitis C virus RNA genome. Cell Host Microbe 4:77–85. doi:10.1016/j.chom.2008.05.013 S1931-3128(08)00173-X [pii]

Jurak I, Griffiths A, Coen DM (2011) Mammalian alphaherpesvirus miRNAs. Biochim Biophys Acta 1809:641–653. doi:10.1016/j.bbagrm.2011.06.010 S1874-9399(11)00102-7 [pii]

Kato T, Furusaka A, Miyamoto M et al (2001) Sequence analysis of hepatitis C virus isolated from a fulminant hepatitis patient. J Med Virol 64:334–339. doi:10.1002/jmv.1055 [pii]

Kelly EJ, Hadac EM, Greiner S, Russell SJ (2008) Engineering microRNA responsiveness to decrease virus pathogenicity. Nat Med 14:1278–1283. doi:10.1038/nm.1776 nm.1776 [pii]

Kieff E, Rickinson A (2007) Epstein-Barr virus and its replication. In: Knipe DM, Howley PM (eds) Fields virology, 5th edn. pp 2603–2654

Kim VN, Han J, Siomi MC (2009) Biogenesis of small RNAs in animals. Nat Rev Mol Cell Biol 10:126–139. doi:10.1038/nrm2632 nrm2632 [pii]

Kincaid RP, Sullivan CS (2012) Virus-encoded microRNAs: an overview and a look to the future. PLoS Pathog 8:e1003018. doi:10.1371/journal.ppat.1003018 PPATHOGENS-D-12-01682 [pii]

Kincaid RP, Burke JM, Sullivan CS (2012) RNA virus microRNA that mimics a B-cell oncomiR. Proc Natl Acad Sci USA 109:3077–3082. doi:10.1073/pnas.1116107109 1116107109 [pii]

Kluiver J, Poppema S, de Jong D et al (2005) BIC and miR 155 are highly expressed in Hodgkin, primary mediastinal and diffuse large B cell lymphomas. J Pathol 207:243–249. doi:10.1002/path.1825

Krutzfeldt J, Rajewsky N, Braich R, Rajeev KG, Tuschl T, Manoharan M, Stoffel M (2005) Silencing of microRNAs in vivo with 'antagomirs'. Nature 438:685–689. doi:10.1038/nature04303 nature04303 [pii]

Lagos D, Pollara G, Henderson S et al (2010) miR-132 regulates antiviral innate immunity through suppression of the p300 transcriptional co-activator. Nat Cell Biol 12:513–519. doi:10.1038/ncb2054 ncb2054 [pii]

Lagos-Quintana M, Rauhut R, Yalcin A, Meyer J, Lendeckel W, Tuschl T (2002) Identification of tissue-specific microRNAs from mouse. Curr Biol 12:735–739 S0960982202008096 [pii]

Lanford RE, Hildebrandt-Eriksen ES, Petri A et al (2010) Therapeutic silencing of microRNA-122 in primates with chronic hepatitis C virus infection. Science 327:198–201. doi:10.1126/science.1178178 science.1178178 [pii]

Langlois RA, Varble A, Chua MA, Garcia-Sastre A, tenOever BR (2012) Hematopoietic-specific targeting of influenza A virus reveals replication requirements for induction of antiviral immune responses. Proc Natl Acad Sci USA 109:12117–12122. doi:10.1073/pnas.1206039109 1206039109 [pii]

Leber MF, Bossow S, Leonard VH et al (2011) MicroRNA-sensitive oncolytic measles viruses for cancer-specific vector tropism. Mol Ther 19:1097–1106. doi:10.1038/mt.2011.55 mt201155 [pii]

Lee SI, Murthy SC, Trimble JJ, Desrosiers RC, Steitz JA (1988) Four novel U RNAs are encoded by a herpesvirus. Cell 54:599–607 S0092-8674(88)80004-7 [pii]

Lemon SM, Walker C, Alter MJ, Yi M (2007) Hepatitis C Virus. In: Knipe DM, Howley PM (eds) Fields virology, pp 1253–1304

Li YP, Gottwein JM, Scheel TK, Jensen TB, Bukh J (2011) MicroRNA-122 antagonism against hepatitis C virus genotypes 1–6 and reduced efficacy by host RNA insertion or mutations in the HCV 5′ UTR. Proc Natl Acad Sci USA 108:4991–4996. doi:10.1073/pnas.1016606108 1016606108 [pii]

Li Y, Masaki T, Yamane D, McGivern DR, Lemon SM (2013) Competing and noncompeting activities of miR-122 and the 5′ exonuclease Xrn1 in regulation of hepatitis C virus replication. Proc Natl Acad Sci USA 110:1881–1886. doi:10.1073/pnas.1213515110 1213515110 [pii]

Libri V, Helwak A, Miesen P et al (2012) Murine cytomegalovirus encodes a miR-27 inhibitor disguised as a target. Proc Natl Acad Sci USA 109:279–284. doi:10.1073/pnas.1114204109 1114204109 [pii]

Lin Z, Flemington EK (2011) miRNAs in the pathogenesis of oncogenic human viruses. Cancer Lett 305:186–199. doi:10.1016/j.canlet.2010.08.018 S0304-3835(10)00426-X [pii]
Lin LT, Noyce RS, Pham TN et al (2010) Replication of subgenomic hepatitis C virus replicons in mouse fibroblasts is facilitated by deletion of interferon regulatory factor 3 and expression of liver-specific microRNA 122. J Virol 84:9170–9180. doi:10.1128/JVI.00559-10 JVI.00559-10 [pii]
Linnstaedt SD, Gottwein E, Skalsky RL, Luftig MA, Cullen BR (2010) Virally induced cellular microRNA miR-155 plays a key role in B-cell immortalization by Epstein-Barr virus. J Virol 84:11670–11678. doi:10.1128/JVI.01248-10 JVI.01248-10 [pii]
Liu J, Carmell MA, Rivas FV et al (2004) Argonaute2 is the catalytic engine of mammalian RNAi. Science 305:1437–1441
Lo AK, To KF, Lo KW, Lung RW, Hui JW, Liao G, Hayward SD (2007) Modulation of LMP1 protein expression by EBV-encoded microRNAs. Proc Natl Acad Sci USA 104:16164–16169. doi:10.1073/pnas.0702896104 0702896104 [pii]
Loeb GB, Khan AA, Canner D et al (2012) Transcriptome-wide miR-155 binding map reveals widespread noncanonical microRNA targeting. Mol Cell 48:760–770. doi:10.1016/j.molcel.2012.10.002 S1097-2765(12)00854-4 [pii]
Lohmann V, Korner F, Koch J, Herian U, Theilmann L, Bartenschlager R (1999) Replication of subgenomic hepatitis C virus RNAs in a hepatoma cell line. Science 285:110–113 7638 [pii]
Lu S, Cullen BR (2004) Adenovirus VA1 noncoding RNA can inhibit small interfering RNA and microRNA biogenesis. J Virol 78:12868–12876. doi:10.1128/JVI.78.23.12868-12876.2004 78/23/12868 [pii]
Machlin ES, Sarnow P, Sagan SM (2011) Masking the 5′ terminal nucleotides of the hepatitis C virus genome by an unconventional microRNA-target RNA complex. Proc Natl Acad Sci USA 108:3193–3198. doi:10.1073/pnas.1012464108 1012464108 [pii]
Machlin ES, Sarnow P, Sagan SM (2012) Combating hepatitis C virus by targeting microRNA-122 using locked nucleic acids. Curr Gene Ther 12:301–306 CGT-12-4-301 [pii]
Manns MP, Wedemeyer H, Cornberg M (2006) Treating viral hepatitis C: efficacy, side effects, and complications. Gut 55:1350–1359. doi:10.1136/gut.2005.076646 55/9/1350 [pii]
Marcinowski L, Tanguy M, Krmpotic A et al (2012) Degradation of cellular mir-27 by a novel, highly abundant viral transcript is important for efficient virus replication in vivo. PLoS Pathog 8:e1002510. doi:10.1371/journal.ppat.1002510 PPATHOGENS-D-11-02267 [pii]
Meister G, Landthaler M, Patkaniowska A, Dorsett Y, Teng G, Tuschl T (2004) Human Argonaute2 mediates RNA cleavage targeted by miRNAs and siRNAs. Mol Cell 15:185–197
Metzler M, Wilda M, Busch K, Viehmann S, Borkhardt A (2004) High expression of precursor microRNA-155/BIC RNA in children with Burkitt lymphoma. Genes Chromosomes Cancer 39:167–169. doi:10.1002/gcc.10316
Mortimer SA, Doudna JA (2013) Unconventional miR-122 binding stabilizes the HCV genome by forming a trimolecular RNA structure. Nucleic Acids Res. doi:10.1093/nar/gkt075 gkt075 [pii]
Motsch N, Pfuhl T, Mrazek J, Barth S, Grasser FA (2007) Epstein-Barr virus-encoded latent membrane protein 1 (LMP1) induces the expression of the cellular microRNA miR-146a. RNA Biol 4:131–137 5206 [pii]
Mrazek J, Kreutmayer SB, Grasser FA, Polacek N, Huttenhofer A (2007) Subtractive hybridization identifies novel differentially expressed ncRNA species in EBV-infected human B cells. Nucleic Acids Res 35:e73. doi:10.1093/nar/gkm244 gkm244 [pii]
Murray CL, Rice CM (2011) Turning hepatitis C into a real virus. Annu Rev Microbiol 65:307–327. doi:10.1146/annurev-micro-090110-102954
Murthy S, Kamine J, Desrosiers RC (1986) Viral-encoded small RNAs in herpes virus saimiri induced tumors. EMBO J 5:1625–1632
Murthy SC, Trimble JJ, Desrosiers RC (1989) Deletion mutants of herpesvirus saimiri define an open reading frame necessary for transformation. J Virol 63:3307–3314
Norman KL, Sarnow P (2010) Modulation of hepatitis C virus RNA abundance and the isoprenoid biosynthesis pathway by microRNA miR-122 involves distinct mechanisms. J Virol 84:666–670. doi:10.1128/JVI.01156-09 JVI.01156-09 [pii]

O'Connell RM, Taganov KD, Boldin MP, Cheng G, Baltimore D (2007) MicroRNA-155 is induced during the macrophage inflammatory response. Proc Natl Acad Sci USA 104:1604–1609. doi:10.1073/pnas.0610731104 0610731104 [pii]

O'Connell RM, Rao DS, Chaudhuri AA et al (2008) Sustained expression of microRNA-155 in hematopoietic stem cells causes a myeloproliferative disorder. J Exp Med 205:585–594. doi:10.1084/jem.20072108 jem.20072108 [pii]

O'Connell RM, Chaudhuri AA, Rao DS, Baltimore D (2009) Inositol phosphatase SHIP1 is a primary target of miR-155. Proc Natl Acad Sci USA 106:7113–7118. doi:10.1073/pnas.0902636106 0902636106 [pii]

Okamura K, Hagen JW, Duan H, Tyler DM, Lai EC (2007) The mirtron pathway generates microRNA-class regulatory RNAs in Drosophila. Cell 130:89–100. doi:10.1016/j.cell.2007.06.028 S0092-8674(07)00795-7 [pii]

Perez JT, Pham AM, Lorini MH, Chua MA, Steel J, tenOever BR (2009) MicroRNA-mediated species-specific attenuation of influenza A virus. Nat Biotechnol 27:572–576. doi:10.1038/nbt.1542 nbt.1542 [pii]

Pfeffer S, Zavolan M, Grasser FA et al (2004) Identification of virus-encoded microRNAs. Science 304:734–736

Pfeffer S, Sewer A, Lagos-Quintana M et al (2005) Identification of microRNAs of the herpesvirus family. Nat Methods 2:269–276

Randall G, Panis M, Cooper JD et al (2007) Cellular cofactors affecting hepatitis C virus infection and replication. Proc Natl Acad Sci USA 104:12884–12889. doi:10.1073/pnas.0704894104 0704894104 [pii]

Riley KJ, Rabinowitz GS, Yario TA, Luna JM, Darnell RB, Steitz JA (2012) EBV and human microRNAs co-target oncogenic and apoptotic viral and human genes during latency. EMBO J 31:2207–2221. doi:10.1038/emboj.2012.63 emboj201263 [pii]

Roberts AP, Lewis AP, Jopling CL (2011) miR-122 activates hepatitis C virus translation by a specialized mechanism requiring particular RNA components. Nucleic Acids Res 39:7716–7729. doi:10.1093/nar/gkr426 gkr426 [pii]

Rodriguez A, Vigorito E, Clare S et al (2007) Requirement of bic/microRNA-155 for normal immune function. Science 316:608–611. doi:10.1126/science.1139253 316/5824/608 [pii]

Ruby JG, Jan CH, Bartel DP (2007) Intronic microRNA precursors that bypass Drosha processing. Nature 448:83–86. doi:10.1038/nature05983 nature05983 [pii]

Russell SJ, Peng KW, Bell JC (2012) Oncolytic virotherapy. Nat Biotechnol 30:658–670. doi:10.1038/nbt.2287 nbt.2287 [pii]

Saetrom P, Heale BS, Snove O Jr, Aagaard L, Alluin J, Rossi JJ (2007) Distance constraints between microRNA target sites dictate efficacy and cooperativity. Nucleic Acids Res 35:2333–2342. doi:10.1093/nar/gkm133 gkm133 [pii]

Saito K, Siomi MC (2010) Small RNA-mediated quiescence of transposable elements in animals. Dev Cell 19:687–697. doi:10.1016/j.devcel.2010.10.011 S1534-5807(10)00464-8 [pii]

Samols MA, Hu J, Skalsky RL, Renne R (2005) Cloning and identification of a microRNA cluster within the latency-associated region of Kaposi's sarcoma-associated herpesvirus. J Virol 79:9301–9305

Sandhu SK, Volinia S, Costinean S et al (2012) miR-155 targets histone deacetylase 4 (HDAC4) and impairs transcriptional activity of B-cell lymphoma 6 (BCL6) in the Emu-miR-155 transgenic mouse model. Proc Natl Acad Sci USA 109:20047–20052. doi:10.1073/pnas.1213764109 1213764109 [pii]

Santanam U, Zanesi N, Efanov A et al (2010) Chronic lymphocytic leukemia modeled in mouse by targeted miR-29 expression. Proc Natl Acad Sci USA 107:12210–12215. doi:10.1073/pnas.1007186107 1007186107 [pii]

Sarrazin C, Hezode C, Zeuzem S, Pawlotsky JM (2012) Antiviral strategies in hepatitis C virus infection. J Hepatol 56(Suppl 1):S88–S100. doi:10.1016/S0168-8278(12)60010-5 S0168-8278(12)60010-5 [pii]

Selbach M, Schwanhausser B, Thierfelder N, Fang Z, Khanin R, Rajewsky N (2008) Widespread changes in protein synthesis induced by microRNAs. Nature 455:58–63. doi:10.1038/nature07228 nature07228 [pii]

Seto E, Moosmann A, Gromminger S, Walz N, Grundhoff A, Hammerschmidt W (2010) Micro RNAs of Epstein-Barr virus promote cell cycle progression and prevent apoptosis of primary human B cells. PLoS Pathog 6:e1001063. doi:10.1371/journal.ppat.1001063
Shapiro JS, Varble A, Pham AM, Tenoever BR (2010) Noncanonical cytoplasmic processing of viral microRNAs. RNA 16:2068–2074. doi:10.1261/rna.2303610 rna.2303610 [pii]
Shapiro JS, Langlois RA, Pham AM, Tenoever BR (2012) Evidence for a cytoplasmic microprocessor of pri-miRNAs. RNA 18:1338–1346. doi:10.1261/rna.032268.112 rna.032268.112 [pii]
Shimakami T, Yamane D, Jangra RK, Kempf BJ, Spaniel C, Barton DJ, Lemon SM (2012a) Stabilization of hepatitis C virus RNA by an Ago2-miR-122 complex. Proc Natl Acad Sci USA 109:941–946. doi:10.1073/pnas.1112263109 1112263109 [pii]
Shimakami T, Yamane D, Welsch C, Hensley L, Jangra RK, Lemon SM (2012b) Base pairing between hepatitis C virus RNA and microRNA 122 3′ of its seed sequence is essential for genome stabilization and production of infectious virus. J Virol 86:7372–7383. doi:10.1128/JVI.00513-12 JVI.00513-12 [pii]
Shin C, Nam JW, Farh KK, Chiang HR, Shkumatava A, Bartel DP (2010) Expanding the microRNA targeting code: functional sites with centered pairing. Mol Cell 38:789–802. doi:10.1016/j.molcel.2010.06.005 S1097-2765(10)00446-6 [pii]
Skalsky RL, Cullen BR (2010) Viruses, microRNAs, and host interactions. Annu Rev Microbiol 64:123–141. doi:10.1146/annurev.micro.112408.134243
Skalsky RL, Samols MA, Plaisance KB et al (2007) Kaposi's sarcoma-associated herpesvirus encodes an ortholog of miR-155. J Virol 81:12836–12845. doi:10.1128/JVI.01804-07 JVI.01804-07 [pii]
Skalsky RL, Corcoran DL, Gottwein E et al (2012) The viral and cellular microRNA targetome in lymphoblastoid cell lines. PLoS Pathog 8:e1002484. doi:10.1371/journal.ppat.1002484 PPATHOGENS-D-11-01612 [pii]
Sullivan CS, Grundhoff AT, Tevethia S, Pipas JM, Ganem D (2005) SV40-encoded microRNAs regulate viral gene expression and reduce susceptibility to cytotoxic T cells. Nature 435:682–686
Sumpter R Jr, Loo YM, Foy E et al (2005) Regulating intracellular antiviral defense and permissiveness to hepatitis C virus RNA replication through a cellular RNA helicase, RIG-I. J Virol 79:2689–2699. doi:10.1128/JVI.79.5.2689-2699.2005 79/5/2689 [pii]
Tam W, Ben-Yehuda D, Hayward WS (1997) bic, a novel gene activated by proviral insertions in avian leukosis virus-induced lymphomas, is likely to function through its noncoding RNA. Mol Cell Biol 17:1490–1502
Thai TH, Calado DP, Casola S et al (2007) Regulation of the germinal center response by microRNA-155. Science 316:604–608. doi:10.1126/science.1141229 316/5824/604 [pii]
Thorley-Lawson DA (2005) EBV persistence and latent infection in vivo. In: Robertson ES (ed) Epstein–Barr virus. Caister Academic Press, Norfolk, pp 309–357
Tuddenham L, Pfeffer S (2011) Roles and regulation of microRNAs in cytomegalovirus infection. Biochim Biophys Acta 1809:613–622. doi:10.1016/j.bbagrm.2011.04.002 S1874-9399(11)00059-9 [pii]
Tuddenham L, Jung JS, Chane-Woon-Ming B, Dolken L, Pfeffer S (2012) Small RNA deep sequencing identifies microRNAs and other small noncoding RNAs from human herpesvirus 6B. J Virol 86:1638–1649. doi:10.1128/JVI.05911-11 JVI.05911-11 [pii]
Umbach JL, Cullen BR (2009) The role of RNAi and microRNAs in animal virus replication and antiviral immunity. Genes Dev 23:1151–1164. doi:10.1101/gad.1793309 23/10/1151 [pii]
Umbach JL, Kramer MF, Jurak I, Karnowski HW, Coen DM, Cullen BR (2008) MicroRNAs expressed by herpes simplex virus 1 during latent infection regulate viral mRNAs. Nature. doi:10.1038/nature07103 nature07103 [pii]
Umbach JL, Nagel MA, Cohrs RJ, Gilden DH, Cullen BR (2009) Analysis of human alphaherpesvirus microRNA expression in latently infected human trigeminal ganglia. J Virol 83:10677–10683. doi:10.1128/JVI.01185-09 JVI.01185-09 [pii]

Vigorito E, Perks KL, Abreu-Goodger C et al (2007) microRNA-155 regulates the generation of immunoglobulin class-switched plasma cells. Immunity 27:847–859. doi:10.1016/j.immuni.2007.10.009 S1074-7613(07)00503-1 [pii]

Villanueva RA, Jangra RK, Yi M, Pyles R, Bourne N, Lemon SM (2010) miR-122 does not modulate the elongation phase of hepatitis C virus RNA synthesis in isolated replicase complexes. Antiviral Res 88:119–123. doi:10.1016/j.antiviral.2010.07.004 S0166-3542(10)00669-8 [pii]

Volinia S, Calin GA, Liu CG et al (2006) A microRNA expression signature of human solid tumors defines cancer gene targets. Proc Natl Acad Sci USA 103:2257–2261. doi:10.1073/pnas.0510565103 0510565103 [pii]

Wakita T, Pietschmann T, Kato T et al (2005) Production of infectious hepatitis C virus in tissue culture from a cloned viral genome. Nat Med 11:791–796. doi:10.1038/nm1268 nm1268 [pii]

Wilson JA, Zhang C, Huys A, Richardson CD (2011) Human Ago2 is required for efficient microRNA 122 regulation of hepatitis C virus RNA accumulation and translation. J Virol 85:2342–2350. doi:10.1128/JVI.02046-10 JVI.02046-10 [pii]

Xu G, Fewell C, Taylor C et al (2010) Transcriptome and targetome analysis in MIR155 expressing cells using RNA-seq. RNA 16:1610–1622. doi:10.1261/rna.2194910 rna.2194910 [pii]

Yekta S, Shih IH, Bartel DP (2004) MicroRNA-directed cleavage of HOXB8 mRNA. Science 304:594–596. doi:10.1126/science.1097434 304/5670/594 [pii]

Yin Q, McBride J, Fewell C et al (2008) MicroRNA-155 is an Epstein-Barr virus-induced gene that modulates Epstein-Barr virus-regulated gene expression pathways. J Virol 82:5295–5306. doi:10.1128/JVI.02380-07 JVI.02380-07 [pii]

Zeng Y, Wagner EJ, Cullen BR (2002) Both natural and designed micro RNAs can inhibit the expression of cognate mRNAs when expressed in human cells. Mol Cell 9:1327–1333

Zeng Y, Yi R, Cullen BR (2003) MicroRNAs and small interfering RNAs can inhibit mRNA expression by similar mechanisms. Proc Natl Acad Sci USA 100:9779–9784

Zhao Y, Yao Y, Xu H et al (2009) A functional MicroRNA-155 ortholog encoded by the oncogenic Marek's disease virus. J Virol 83:489–492. doi:10.1128/JVI.01166-08 JVI.01166-08 [pii]

Zhao Y, Xu H, Yao Y et al (2011) Critical role of the virus-encoded microRNA-155 ortholog in the induction of Marek's disease lymphomas. PLoS Pathog 7:e1001305. doi:10.1371/journal.ppat.1001305

Zhong J, Gastaminza P, Cheng G et al (2005) Robust hepatitis C virus infection in vitro. Proc Natl Acad Sci USA 102:9294–9299. doi:10.1073/pnas.0503596102 0503596102 [pii]

Zhu JY, Pfuhl T, Motsch N, Barth S, Nicholls J, Grasser F, Meister G (2009) Identification of novel Epstein-Barr virus microRNA genes from nasopharyngeal carcinomas. J Virol 83:3333–3341. doi:10.1128/JVI.01689-08 JVI.01689-08 [pii]

Interplay Between DNA Tumor Viruses and the Host DNA Damage Response

Karyn McFadden and Micah A. Luftig

Abstract Viruses encounter many challenges within host cells in order to replicate their nucleic acid. In the case of DNA viruses, one challenge that must be overcome is recognition of viral DNA structures by the host DNA damage response (DDR) machinery. This is accomplished in elegant and unique ways by different viruses as each has specific needs and sensitivities dependent on its life cycle. In this review, we focus on three DNA tumor viruses and their interactions with the DDR. The viruses Epstein-Barr virus (EBV), Kaposi's sarcoma-associated herpesvirus (KSHV), and human papillomavirus (HPV) account for nearly all of the virus-associated human cancers worldwide. These viruses have also been excellent models for the study of oncogenic virus-mediated cell transformation. In this review, we will discuss how each of these viruses engage and subvert aspects of the host DDR. The first level of DDR engagement is a result of the genetic linkage between the oncogenic potential of these viruses and their ability to replicate. Namely, the promotion of cells from quiescence into the cell cycle to facilitate virus replication can be sensed through aberrant cellular DNA replication structures which activate the DDR and hinder cell transformation. DNA tumor viruses subvert this growth-suppressive DDR through changes in viral oncoprotein expression which ultimately facilitate virus replication. An additional level of DDR engagement is through direct detection of replicating viral DNA. These interactions parallel those observed in other DNA virus systems in that the need to subvert these intrinsic sensors of aberrant DNA structure in order to replicate must be in place. DNA tumor viruses are no exception. This review will cover the molecular features of DNA tumor virus interactions with the host DDR and the consequences for virus replication.

K. McFadden · M. A. Luftig (✉)
Department of Molecular Genetics and Microbiology, Center for Virology, Duke University Medical Center, Durham, NC 27710, USA
e-mail: micah.luftig@duke.edu

B. R. Cullen (ed.), *Intrinsic Immunity*, Current Topics in Microbiology and Immunology 371, DOI: 10.1007/978-3-642-37765-5_9,

Contents

Abbreviations

DDR	DNA damage response
EBV	Epstein-Barr virus
KSHV	Kaposi's sarcoma-associated herpesvirus
HPV	Human papillomavirus
ATM	Ataxia-telangiectasia mutated
ATR	Ataxia-telangiectasia and RAD3-related
MRN	Mre11-Rad50-Nbs1
BL	Burkitt's lymphoma
LCL	Lymphoblastoid cell line
DSB	Double-stranded break
ssDNA	Single-stranded DNA
EBNA	Epstein-Barr virus nuclear antigen
LMP	(EBV) latent membrane protein
vFLIP	(KSHV) viral FLICE (FADD-like IL-1β-converting enzyme) inhibitory protein

1 Introduction

This review will focus on the interplay between DNA tumor viruses and the host DNA damage response (DDR). We will first discuss the components of the DDR sensing machinery and signaling pathways, then the interplay between viruses and the DDR. The role of the DDR in regulating replication of each specific virus, Epstein-Barr virus (EBV), Kaposi's sarcoma-associated herpesvirus (KSHV), and human papillomaviruses (HPV), will then be discussed individually. Finally, we will address emerging themes in the interactions between viral proteins and DNA with the DDR and convergent roles of the DDR in regulating latent versus lytic virus replication.

1.1 The DNA Damage Response

Our DNA must be replicated faithfully in the face of continuous damage by both endogenous and exogenous agents. Damaged DNA that is not repaired during the cell cycle has the potential to cause genomic instability which could ultimately result in cancer. The DDR is a signal transduction pathway that senses and recruits repair complexes to the site of damage and can stall cell cycle progression until the repair is completed [reviewed in (Bartek et al. 2007)]. The presence of damaged DNA triggers a response mediated by members of the phosphoinositide-3-like kinase (PIKK) family of serine/threonine kinases: ataxia-telangiectasia mutated (ATM), ataxia-telangiectasia and RAD3-related (ATR), or DNA-dependent protein kinase (DNA-PK) (Smith et al. 2010). This leads to the activation via phosphorylation of a number of downstream factors leading to repair or in the face of irreparable damage, senescence, or apoptosis.

The most detrimental form of DNA damage is a double-stranded break (DSB) which is sensed by the ATM arm of the DDR (Fig. 1a and reviewed in Derheimer and Kastan 2010). In response to DSBs, the Mre11-Rad50-Nbs1 (MRN) complex binds to the site of DNA damage and ATM is recruited to the damaged DNA, where it interacts with the MRN complex (Derheimer and Kastan 2010; Falck et al. 2005). ATM then undergoes an activating autophosphorylation at serine 1981, which causes it to transition from a dimer to a monomer (So et al. 2009; Bakkenist and Kastan 2003). ATM is further activated through additional phosphorylations as well as acetylation of lysine 3016 by TIP60 (Sun et al. 2005). Recent evidence also implicates the modulation of phosphorylation by protein phosphatase 2A (Goodarzi et al. 2004) and WIP1 phosphatases in the activation of ATM (Shreeram et al. 2006). ATM then phosphorylates H2AX on S139 (known as γH2AX) (Burma et al. 2001) which recruits the critical adaptor protein MDC1 (Stewart et al. 2003). The ubiquitin ligase proteins RNF8 and RNF168 bind to ATM-phosphorylated MDC1 leading to the ubiquitination of γH2AX, which scaffolds the interactions with 53BP1 and Brca1 (Huen et al. 2007; Mailand et al. 2007; Kolas et al. 2007; Doil et al. 2009). These events nucleate and retain a host of DNA repair factors at sites of DNA damage as well as enabling the phosphorylation of downstream targets of ATM including Chk2, p53, and Cdc25 family members that mediate cell cycle arrest and apoptosis (Smith et al. 2010; Matsuoka et al. 2007).

ATR is essential in cells for DNA replication as it responds to single-stranded DNA (ssDNA) exposure following replication fork collapse and the uncoupling of

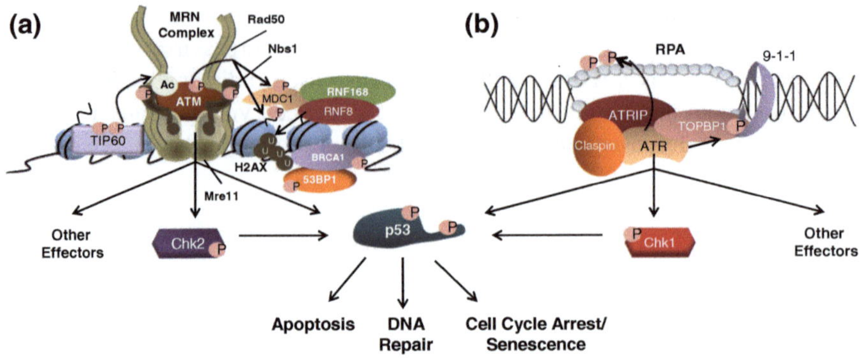

Fig. 1 Schematic diagram of the ATM and ATR DNA damage response signaling pathways. **a** ATM activation responding to a DNA double-stranded break (*DSB*). The initial complex of Mre11, Rad50, and Nbs1 targets the DSB and recruits ATM to activate its kinase domain. This leads to phosphorylation and monomerization, which then triggers a cascade of target phosphorylations. H2AX, the histone H2A variant, is first phosphorylated and provides a scaffold for MDC1, which recruits the ring-finger Ub ligases RNF8 and RNF168 to ubiquitinate H2AX. Concomitantly, enzymes such as Tip60 are recruited that can facilitate both ATM activation and chromatin de-compaction through acetylation of ATM directly and histone tails, respectively. The ubiquitination of H2AX further promotes recruitment and retention of factors such as 53BP1 and Brca1 that establish stable DDR foci propagating ATM signaling to effector molecules such as the checkpoint kinase Chk2, the transcription factor p53, and others. **b** ATR activation responding to ssDNA exposure following DNA replicative stress. Following replication fork collapse, ssDNA exposure is sensed by the protein RPA that coats these regions and serves as a scaffold to recruit the kinase ATR through its co-factor ATRIP. The critical ATR-activating factor TopBP1 is recruited through the Rad9-Rad1-Hus1 (9-1-1) complex at the replication fork. Finally, Claspin recruitment to the complex enables ATR phosphorylation of Chk1 and many other downstream effectors. A common downstream target is p53, which can be phosphorylated by both ATM and ATR leading to phenotypic outcomes ranging from DNA repair to apoptosis and senescence depending on the strength and duration of the DNA-damaging signal

the replication complex (Fig. 1b and reviewed in Nam and Cortez 2011). Replication protein A (RPA) coats the ssDNA and recruits ATR through its partner ATR-interacting protein (ATRIP) (Zou and Elledge 2003). Topoisomerase-binding protein 1 (TopBP1) is recruited to the ssDNA through the Rad9-Rad1-Hus1 (9-1-1) complex and then binds to both ATR and ATRIP. This stimulates ATR kinase activity leading to the phosphorylation of downstream effectors (Kumagai et al. 2006; Matsuoka et al. 2007) including the critical mediator Claspin, which recruits Chk1 to ssDNA, thus enabling ATR to phosphorylate and activate its kinase domain to promote checkpoint activation (Jeong et al. 2003; Kumagai and Dunphy 2003; Smith et al. 2010).

Chk1 is the primary effector of the intra-S and G2/M checkpoints while Chk2 exerts influence in G1/S and intra-S (Bartek and Lukas 2003; Stracker et al. 2009). Both Chk1 and Chk2 phosphorylate the Cdc25 family of phosphatases that function to activate cyclin-dependent kinases (CDKs) by removing inhibitory phosphorylations (Smith et al. 2010). Chk2 also modulates cell cycle progression by phosphorylating E2F, and both Chk1 and Chk2 can phosphorylate p53 with downstream consequences on G1/S-phase progression and apoptosis. Chk1 and Chk2 also play a role in DNA repair through the phosphorylation and activation of components of the homologous recombination (HR) pathway (Stracker et al. 2009). Despite the apparent dissection of function between the ATM and ATR pathway signaling to Chk2 and Chk1, respectively, these pathways are often functionally redundant where common targets are temporally offset in being phosphorylated by one kinase or the other (Stiff et al. 2006). Nevertheless, the engagement of the DDR leads to a rapid accumulation of factors with characteristic activating phosphorylations at sites of DNA damage. These hallmarks are useful measures of DDR activity, but detailed genetic analysis of individual signaling molecules is necessary to get an accurate picture of how these complex signaling networks are built.

1.2 Viruses and the DDR

Viruses face extraordinary challenges upon infection of a host cell. Both RNA and DNA viruses must evade innate sensors of foreign nucleic acids and other pattern recognition receptors to begin protein production, genome replication, and new particle formation. In the case of DNA viruses, a major challenge involves a specific conflict setup between the viral and host DNA genomes. The viral DNA genome may be recognized as damaged DNA and initiates a DDR that could have beneficial or catastrophic consequences for virus replication. For instance, adenovirus DNA structures are recognized by the host DDR to prevent cleavage and packaging of progeny DNA. An adenovirus protein E3 ubiquitin ligase complex (E4orf3/E1B55k) specifically degrades the components of the host DNA damage–sensing Mre11/Rad50/Nbs1 (MRN) complex to allow faithful DNA processing and prevent concatemerization (Stracker et al. 2002). This landmark paper highlighted the complexity of the conflict between DNA viruses and the host DDR. Subsequent work in other DNA viruses has indicated both a positive role for the DDR in some aspects of virus replication as well as a limiting role in others (Weitzman et al. 2010).

In the specific case of DNA tumor viruses, several aspects of virus replication and virus-induced cell transformation involve the DDR (Nikitin and Luftig 2011, 2012). Viral oncoproteins are required to drive infected cells from quiescence into the cell cycle in order to promote virus replication. The consequence of viral oncoprotein expression is cellular DNA replicative stress that can activate a growth-suppressive DDR. In the setting of viral latency or during virus replication, viral proteins can directly engage and either activate or inactivate components of the host DDR. This engagement can engender the specificity necessary to either promote certain aspects of the DDR, such as repair factor recruitment, or negate other activities, such as downstream activation of apoptosis. A final mechanism by which DNA tumor viruses can engage the DDR is through direct recognition of aberrant viral DNA structures. Under certain circumstances, such as latent DNA replication of large DNA tumor viruses (e.g., gamma-herpesviruses), DDR factor recruitment is important in maintaining episomal DNA. Similarly, during lytic replication of these viruses, DDR factors assemble at sites of viral DNA replication to recruit factors likely important for recombination and repair of viral DNA prior to packaging. However, robust DDR signaling can activate checkpoints and apoptosis that must be quenched downstream in order to allow virus replication to proceed. The coordination of these processes is an area of intense study and DNA tumor viruses provide an excellent model with which to strengthen our understanding of these responses. In this review, we will focus on the mechanisms by which human DNA tumor viruses engage and subvert the host DDR. The three representative viruses discussed will be EBV, KSHV, and the HPV.

2 Epstein-Barr Virus and the DNA Damage Response

2.1 Epstein-Barr Virus

EBV is a human gamma-herpesvirus that is found ubiquitously in the adult population, where it establishes a lifelong latency with no discernible clinical manifestations in the majority of those infected (Rickinson and Kieff 2007). Infection with EBV generally occurs early in life and the virus establishes a lifelong latency in B lymphocytes. EBV has been linked to a number of benign and malignant diseases such as infectious mononucleosis, endemic Burkitt's lymphoma (eBL), and the epithelial cell-derived nasopharyngeal carcinomas (NPC) (Rickinson and Kieff 2007). EBV is a concern in immunocompromised patients, where it has been linked to a large subset of HIV-1 associated B-cell lymphomas and post-transplant lymphoproliferative disorders (Rickinson and Kieff 2007).

Initial EBV infection of epithelial cells typically results in lytic replication, which is followed by infection of resting B cells where long-term latency is established. Most EBV-induced malignancies are associated with the latent stage of the viral replicative cycle, during which only a limited set of viral proteins and small non-coding RNAs are produced. The proteins expressed during latent infection include the EBV nuclear antigens (EBNAs 1, 2, 3A, 3B, 3C, and LP) and the latent membrane proteins (LMP1 and LMP2) (Yates et al. 1985). Akin to many oncogenic viruses, EBV alters the host cell to ensure long-term proliferation and survival. This includes induction of the cell cycle and evasion of innate host defenses that would result in long-term senescence or apoptosis. The first viral proteins expressed in the infected cell are EBNA-LP and EBNA2, which transcriptionally up-regulate the expression of a number of viral and cellular genes, thereby inducing a transition of resting B cells into the cell cycle (Kaiser et al. 1999; Sinclair et al. 1994). These genes include the viral EBNA1 and EBNA3A, 3B, and 3C as well as cellular S-phase-initiating proto-oncogenes such as c-Myc, E2F1, and cyclin D2 (Kaiser et al. 1999) (Fig. 2a, bottom). The EBNA3 proteins modulate EBNA2-dependent gene expression in addition to regulating a number of other cellular targets (Johannsen et al. 1996; Zhao et al. 1996). EBNA2 and EBNA3C then co-activate the viral LMP1, a constitutively active mimic of the TNF receptor

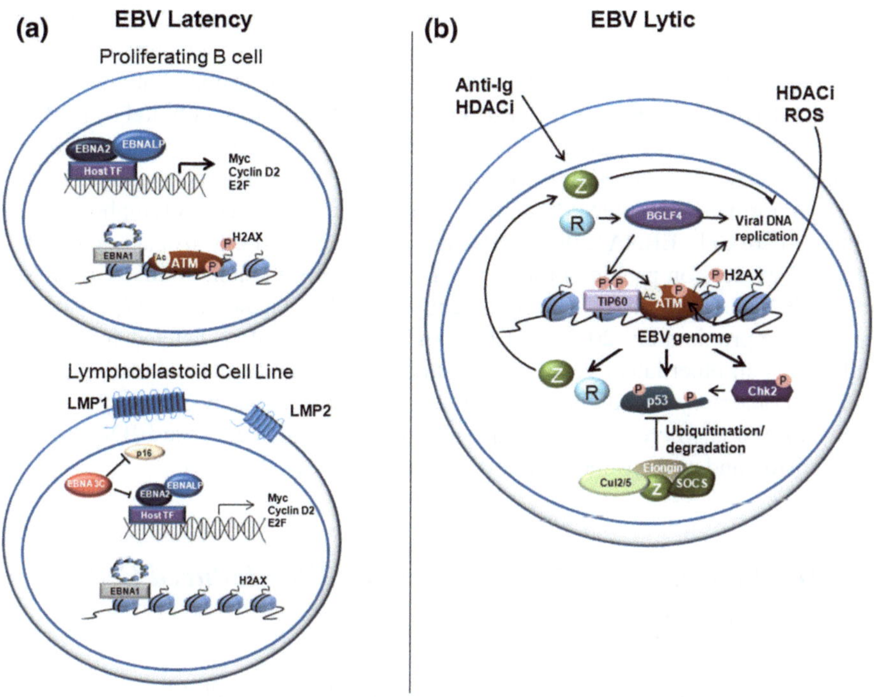

Fig. 2 Epstein-Barr virus and the DNA damage response. **a** DDR activation during EBV latency. *Top*, Early after B-cell infection with EBV, cells hyper-proliferate dependent on the expression of the viral EBNA2 and EBNA-LP proteins driving c-Myc and other S-phase-promoting genes. This results in an activated DDR including ATM and H2AX phosphorylation. *Bottom*, in later divisions after infection, and also in immortalized LCLs, the expression of the EBNA3C protein mitigates the DDR, presumably through attenuating the CDK inhibitor p16. Additional viral proteins are expressed such as LMP1 and LMP2 that contribute to the LCL survival phenotype. **b** DDR activation during EBV lytic infection. Reactivation from diverse stimuli converge on expression of the critical viral trans-activator Z. In the case of anti-Ig treatment of latently infected B cells or pan-HDAC inhibition, these effects are direct on the Z promoter (*top left*). However, other molecules such as DNA-damaging agents like reactive oxygen species (*ROS*) can directly promote ATM activation. The consequence of this activation is also Z expression through the ATM-dependent regulation of viral lytic promoters and Tip60-mediated acetylation of viral chromatin. Expression of the viral kinase BGLF4 further supports this mechanism through direct activation of Tip60, ATM, and H2AX. Ultimately, the downstream signaling from ATM can lead to the deleterious effect of p53-mediated apoptosis. However, the viral Z protein can form an E3 ubiquitin ligase complex to promote p53 degradation at late stages during viral lytic replication

and B-cell co-stimulatory molecule CD40 (Lin et al. 2002; Cahir McFarland et al. 1999). This triggers NFκB, p38, and JNK signaling pathways resulting in the modulation in gene expression patterns to allow for long-term cellular proliferation and survival. Additionally, LMP2A acts as a mimic of the B-cell receptor, leading to constitutive, high-level expression of PI3K signaling and pro-survival pathways (Miller et al. 1995). EBNA1 is a DNA-binding transcription factor that primarily functions to replicate and segregate the EBV episome (Yates et al. 1985).

EBV encodes three distinct latency programs which are tightly linked to the differentiation state of the infected B cell and exhibit distinct patterns of gene expression (Thorley-Lawson 2001). Infection of resting primary human peripheral blood B cells with EBV in vitro transforms them into indefinitely proliferating non-tumorigenic lymphoblastoid cell lines (LCLs) which express the full set of latency genes (latency III). In contrast, the more restricted latency II program is found in normal EBV-infected germinal center B cells, Hodgkin's lymphomas, and NPCs, where only EBNA1, the LMPs, and viral non-coding RNAs are detected. A latency I expression profile is found in BL where only the non-coding viral RNAs and EBNA1 are expressed. A more recently described latency type, Wp-restricted, found in approximately 20 % of EBV-positive BLs, is often deleted for EBNA2 and retains all other latent genes (Kelly et al. 2002). Finally, normal resting peripheral blood memory B cells in asymptomatic individuals display a latency 0 type of gene expression where essentially all viral genes are silenced, save EBNA1 during cell division and perhaps the viral miRNAs (Babcock et al. 2000).

2.2 *The Role of the DDR in Suppressing EBV-Induced Cell Proliferation*

Viral induction of proliferation and survival pathways can result in oncogenic transformation, as evidenced by the tight association of latent EBV infection with many types of malignancies. However, these malignancies are rare despite the greater than 90 % prevalence of the virus in the adult population. In support of this, the efficiency of B-cell transformation is low, with only ~1 % of infected cells becoming immortalized (Henderson et al. 1977; Sugden and Mark 1977). This implies that the host cell must be able to restrict virally induced proliferation. This was addressed by work by Nikitin, et al. who demonstrated that at least 10 % of this restriction is due to the induction of the DDR (Nikitin et al. 2010). Infection of primary B cells with EBV leads to a transient period of hyper-proliferation that peaks 4–6 days after infection and eventually slows to the rate observed in LCLs. This hyper-proliferative phase coincides with the activation of the ATM-dependent DDR, as indicated by the phosphorylation of H2AX, ATM, Chk2 and the formation of 53BP1 nuclear foci (Fig. 2a). These cells were demonstrated to be nearly 100 % latently infected, and the DDR markers did not co-localize with viral episomes. This indicates that the DDR was activated in response to host DNA damage and not to the replication or presence of the viral DNA. Inhibition of ATM or Chk2 enhanced B-cell transformation, implying that the DDR is an innate mechanism that the cell uses to restrict viral-mediated cellular proliferation. The enhanced transformation observed in the presence of inhibitors of the DDR was only effective during the hyper-proliferative period. As the infected B cells transitioned to become immortalized LCLs, there was a reduction in both DDR markers and sensitivity to the DDR.

The differences in host gene expression between resting B cells, transformed LCLs, and this hyper-proliferative population could reveal how the host cell can limit transformation. Indeed, gene expression microarrays indicated that there are distinct transcriptomic changes between the hyper-proliferating cells in comparison to either LCLs or uninfected B cells (Nikitin et al. 2010; Price et al. 2012; Forte et al. 2012). These include enrichment for genes that are involved in cell cycle progression and the DDR. This also corresponded to a transition from a period of elevated EBNA-LP and EBNA2 activity to an accumulation of EBNA3 proteins. As previously noted, EBNA-LP and EBNA2 up-regulate expression of a number of genes involved in cell cycle progression, and this activity is repressed though the action of the EBNA3 proteins (Kieff and Rickinson 2007). This could indicate a model in which initial infection causes a transient period of hyper-proliferation induced by EBNA2-regulated genes such as c-Myc, which results in replicative stress and induction of the DDR-mediated growth arrest.

In order to establish latency, the virus must overcome the DDR-induced cell cycle arrest. Loss of EBNA3C prevents EBV-mediated B-cell transformation and also results in a dramatic increase in the activation of DDR markers (Nikitin et al. 2010). Previous work has demonstrated that EBNA3C represses the MDM2 antagonist p14ARF and the cyclin-dependent kinase inhibitor p16INK4a through epigenetic silencing of their promoter (Skalska et al. 2010; Maruo et al. 2011). Given recent evidence suggesting that persistent DDR signaling at sites of irreparable DNA damage leading to senescence can be exacerbated by over-expression of p16 (Rodier et al. 2011), the role of EBNA3C in suppressing p16 may, in fact, be critical for its ability to attenuate the EBV-induced DDR.

EBNA3C has also been demonstrated to interact with a number of host factors including RBPJ, CtBP, NCoR, HDACs, and others (Bajaj et al. 2008; Rosendorff et al. 2004; Subramanian et al. 2002; Touitou et al. 2005). Notably, EBNA3C was shown to directly interact with Chk2, and this interaction relieved a G2/M cell cycle block (Choudhuri et al. 2007). EBNA3C has also been linked to S-phase entry through the stabilization of the cyclin D1/CDK6 complex (Saha et al. 2011) and recruitment of SCFSkp2 E3 ligase for ubiquitin-mediated degradation of pRB (Knight and Robertson 2004). Therefore, EBNA3C is a key viral factor that is likely to regulate the host DDR through multiple mechanisms throughout the cell cycle.

2.3 EBV and Genomic Instability

DNA damage that is not repaired could ultimately lead to genomic instability and cancer. Unlike the karyotypically stable LCLs (Lacoste et al. 2010), chromosomal aberrations have been observed in EBV-associated Burkitt's lymphomas (BLs) and other cancers [reviewed in (Gruhne et al. 2009a)]. A direct link between EBV viral proteins and genomic instability was uncovered by the Masucci group in studies looking at paired EBV-negative and EBV-positive BL lines. The presence of EBV caused a 3–10-fold increase in chromosomal abnormalities including dicentric chromosomes as well as the induction of a DDR response as indicated by the presence of γH2AX (Kamranvar et al. 2007). Further analysis revealed that EBNA1 alone induced the chromosomal dysfunction and up-regulation of the DDR (Gruhne et al. 2009b). EBNA1 increased expression of the transcription factor Nox2, which in turn increased levels of reactive oxygen species (ROS) resulting in DNA damage and aberrant chromosomal structures. In a follow-up study, the authors found telomere dysfunction in a number of EBV-positive BL cell lines while the corresponding EBV-negative lines were karyotypically normal (Kamranvar and Masucci 2011). BL cells that express EBNA1 exhibit increased telomere dysfunction such as extra-chromosomal telomere signals, the accumulation of telomere dysfunction–induced foci (TIFs), and telomeric-sister chromatid exchange. While EBNA1 had no effect on the overall expression of the shelterin components, it did reduce the association of TRF2 with telomeres. EBNA1 appears to induce this damage through ROS as compounds that act as ROS scavengers were able to inhibit the appearance of DDR markers and reconstitute TRF2-protected telomeres (Kamranvar and Masucci 2011).

EBNA1 is not the only viral protein implicating in inducing genomic instability (Gruhne et al. 2009a). In an EBV-negative B-lymphoma cell line, BJAB, ectopic expression of LMP1 or EBNA3C also promoted genomic instability as detected by chromosomal aberrations and the up-regulation of DDR markers (Gruhne et al. 2009c). The three viral proteins appear to induce this instability through different mechanisms. Studies in epithelial cells initially indicated that LMP1 could inhibit DNA repair by inducing the cytoplasmic accumulation and thus inhibition of Foxo3, a transcription factor implicating in cell cycle arrest and DNA repair mechanisms (Chen et al. 2008). Furthermore, LMP1-expressing B cells had a reduced capacity to repair damaged DNA and a decrease in ATM response (Gruhne et al. 2009c). In contrast, EBNA3C over-expression promoted aneuploidy through a mitotic spindle checkpoint failure. More specifically, EBNA3C led to a decrease in the expression of the checkpoint protein BubR1 which binds cdc2 and prevents anaphase progression (Gruhne et al. 2009c). Additionally, the interaction of EBNA3C with Chk2 and other cell cycle components discussed above could be involved in promoting genomic instability through aberrant mitotic progression (Choudhuri et al. 2007; Knight et al. 2005; Saha et al. 2009). Collectively, these studies indicate that the aberrant expression of EBV latent oncoproteins may contribute to oncogenesis through disrupting normal controls of DNA repair and cell cycle checkpoints.

2.4 Latent EBV DNA Replication Requires the DDR

During viral latency, each EBV episome is replicated once per cell cycle by EBNA1 [(Yates et al. 1985); reviewed in (Hammerschmidt and Sugden 2013)]. EBNA1 binds to specific regions of the EBV genome, namely oriP, the origin of plasmid replication, and recruits the cellular replication machinery. EBNA1 binds to a family of repeats (FR) and a dyad symmetry (DS) element within oriP (Rawlins et al. 1985). Host proteins including the cellular origin replication complex (ORC) and mini chromosome maintenance (MCM) complex are enriched in association with EBNA1 at the DS elements during G1 and early S phase (Ritzi et al. 2003). Interestingly, there are also repeat elements adjacent to EBNA1-binding sites within the DS that are very similar to telomere repeat elements (Deng et al. 2002, 2003). In fact, these sites bind the shelterin components TRF1, TRF2, and hRap1. Shelterin components normally function to protect telomeres from DDR sensing [reviewed in (Palm and de Lange 2008)]. Through these sites, EBNA1 and TRF2 interact and recruit ORC to the viral episome (Norseen et al. 2008; Tatsumi et al. 2008). TRF2 and hRap1 binding are both critical for EBNA1-mediated replication while TRF1 appears to be inhibitory (Deng et al. 2003). Thus, EBV co-opts the normal cellular DNA machinery of telomeric DNA repeats to recruit shelterin components to facilitate replication of the viral genome. This is likely important to suppress hyperactivation of the DDR at latent EBV genomes, as occurs at protected telomeric ends due to the activity of TRF2 (Palm and de Lange 2008).

The firing of oriP is also delayed in early S phase by a host checkpoint regulation that is dependent on the recruitment of HDACs by TRF2 (Zhou et al. 2009). In early S phase, the nucleosomes adjacent to the TRF2-binding sites are de-acetylated. As S-phase progresses, the nucleosomes become acetylated and oriP replication ensues (Zhou et al. 2009). The activity of the DDR protein Chk2 is down-regulated after interaction with TRF2 (Buscemi et al. 2009) in a cell cycle-dependent manner with enrichment in S phase (Zhou et al. 2010). Chk2 depletion by siRNA or Chk2 kinase inhibitor inhibits oriP replication and episome maintenance and advances the replication timing of OriP (Zhou et al. 2010). Mechanistically, the depletion of Chk2 alters histone de-acetylation at DS sites within oriP. In vitro studies indicate that Chk2 phosphorylates TRF2 on its amino-terminal domain and a phospho-mimetic TRF2 mutant precludes ORC recruitment to oriP.

In addition to shelterin components, EBNA1 also recruits the DNA damage–sensing MRN complex to oriP during S phase (Dheekollu et al. 2007). In the absence of Nbs1, similar to loss of Chk2, oriP maintenance and EBV genome maintenance are suppressed, indicating an important role for MRN recruitment to the oriP. Therefore, the oriP DNA element serves as a unique scaffold for DDR factors and signaling that is required to orchestrate the faithful replication of EBV DNA during latency as well as presumably to promote the proper resolution of Holliday junction structures formed late during each S phase (Dheekollu et al. 2007).

2.5 The Role of the DDR in Facilitating EBV Lytic Replication

EBV latently infected B cells can switch into a lytic phase of infection, generating progeny virions that can infect a new host. Lytic reactivation is a highly regulated process that begins with expression of two immediate early viral genes, BZLF1 and BRLF1, which encode the proteins Z and R, respectively (Kieff and Rickinson 2007). This expression results in the transactivation of early lytic genes, which encode viral proteins responsible for replication. After viral replication, the late genes are transcribed, resulting in the production of the structural proteins needed to produce new virions. EBV, like many viruses, usurps host cell cycle controls in order to provide an S-phase-like environment for efficient replication.

The best-characterized viral protein involved in this process is Z which can interact with and alter components of the cell cycle checkpoint in addition to its role as a transcription factor. The link between Z and the DDR was first observed by Kudoh et al. (2005) using a tetracycline inducible BZLF1 (Z) construct to induce lytic replication. Induction of Z correlated with the phosphorylation and activation of ATM, Chk2, H2AX, and p53 as well as the co-localization of ATM-S1981 and the MRN complex with EBV replication centers. The activation of these proteins would normally lead to cell cycle arrest and apoptosis. However, the authors also observed hyper-phosphorylation of Rb and the subsequent release and activation of E2F1 and its downstream targets, such as cyclins A and E, indicating cell cycle progression.

One of the ways that Z regulates cell cycle progression is through its interaction with p53 (Zhang et al. 1994; Mauser et al. 2002; Kudoh et al. 2005). Over-expression of Z in B958 cells induces the activation of p53 but without the subsequent increase in protein levels. Additionally, the expression of downstream targets such as MDM2 and p21CIP1 was not elevated (Kudoh et al. 2005; Mauser et al. 2002) although p53 retained its ability to bind to DNA and had increased post-translational modifications (Mauser et al. 2002). Subsequent studies revealed that Z can function as an adaptor of the Elongin B/C-cul2/5-SOCS-(ECS) box ubiquitin ligase complex through direct interaction with Cul2 and Cul5 (Kato et al. 2009). The targeting and subsequent degradation of p53 by Z was enhanced by phosphorylation of p53 on S15 and S20 as a result of the activation of the ATM pathway.

To further delineate the interplay between the DDR and lytic replication, the Hayward group identified host substrates phosphorylated by a conserved herpesvirus kinase (Li et al. 2011). The EBV orthologue, BGLF4, is a mimic of host CDKs and plays a key role in the establishment of an S-phase-like environment in infected cells. Using a protein microarray to identify BGLF4 substrates, Li et al. (2011) observed enrichment for proteins involved in the ATM arm of the DDR, including TIP60, a chromatin remodeling protein that is involved in both DNA repair and transcription. BGLF4 interaction with TIP60 results in its phosphorylation and the activation of its histone and protein acetylation activity. TIP60 has previously been shown to activate ATM in response to DNA damage (Sun et al. 2005), and its activation by BGLF4 also results in the phosphorylation of ATM and its downstream effectors (Fig. 2b). Critically, loss of either TIP60 or ATM suppressed EBV DNA replication after lytic induction (Li et al. 2011). Mechanistically, BGLF4 and TIP60 are recruited to the EBV lytic replication origin (OriLyt) and other viral promoters stimulating both γH2AX and acetylated H2AK5, thereby leading to increased viral lytic gene expression and DNA replication.

The importance of the DDR for lytic replication was further confirmed by the Kenney group by exploring the importance of ATM in response to induction by a variety of stimuli (Hagemeier et al. 2012). Treatment of Akata BL cells with the ATM inhibitor KU55933 suppressed viral reactivation after induction with paclitaxel, a chemotherapeutic agent; sodium butyrate, an HDAC inhibitor; activation of the B-cell receptor by anti-IgG treatment; cytokine treatment with TGFβ or treatment with the proteasome inhibitor bortezomib. The importance of ATM for lytic reactivation was also seen in both EBV-positive BL lines as well as in epithelial cells. Of note, not all of the BL lines used in this study contained functional p53, indicating that the role of ATM in lytic reactivation is at least partially p53 independent. Since ROS was recently shown to activate ATM kinase activity (Guo et al. 2010), the authors speculated that it also might lead to lytic reactivation. Indeed, ROS did lead to lytic reactivation, which was blocked by subsequent treatment with an ATM inhibitor or by knocking down ATM with shRNAs.

Viruses subvert host ubiquitination pathways in most phases of their life cycle in order to evade innate immune responses and to degrade proteins that would hinder their replication. Similarly, ubiquitin specific proteases are also targeted by viral proteins such as EBNA1 to alter the turnover of proteins needed for replication (Saridakis et al. 2005). Recent work by the Masucci group has demonstrated that the tegument protein BPLF1 is a de-neddylase that regulates Cullin-ring ligases (CRL) to alter S-phase progression (Gastaldello et al. 2010). BPLF1 is able to cleave both ubiquitin adducts as well as NEDD8 conjugates in transfected cells, although evidence points to its de-neddylation activity as being more functionally relevant. BPLF1 directly interacts with CRLs and modulates their activity as seen through an accumulation of substrates such as Cdc25A, p21CIP1, and Cdt1, an S-phase licensing factor. The stabilization of Cdt1 through its de-neddylation promotes aberrant S-phase entry and leads to the accumulation of phosphorylated ATM and Chk2. Inhibition of BPLF1 using shRNAs reduced EBV replication to a level similar to that seen with inhibition of Z (Gastaldello et al. 2010).

While S-phase entry is a critical aspect of EBV lytic DNA replication, the consequences on cell cycle arrest through aberrant entry via Cdt1 stabilization could be catastrophic for the virus. Interestingly, recent work by the Pagano laboratory indicates that BPLF1 can de-ubiquitinate the major host DNA replication processivity factor PCNA (Whitehurst et al. 2012). Since mono-ubiquitinated PCNA (PCNA-Ub) recruits trans-lesion synthesis (TLS) polymerase to sites of DNA damage (Yang et al. 2010) and replaces the replicative polymerase, thus allowing it to bypass the lesion, the authors speculate that stalling of cellular DNA replication may free cellular factors that are needed for viral DNA replication (Whitehurst et al. 2012). Additionally, the downstream signaling consequences of PCNA-Ub, such as strong checkpoint activation and apoptosis, would be ameliorated by BPLF1. Thus, BPLF1 simultaneously functions to promote pseudo-S-phase maintenance through Cdt1 stabilization while cutting off downstream signaling through inhibiting PCNA-Ub.

A model can then be considered in which DNA damage or ROS leads to the activation of ATM (Fig. 2b). This leads to the activation of the Z and R promoters and the expression of the immediate early genes. Data from the Tsurumi, Kenney, and other groups indicate that p53 may be important at this early stage in lytic replication (Sato and Tsurumi 2010). The combined activity of Z and R would then lead to the up-regulation of BGLF4, which phosphorylates TIP60 resulting in an increase in its HAT activity. TIP60 then further activates ATM as well as the

OriLyt promoter, leading to lytic reactivation. Viral replication is aided at this late stage by the inhibition of p53-mediated apoptosis through Z-dependent ubiquitination as well as by the regulation of DDR signaling by BPLF1.

3 Kaposi's Sarcoma-Associated Herpesvirus and the DDR

3.1 Kaposi's Sarcoma-Associated Herpesvirus

KSHV is a large, double-stranded DNA virus of the gamma-herpesvirus family that was originally identified in the lesions of patients suffering from Kaposi's sarcoma (Chang et al. 1994). KSHV has been found in all forms of KS as well as primary effusion lymphoma (PEL) and multi-centric Castleman's disease [reviewed in (Mesri et al. 2010)]. Like all members of the herpesvirus family, KSHV can replicate in both lytic and latent stages. During latency, the virus is maintained as an extra-chromosomal episome that is linked to the host chromosome through the virally encoded protein latency-associated nuclear antigen (LANA) (Ballestas et al. 1999).

The major viral transcripts expressed during latency function to keep the cell alive and replicating. The proteins include LANA, viral cyclin (v-cyclin), viral FLICE inhibitory protein (vFLIP), and the kaposins, and KSHV also encoded many miRNAs (Mesri et al. 2010). LANA plays a role in abrogating host tumor suppressor pathways by inhibiting p53 and Rb-E2F (Friborg et al. 1999; Si and Robertson 2006; Radkov et al. 2000; Mesri et al. 2010) in addition to its critical role in episome maintenance (Ballestas et al. 1999). LANA also activates telomerase reverse transcriptase (TERT) expression further promoting infected-cell survival (Verma et al. 2004; Mesri et al. 2010).

KSHV v-cyclin is structurally homologous to cellular D-type cyclins and can bind to and activate cellular CDK6 [reviewed in (Mittnacht and Boshoff 2000; Verschuren et al. 2004)]. The v-cyclin/CDK6 complex has a broader substrate range than its cellular counterpart and is resistant to inhibition by CDK inhibitors. vFLIP can activate NFκB by binding to the activating IKK complex, which results in the expression of anti-apoptotic genes and the induction of cytokine secretion (Guasparri et al. 2004; Mesri et al. 2010). Importantly, vFLIP also serves as a key negative regulator of autophagy in latently infected cells (Lee et al. 2009).

3.2 The Role of the DDR in Suppressing KSHV-Induced Cell Proliferation

The first link between latent KSHV infection and the DDR came from the Ojala group exploring the effect of v-cyclin in telomerase immortalized endothelial cells (Koopal et al. 2007) (Fig. 3a). Over-expression of v-cyclin decreased cellular

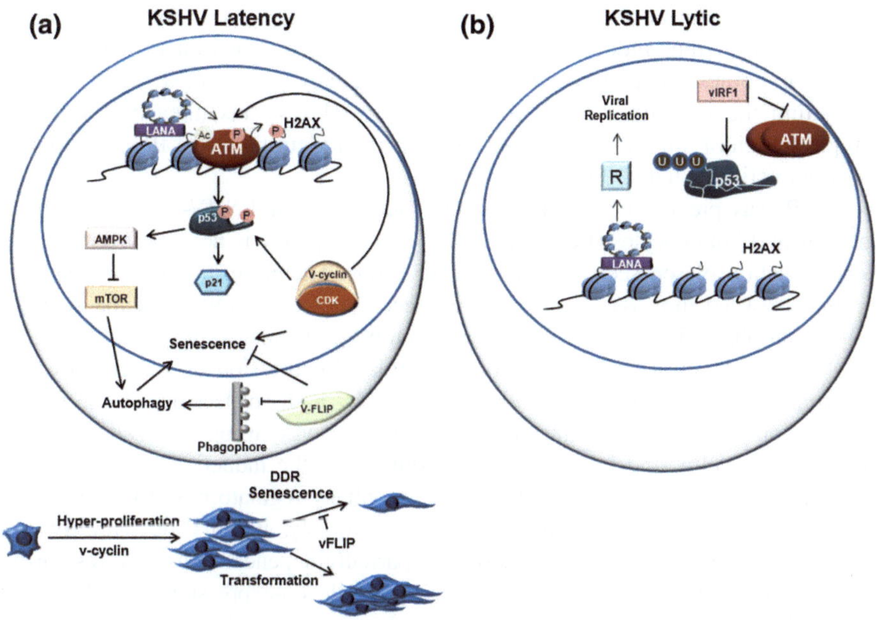

Fig. 3 Kaposi's sarcoma-associated herpesvirus and the DNA damage response. **a** DDR activation during KSHV latency. The induction of cell cycle progression by the viral cyclin (v-cyclin) in complex with CDK6 leads to the activation of an ATM-mediated DDR. The consequence of this activation is p53 induction, oncogene-induced senescence (*OIS*), and autophagy. However, co-expression of the KSHV v-FLIP protein overrides v-cyclin-mediated OIS by suppressing autophagy. **b** KSHV and the DDR during lytic reactivation. Little is understood about KSHV lytic viral DNA replication and the DDR. However, the expression of the lytic cycle vIRF1 gene is capable of suppressing p53 activity and, in particular, the ability of ATM to activate p53. This suggests that if ATM were to be activated by KSHV lytic DNA replication (as occurs in EBV, HPV, and MHV68 infection), then a plausible mechanism to overcome checkpoint activation or apoptosis would be in place

proliferation through the induction of an S-phase arrest. They also demonstrated that viral protein expression could lead to cellular senescence, as has also been observed by over-expression of classical cellular oncogenes. Both primary and immortalized endothelial cells had a strong induction of DDR markers, such as phosphorylated H2AX, ATM, Chk2, and p53. Treatment of v-cyclin-expressing cells with an inhibitor of the ATM kinase (KU-55933) led to a reduction in DDR markers and a release of the S-phase arrest, which resulted in aberrant mitosis (Koopal et al. 2007). This could be due to v-cyclin-induced centrosome amplification, which can lead to aberrant mitosis in the absence of a functional DDR (Verschuren et al. 2002). The DDR pathway was activated in KSHV-infected endothelial cells after they overcame an early period of arrest. The effect of the DDR could be bypassed by ablation of p53 or Rb signaling (Koopal et al. 2007).

This work was followed up by Leidal, et al, who demonstrated that KSHV-infected primary human fibroblasts activated the DDR but did not undergo senescence (Leidal et al. 2012). This is in contrast to v-cyclin-expressing cells, which had a greater induction of the DDR as well as senescence markers. The authors hypothesized that one of the KSHV latency genes must be capable of abrogating the v-cyclin-induced senescence. To address this, they performed an ectopic screen of a number of latency proteins and identified v-FLIP as a suppressor of v-cyclin-mediated senescence.

vFLIP was previously shown to inhibit autophagy in KSHV-infected endothelial cells and lymphocytes (Lee et al. 2009), and autophagy is necessary for the efficient induction of senescence (Young et al. 2009). Therefore, the authors explored the link between the loss of oncogene-induced senescence (OIS) and the suppression of autophagy. KSHV v-cyclin induced autophagy similar to the potent oncogene HrasV12 (Leidal et al. 2012). Temporally regulated expression of v-cyclin revealed an initial period of hyper-proliferation followed by a sharp decrease in the rate of proliferation coinciding with the onset of OIS. p53 was strongly activated in cells expressing v-cyclin, and its activity was required for the induction of OIS through the transcriptional up-regulation of the key autophagy regulators Sestrin1 and damage-regulated autophagy modulator (DRAM). Notably, the mTOR pathway was inactivated concomitant with OIS induction, partially dependent on AMP-responsive protein kinase (AMPK), a target of Sestrin1. While co-expression of vFLIP with v-cyclin had no effect on DDR markers, the induction of autophagy and OIS was dramatically reduced. Surprisingly, in contrast to its role in cell survival through NFκB activity (Chaudhary et al. 1999; Liu et al. 2002), the specific antagonism of autophagy by vFLIP was responsible for its ability to override v-cyclin-mediated OIS. This cooperativity likely contributes significantly to KSHV-mediated tumorigenesis (Fig. 3a).

3.3 The Role of the DDR in KSHV Lytic Replication

While the default pathway for KSHV infection is latency, it can be induced to reactivate by a variety of stimuli, resulting in the generation of a full set of viral proteins and the production of new virions. The switch between latency and lytic replication is primarily mediated by the protein RTA, encoded by ORF5 (Ganem 2006). The expression of RTA and other immediate early genes is followed by the expression of the delayed-early genes. These genes include a viral DNA polymerase, primase, helicase, and single-stranded binding proteins. These proteins can form a stable replication complex on the OriLyt promoter in the host cell nucleus. The final stage of lytic reactivation is the induction of late genes, which largely encode structural proteins necessary for viral particle formation (Ganem 2006).

As with all of the DNA viruses discussed, KSHV needs to modulate the host cell to ensure that there is an appropriate S-phase-like environment for replication as well as the evasion of host cell defenses. KSHV evades host innate immunity by encoding several homologs of host genes, such as viral interleukin-6 (vIL-6), viral Bcl2, and four homologs of interferon regulatory factors (vIRF1-4) (reviewed in Jacobs and Damania 2011). vIRF1 was initially reported to interact with and inactivate p53 by suppressing its acetylation and its transcriptional activation potential (Seo et al. 2001). This was further explored by Shin et al. (2006), who used ectopic expression of a tetracycline-regulated vIRF1 in a KSHV-null B-lymphoma line, BJAB. Expression of vIRF1 resulted in an increase in p53 ubiquitination and the inhibition of the activating phosphorylation of S15. Further analysis revealed that vIRF1 interacted with ATM through its C-terminal domain and that this lead to a decrease in its activation, as revealed by phosphorylation of S1981. The activation of downstream targets of ATM such as Chk2, H2AX, and p53 was also decreased by vIRF1, while the expression of the cell cycle regulatory protein, Cdc25, was increased, consistent with a role for vIRF1 in suppressing downstream checkpoint activation by ATM. Therefore, these studies suggest a possible mechanism by which KSHV oncoproteins subvert downstream DDR signaling during lytic reactivation (Fig. 3b). It will be of great interest in the future to discern whether, as in the case of EBV, KSHV also utilizes aspects of ATM signaling to promote lytic viral gene expression and DNA replication. Indeed, evidence from the KSHV-related murine gamma-herpesvirus MHV68 suggests a role for ATM and H2AX in virus replication (Tarakanova et al. 2007).

4 Human Papillomaviruses and the DNA Damage Response

4.1 Human Papillomaviruses

HPVs are small double-stranded DNA viruses that generally cause benign skin lesions. Over 120 serotypes of HPV have been identified that can be further classified as low or high risk according to their association with benign or malignant lesions, respectively. Infection with a high-risk variant, such as HPV 16 or 18, has been associated with 99 % of cervical cancers (Walboomers et al. 1999) and ~25 % of head and neck cancers (Gillison 2004).

The HPV lifecycle is closely linked to the differentiation state of epithelial cells [reviewed in (Moody and Laimins 2010; McBride et al. 2012)]. Initially, the virus infects the basal layer of squamous epithelia and establishes itself as a low copy number extra-chromosomal episome. These cells represent the only actively replicating cells within the epithelia and are exposed to infection through abrasions or wounding. Infection leads to an initial burst of viral DNA replication that is followed by a period where the virus DNA replicates in concert with the host chromosome. As uninfected epithelial cells divide, they leave the basement membrane layer and undergo terminal differentiation until they are eventually shed as dead and dying cells. HPV undergoes its final replication cycle, called vegetative replication, in the differentiated cell layer. Newly generated virions can then be shed and spread to new targets.

Restricting its replicative life cycle to differentiated cells allows the virus to remain undetected by the immune system for extended periods of time. However, terminal differentiation entails an exit from the cell cycle which would deprive the virus of the host replicative machinery. To circumvent this, HPV encodes several proteins (E5, E6 and E7) that promote proliferation by delaying exit from the cell cycle. E6 and E7 are the primary transforming genes of HPV through their antagonism of p53 and Rb, respectively (Dyson et al. 1989; Munger et al. 1989a, b). The E1 and E2 proteins are essential for viral DNA replication and episome maintenance and may play a role in promoting genomic instability and cancer progression, as will be discussed below (Amin et al. 2000; Androphy et al. 1987).

4.2 The Role of the DDR in Suppressing HPV-Induced Cell Proliferation

Despite the high frequency of HPV infection, the rate of cancer progression is low and only occurs if the virus has not been cleared over an extended period of time. One striking observation is that a majority of the cervical cancer biopsies show integration of the high-risk HPVs (Moody and Laimins 2010). While this appears to occur randomly there does seem to be a preference for common fragile sites within the genome. One consequence of HPV genomic integration is the loss of the E2 protein, which is responsible for repressing viral gene expression in addition to its role as a replication factor. This leads to uncontrolled expression of E6 and E7 (Schwarz et al. 1985) which can independently transform cells and lead to genomic instability through mitotic defects (Duensing et al. 2000).

The Ustav group demonstrated that integrated HPV genomes in HeLa and SiHa cells undergo an "onion skip" type of replication in the presence of exogenous E1 (Kadaja et al. 2009). This leads to DNA replication structures that are recognized by the ATM arm of the DDR (Fig. 4a). This ultimately leads to repair by either HR or NHEJ, the latter of which can produce mutations or chromosomal translocations. In fact, genomic instability was observed in SiHa cells transfected with E1. In other studies, over-expression of E1 from diverse HPV strains induced cell cycle arrest and a DDR in undifferentiated keratinocytes in the absence of viral DNA (Fradet-Turcotte et al. 2011; Sakakibara et al. 2011; Reinson et al. 2013). Expression of E2 alone, or coupled with the viral origin, was able to partially suppress this effect in one study (Fradet-Turcotte et al. 2011) but not in another (Sakakibara et al. 2011). The ability of E1 to induce a DDR was seen even with basal-level expression of the protein, indicating that it is not simply an artifact of over-expression. The DDR foci partially co-localized with E1 and E2 as well as sites of DNA synthesis (Sakakibara et al. 2011). The level of nucleotide incorporation detected was below what would normally be detected for S-phase progression and could indicate regions of DNA repair. Consistent with this, the DDR foci were also TUNEL positive indicating regions of fragmented DNA (Sakakibara et al. 2011) and had DSBs detected by alkaline comet assay (Reinson et al. 2013).

High-risk HPV E6 and E7 expression in undifferentiated keratinocytes is sufficient to induce replicative stress, which could result in collapsed replication forks and the activation of ATR and ATM pathways (Moody and Laimins 2009; Bester et al. 2011). While HPV oncoprotein expression drives aberrant S phase, promoting a DDR,

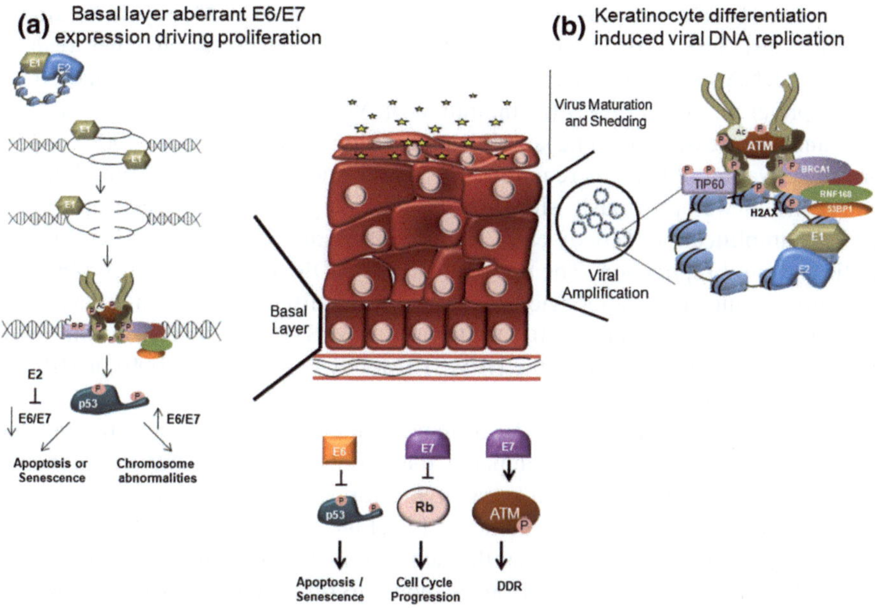

Fig. 4 Human papillomavirus and the DNA damage response. **a** HPV infection of basal layer epithelial cells and DDR activation. The infection of basal epithelial cells by HPV leads to a constitutive ATM-dependent DDR. This may be due to the role of HPV E1 in promoting replication of integrated HPV genomes or perhaps promoting DNA damage at other sites through the host genome (*left*). Normally, in this layer, E2 protein suppresses the expression of E6 and E7. If E2 is abrogated due to integration, then E7 up-regulation could lead to increased S-phase entry promoting the DDR or through direct engagement and activation of ATM (*bottom, right*). **b** During vegetative DNA replication upon differentiation of the basal keratinocyte layer, the ATM pathway is activated and DDR factors co-localize with viral genomes (*right*). ATM and Chk2 activity are important for viral DNA replication in this phase of the HPV life cycle. The uppermost layer of keratinocytes in the figure indicates virion release (*stars*)

the potential for senescence or apoptosis is strongly mitigated by the ability of E6 to leverage E6AP to degrade p53 (Scheffner et al. 1993). Therefore, HPV infection in undifferentiated keratinocytes can drive an oncogenic stress, but cell proliferation in the context of viral genome integration is maintained due to potent anti-apoptotic activities. The role of the DDR in suppressing HPV tumorigenesis may then be restricted to its functions as a regulator of viral DNA replication, as discussed below.

4.3 The Role of the DDR in Facilitating HPV Vegetative DNA Replication

Emerging evidence suggests that the DDR plays a critical role in HPV amplification in differentiated keratinocytes. Keratinocytes containing the HPV-31 episome differentiate upon calcium treatment. Differentiation increases markers of the canonical ATM pathway, which localize to nuclear foci (Moody and Laimins 2009). These foci also contain the HPV episome as well as enzymes responsible for DNA repair and HR such as PCNA, pRPAS32-S33, Rad51, and Brca1 (Gillespie et al. 2012) (Fig. 4b). Inhibition of ATM led to diminished replication foci and suppressed episome amplification (Moody and Laimins 2009). ATM-driven activation of Chk2 is required for caspase 7 activation (Moody and Laimins 2009) which was previously shown to be necessary for the cleavage of E1 and genome amplification (Moody et al. 2007), further solidifying the link between the DDR and viral replication. The up-regulation of DDR markers was also observed in primary human keratinocytes that underwent differentiation in organotypic raft cultures (Banerjee et al. 2011). Strong pATM-S1981 was seen in differentiated cells that expressed either the HPV-31 (Moody and Laimins 2009) or HPV-18 genome (Banerjee et al. 2011).

DDR activation appears to be primarily induced by E7 in differentiated cells. Banerjee et al. (2011) compared the expression of the HPV-18 genome, E7 alone, or vector controls in organotypic raft cultures using primary human keratinocytes. The full viral genome or E7 alone were able to induce up-regulation of pATM-S1981 as well as the activation of Chk2 and the ATR effector Chk1. Additionally, E7 was previously shown to directly interact with ATM and induce the activation of Chk2 (Moody and Laimins 2009). Cultures expressing E7 alone or the HPV-18 genome exhibited a prolonged G2 phase (Wang et al. 2009; Banerjee et al. 2011) as evidenced by the accumulation of inactive Cdk1 and Cdc25 in the cytoplasm and the lack of mitotic cells in culture. Active Chk1, Chk2, and JNK are known to inactivate Cdc25 and therefore the activation of the DDR could lead to a prolonged G2 phase, which can enhance viral amplification (Banerjee et al. 2011).

A model is starting to emerge in which HPV is able to exploit the intrinsic DDR to amplify its genome. E7 can interact with and activate ATM in differentiated cells leading to the activation of its downstream effectors such as Chk2 and γH2AX (Fig. 4). This results in the co-localization of many repair proteins at the viral genome. Chk2 can then activate caspase 7, resulting in the cleavage of E1, an essential replication factor. The activated DDR would then induce a G2 checkpoint by inactivating Cdk1 and Cdc25 to prevent entry into mitosis. This allows the virus to amplify its genome out of sync with normal cellular S phase, thus preventing competition for replication factors. The inhibition of p53 by E6 prevents the apoptosis normally induced by the activation of the DDR and thereby facilitates viral DNA replication.

5 Conclusions

DNA tumor viruses activate the host DDR in multiple ways. Viral oncoproteins drive cells from quiescence into cell cycle and often lead to aberrant cellular DNA replication that can provoke the host DDR. This process may occur during viral latency, as exemplified by the oncogenic herpesviruses, or during acute infection. Furthermore, viral DNA replication, either latent or lytic, can produce structures that directly activate the DDR. Once activated, the DDR can promote DNA repair and recombination as well as activate cell cycle checkpoints, senescence, or apoptosis. In the case of latent viral oncoprotein-driven cellular DNA replicative stress leading to DDR activation, the consequence is a decrease in cellular proliferation which is ultimately detrimental for the virus. In contrast, during lytic replication, the activation of the DDR through viral DNA replication structures is beneficial for the virus. Therefore, the host DDR plays divergent roles in the replication of human DNA tumor viruses depending on the mode of virus replication.

DNA tumor viruses drive cellular transformation under circumstances where virus replication and the ability to produce progeny virions are inhibited. For example, in the case of high-risk HPVs, unintentional integration of viral DNA into the host genome leads to the aberrantly high expression of the oncoproteins E6 and E7, which can lead to cell transformation and contribute to tumorigenesis. Early expression of these oncoproteins in basal keratinocytes may mimic aspects of this process. As shown by Bester et al. (2011), HPV-16 E6/E7 expression can activate a growth-suppressive DDR due to increased cellular DNA replicative stress. Similarly, in the case of EBV and KSHV, viral latent oncoproteins drive aberrant cellular DNA proliferation and a growth-suppressive DDR (Nikitin et al. 2010; Koopal et al. 2007). Interestingly, recent work by the McCormick lab has shown that KSHV vFLIP is capable of overcoming the DDR-dependent senescence mediated by the v-cyclin protein in KSHV-infected endothelial cells (Leidal et al. 2012). This activity depends on induction of autophagy by vFLIP and mitigates the far downstream consequences of the DDR. Similarly, in the context of EBV infection of primary B cells, the cell proliferation rate is transiently high, concomitant with DDR activation, but then wanes through immortalization dependent on expression of the viral EBNA3C protein (Nikitin et al. 2010). Therefore, DDR activation in latently infected cells (in the case of the gammaherpesivruses) leads to growth suppression that ultimately compromises viral infection. However, viral oncoproteins are able to mitigate these pathways either downstream through modulating autophagy or upstream through attenuating the strength of the oncogenic signal.

The nature of the molecular signal downstream of oncogenic stress that triggers the DDR in these settings remains controversial. Evidence for increased DNA replicative stress (Bartkova et al. 2006; Di Micco et al. 2006), increased fragile sites (Di Micco et al. 2006), increased irreparable DNA damage at telomeres (Fumagalli et al. 2012), and telomere dysfunction (Suram et al. 2012), as well as changes in chromatin structure associated with hyper-activation of transcription at certain loci, has all been implicated in DDR activation downstream of activated oncogenes. The careful genetic dissection of the critical sensors and mediators of each of these aberrant DNA structures in the context of viral oncogene expression will be needed to discern precisely how the DDR is activated and whether these signals engender specificity in the downstream signaling pathways following oncogenic stress.

During lytic viral DNA replication, two mechanisms can give rise to an activated DDR. First, the set of proteins expressed early during infection or reactivation can generate a pseudo-S-phase environment important for viral DNA replication. Often the consequence of this aberrant cell cycle promotion is the activation of the DDR. However, the downstream consequences of the DDR are mitigated, enabling virus replication. In the case of EBV, for example, the promotion of S-phase entry through the BPLF1 protein leads to an activated DDR including ATM phosphorylation. However, the downstream signaling at sites of cellular DNA damage is mitigated in part through the activity of BPLF1 in preventing PCNA mono-ubuiqitination. Furthermore, the viral Z protein is capable of degrading p53 once activated at late stages in lytic reactivation. Thus, as often observed in the context of virus infection, a multi-faceted approach to ensuring cell survival is used to facilitate DNA replication in the context of an activated DDR.

A second signal for DDR activation during lytic replication is the replication of viral DNA. In both the EBV and HPV systems, strong co-localization of DDR factors with sites of viral DNA replication corroborates their induction by aberrant viral DNA structures sensed as damaged DNA. Interestingly, DDR signaling, and ATM in particular, is required for efficient DNA replication and progeny virion formation. It is likely that the recruitment of factors necessary for viral DNA repair and/or recombination is critical in the process of faithful DNA replication and packaging necessary to produce competent infectious progeny. The precise set of ATM targets important for DNA tumor virus replication is an active and exciting area of current research. The mechanistic delineation of the beneficial DDR in virus replication has the potential to lead to new and interesting therapeutic targets in these and other viral diseases. Moreover, it is likely that using these viruses to probe the specificities of DDR activation will lead to new insights into the recognition, repair, and signaling from cellular sites of DNA damage as well.

Acknowledgments This publication resulted (in part) from research supported by the Duke University Center for AIDS Research (CFAR), an NIH funded program (5P30 AI064518); as well as an NIH funded R01 (R01-CA140337); a Viral Oncology Training grant (VOTG: T32-CA009111) and IRTPA Training grant (T32-AI007392). We would also like to thank members of the Luftig lab for helpful discussions.

References

Amin AA, Titolo S, Pelletier A, Fink D, Cordingley MG, Archambault J (2000) Identification of domains of the HPV11 E1 protein required for DNA replication in vitro. Virology 272:137–150

Androphy EJ, Lowy DR, Schiller JT (1987) Bovine papillomavirus E2 trans-activating gene product binds to specific sites in papillomavirus DNA. Nature 325:70–73

Babcock GJ, Hochberg D, Thorley-Lawson AD (2000) The expression pattern of Epstein-Barr virus latent genes in vivo is dependent upon the differentiation stage of the infected B cell. Immunity 13:497–506

Bajaj BG, Murakami M, Cai Q, Verma SC, Lan K, Robertson ES (2008) Epstein-Barr virus nuclear antigen 3C interacts with and enhances the stability of the c-Myc oncoprotein. J Virol 82:4082–4090

Bakkenist CJ, Kastan MB (2003) DNA damage activates ATM through intermolecular autophosphorylation and dimer dissociation. Nature 421:499–506

Ballestas ME, Chatis PA, Kaye KM (1999) Efficient persistence of extrachromosomal KSHV DNA mediated by latency-associated nuclear antigen. Science 284:641–644

Banerjee NS, Wang HK, Broker TR, Chow LT (2011) Human papillomavirus (HPV) E7 induces prolonged G2 following S phase reentry in differentiated human keratinocytes. J Biol Chem 286:15473–15482

Bartek J, Lukas J (2003) Chk1 and Chk2 kinases in checkpoint control and cancer. Cancer Cell 3:421–429

Bartek J, Bartkova J, Lukas J (2007) DNA damage signalling guards against activated oncogenes and tumour progression. Oncogene 26:7773–7779

Bartkova J, Rezaei N, Liontos M, Karakaidos P, Kletsas D, Issaeva N, Vassiliou LV, Kolettas E, Niforou K, Zoumpourlis VC, Takaoka M, Nakagawa H, Tort F, Fugger K, Johansson F, Sehested M, Andersen CL, Dyrskjot L, Orntoft T, Lukas J, Kittas C, Helleday T, Halazonetis TD, Bartek J, Gorgoulis VG (2006) Oncogene-induced senescence is part of the tumorigenesis barrier imposed by DNA damage checkpoints. Nature 444:633–637

Bester AC, Roniger M, Oren YS, Im MM, Sarni D, Chaoat M, Bensimon A, Zamir G, Shewach DS, Kerem B (2011) Nucleotide deficiency promotes genomic instability in early stages of cancer development. Cell 145:435–446

Burma S, Chen BP, Murphy M, Kurimasa A, Chen DJ (2001) ATM phosphorylates histone H2AX in response to DNA double-strand breaks. J Biol Chem 276:42462–42467

Buscemi G, Zannini L, Fontanella E, Lecis D, Lisanti S, Delia D (2009) The shelterin protein TRF2 inhibits Chk2 activity at telomeres in the absence of DNA damage. Curr Biol 19:874–879

Cahir McFarland ED, Izumi KM, Mosialos G (1999) Epstein-Barr virus transformation: involvement of latent membrane protein 1-mediated activation of NF-kappaB. Oncogene 18:6959–6964

Chang Y, Cesarman E, Pessin MS, Lee F, Culpepper J, Knowles DM, Moore PS (1994) Identification of herpesvirus-like DNA sequences in AIDS-associated Kaposi's sarcoma. Science 266:1865–1869

Chaudhary PM, Jasmin A, Eby MT, Hood L (1999) Modulation of the NF-kappa B pathway by virally encoded death effector domains-containing proteins. Oncogene 18:5738–5746

Chen YR, Liu MT, Chang YT, Wu CC, Hu CY, Chen JY (2008) Epstein-Barr virus latent membrane protein 1 represses DNA repair through the PI3K/Akt/FOXO3a pathway in human epithelial cells. J Virol 82:8124–8137

Choudhuri T, Verma SC, Lan K, Murakami M, Robertson ES (2007) The ATM/ATR signaling effector Chk2 is targeted by Epstein-Barr virus nuclear antigen 3C to release the G2/M cell cycle block. J Virol 81:6718–6730

Deng Z, Lezina L, Chen CJ, Shtivelband S, So W, Lieberman PM (2002) Telomeric proteins regulate episomal maintenance of Epstein-Barr virus origin of plasmid replication. Mol Cell 9:493–503

Deng Z, Atanasiu C, Burg JS, Broccoli D, Lieberman PM (2003) Telomere repeat binding factors TRF1, TRF2, and hRAP1 modulate replication of Epstein-Barr virus OriP. J Virol 77:11992–12001

Derheimer FA, Kastan MB (2010) Multiple roles of ATM in monitoring and maintaining DNA integrity. FEBS Lett 584:3675–3681

Dheekollu J, Deng Z, Wiedmer A, Weitzman MD, Lieberman PM (2007) A role for MRE11, NBS1, and recombination junctions in replication and stable maintenance of EBV episomes. PLoS ONE 2:e1257

Di Micco R, Fumagalli M, Cicalese A, Piccinin S, Gasparini P, Luise C, Schurra C, Garre M, Nuciforo PG, Bensimon A, Maestro R, Pelicci PG, d'Adda di Fagagna F (2006) Oncogene-induced senescence is a DNA damage response triggered by DNA hyper-replication. Nature 444:638–642

Doil C, Mailand N, Bekker-Jensen S, Menard P, Larsen DH, Pepperkok R, Ellenberg J, Panier S, Durocher D, Bartek J, Lukas J, Lukas C (2009) RNF168 binds and amplifies ubiquitin conjugates on damaged chromosomes to allow accumulation of repair proteins. Cell 136:435–446

Duensing S, Lee LY, Duensing A, Basile J, Piboonniyom S, Gonzalez S, Crum CP, Munger K (2000) The human papillomavirus type 16 E6 and E7 oncoproteins cooperate to induce mitotic defects and genomic instability by uncoupling centrosome duplication from the cell division cycle. Proc Natl Acad Sci U S A 97:10002–10007

Dyson N, Howley PM, Munger K, Harlow E (1989) The human papilloma virus-16 E7 oncoprotein is able to bind to the retinoblastoma gene product. Science 243:934–937

Falck J, Coates J, Jackson SP (2005) Conserved modes of recruitment of ATM, ATR and DNA-PKcs to sites of DNA damage. Nature 434:605–611

Forte E, Salinas R, Chang C, Zhou T, Linnstaedt SD, Gottwein E, Jacobs C, Jima D, Li QJ, Dave SS, Luftig MA (2012) The Epstein-Barr virus induced tumor suppressor miR-34a is growth promoting in EBV-infected B cells. J Virol 86:6889–6898

Fradet-Turcotte A, Bergeron-Labrecque F, Moody CA, Lehoux M, Laimins LA, Archambault J (2011) Nuclear accumulation of the papillomavirus E1 helicase blocks S-phase progression and triggers an ATM-dependent DNA damage response. J Virol 85:8996–9012

Friborg J Jr, Kong W, Hottiger MO, Nabel GJ (1999) p53 inhibition by the LANA protein of KSHV protects against cell death. Nature 402:889–894

Fumagalli M, Rossiello F, Clerici M, Barozzi S, Cittaro D, Kaplunov JM, Bucci G, Dobreva M, Matti V, Beausejour CM, Herbig U, Longhese MP, d'Adda di Fagagna F (2012) Telomeric DNA damage is irreparable and causes persistent DNA-damage-response activation. Nat Cell Biol 14:355–365

Ganem D (2006) KSHV infection and the pathogenesis of Kaposi's sarcoma. Annu Rev Pathol 1:273–296

Gastaldello S, Hildebrand S, Faridani O, Callegari S, Palmkvist M, Di Guglielmo C, Masucci MG (2010) A deneddylase encoded by Epstein-Barr virus promotes viral DNA replication by regulating the activity of cullin-RING ligases. Nat Cell Biol 12:351–361

Gillespie KA, Mehta KP, Laimins LA, Moody CA (2012) Human papillomaviruses recruit cellular DNA repair and homologous recombination factors to viral replication centers. J Virol 86:9520–9526

Gillison ML (2004) Human papillomavirus-associated head and neck cancer is a distinct epidemiologic, clinical, and molecular entity. Semin Oncol 31:744–754

Goodarzi AA, Jonnalagadda JC, Douglas P, Young D, Ye R, Moorhead GB, Lees-Miller SP, Khanna KK (2004) Autophosphorylation of ataxia-telangiectasia mutated is regulated by protein phosphatase 2A. EMBO J 23:4451–4461

Gruhne B, Kamranvar SA, Masucci MG, Sompallae R (2009a) EBV and genomic instability–a new look at the role of the virus in the pathogenesis of Burkitt's lymphoma. Semin Cancer Biol 19:394–400

Gruhne B, Sompallae R, Marescotti D, Kamranvar SA, Gastaldello S, Masucci MG (2009b) The Epstein-Barr virus nuclear antigen-1 promotes genomic instability via induction of reactive oxygen species. Proc Natl Acad Sci U S A 106:2313–2318

Gruhne B, Sompallae R, Masucci MG (2009c) Three Epstein-Barr virus latency proteins independently promote genomic instability by inducing DNA damage, inhibiting DNA repair and inactivating cell cycle checkpoints. Oncogene 28:3997–4008

Guasparri I, Keller SA, Cesarman E (2004) KSHV vFLIP is essential for the survival of infected lymphoma cells. J Exp Med 199:993–1003

Guo Z, Kozlov S, Lavin MF, Person MD, Paull TT (2010) ATM activation by oxidative stress. Science 330:517–521

Hagemeier SR, Barlow EA, Meng Q, Kenney SC (2012) The cellular ataxia telangiectasia-mutated kinase promotes Epstein-Barr virus lytic reactivation in response to multiple different types of lytic reactivation-inducing stimuli. J Virol 86:13360–13370

Hammerschmidt W, Sugden B (2013) Replication of epstein-barr viral DNA. Cold Spring Harb Perspect Biol 5:a013029

Henderson E, Miller G, Robinson J, Heston L (1977) Efficiency of transformation of lymphocytes by Epstein-Barr virus. Virology 76:152–163

Huen MS, Grant R, Manke I, Minn K, Yu X, Yaffe MB, Chen J (2007) RNF8 transduces the DNA-damage signal via histone ubiquitylation and checkpoint protein assembly. Cell 131:901–914

Jacobs SR, Damania B (2011) The viral interferon regulatory factors of KSHV: immunosuppressors or oncogenes? Front Immunol 2:19

Jeong SY, Kumagai A, Lee J, Dunphy WG (2003) Phosphorylated claspin interacts with a phosphate-binding site in the kinase domain of Chk1 during ATR-mediated activation. J Biol Chem 278:46782–46788

Johannsen E, Miller CL, Grossman SR, Kieff E (1996) EBNA-2 and EBNA-3C extensively and mutually exclusively associate with RBPJkappa in Epstein-Barr virus-transformed B lymphocytes. J Virol 70:4179–4183

Kadaja M, Isok-Paas H, Laos T, Ustav E, Ustav M (2009) Mechanism of genomic instability in cells infected with the high-risk human papillomaviruses. PLoS Pathog 5:e1000397

Kaiser C, Laux G, Eick D, Jochner N, Bornkamm GW, Kempkes B (1999) The proto-oncogene c-myc is a direct target gene of Epstein-Barr virus nuclear antigen 2. J Virol 73:4481–4484

Kamranvar SA, Masucci MG (2011) The Epstein-Barr virus nuclear antigen-1 promotes telomere dysfunction via induction of oxidative stress. Leukemia 25:1017–1025

Kamranvar SA, Gruhne B, Szeles A, Masucci MG (2007) Epstein-Barr virus promotes genomic instability in Burkitt's lymphoma. Oncogene 26:5115–5123

Kato M, Sanada M, Kato I, Sato Y, Takita J, Takeuchi K, Niwa A, Chen Y, Nakazaki K, Nomoto J, Asakura Y, Muto S, Tamura A, Iio M, Akatsuka Y, Hayashi Y, Mori H, Igarashi T, Kurokawa M, Chiba S, Mori S, Ishikawa Y, Okamoto K, Tobinai K, Nakagama H, Nakahata T, Yoshino T, Kobayashi Y, Ogawa S (2009) Frequent inactivation of A20 in B-cell lymphomas. Nature 459:712–716

Kelly G, Bell A, Rickinson A (2002) Epstein-Barr virus-associated Burkitt lymphomagenesis selects for downregulation of the nuclear antigen EBNA2. Nat Med 8:1098–1104

Kieff E, Rickinson A (2007) Epstein-Barr virus and its replication. In: Knipe DM, Howley PM (eds) Fields virology, 5th edn, vol 2. pp 2603–2654

Knight JS, Robertson ES (2004) Epstein-Barr virus nuclear antigen 3C regulates cyclin A/p27 complexes and enhances cyclin A-dependent kinase activity. J Virol 78:1981–1991

Knight JS, Sharma N, Robertson ES (2005) Epstein-Barr virus latent antigen 3C can mediate the degradation of the retinoblastoma protein through an SCF cellular ubiquitin ligase. Proc Natl Acad Sci U S A 102:18562–18566

Kolas NK, Chapman JR, Nakada S, Ylanko J, Chahwan R, Sweeney FD, Panier S, Mendez M, Wildenhain J, Thomson TM, Pelletier L, Jackson SP, Durocher D (2007) Orchestration of the DNA-damage response by the RNF8 ubiquitin ligase. Science 318:1637–1640

Koopal S, Furuhjelm JH, Jarviluoma A, Jaamaa S, Pyakurel P, Pussinen C, Wirzenius M, Biberfeld P, Alitalo K, Laiho M, Ojala PM (2007) Viral oncogene-induced DNA damage response is activated in Kaposi sarcoma tumorigenesis. PLoS Pathog 3:1348–1360

Kudoh A, Fujita M, Zhang L, Shirata N, Daikoku T, Sugaya Y, Isomura H, Nishiyama Y, Tsurumi T (2005) Epstein-Barr virus lytic replication elicits ATM checkpoint signal transduction while providing an S-phase-like cellular environment. J Biol Chem 280:8156–8163

Kumagai A, Dunphy WG (2003) Repeated phosphopeptide motifs in claspin mediate the regulated binding of Chk1. Nat Cell Biol 5:161–165

Kumagai A, Lee J, Yoo HY, Dunphy WG (2006) TopBP1 activates the ATR-ATRIP complex. Cell 124:943–955

Lacoste S, Wiechec E, Dos Santos Silva AG, Guffei A, Williams G, Lowbeer M, Benedek K, Henriksson M, Klein G, Mai S (2010) Chromosomal rearrangements after ex vivo Epstein-Barr virus (EBV) infection of human B cells. Oncogene 29:503–515

Lee JS, Li Q, Lee JY, Lee SH, Jeong JH, Lee HR, Chang H, Zhou FC, Gao SJ, Liang C, Jung JU (2009) FLIP-mediated autophagy regulation in cell death control. Nat Cell Biol 11:1355–1362

Leidal AM, Cyr DP, Hill RJ, Lee PW, McCormick C (2012) Subversion of autophagy by Kaposi's sarcoma-associated herpesvirus impairs oncogene-induced senescence. Cell Host Microbe 11:167–180

Li R, Zhu J, Xie Z, Liao G, Liu J, Chen MR, Hu S, Woodard C, Lin J, Taverna SD, Desai P, Ambinder RF, Hayward GS, Qian J, Zhu H, Hayward SD (2011) Conserved herpesvirus kinases target the DNA damage response pathway and TIP60 histone acetyltransferase to promote virus replication. Cell Host Microbe 10:390–400

Lin J, Johannsen E, Robertson E, Kieff E (2002) Epstein-Barr virus nuclear antigen 3C putative repression domain mediates coactivation of the LMP1 promoter with EBNA-2. J Virol 76:232–242

Liu L, Eby MT, Rathore N, Sinha SK, Kumar A, Chaudhary PM (2002) The human herpes virus 8-encoded viral FLICE inhibitory protein physically associates with and persistently activates the Ikappa B kinase complex. J Biol Chem 277:13745–13751

Mailand N, Bekker-Jensen S, Faustrup H, Melander F, Bartek J, Lukas C, Lukas J (2007) RNF8 ubiquitylates histones at DNA double-strand breaks and promotes assembly of repair proteins. Cell 131:887–900

Maruo S, Zhao B, Johannsen E, Kieff E, Zou J, Takada K (2011) Epstein-Barr virus nuclear antigens 3C and 3A maintain lymphoblastoid cell growth by repressing p16INK4A and p14ARF expression. Proc Natl Acad Sci U S A 108:1919–1924

Matsuoka S, Ballif BA, Smogorzewska A, McDonald ER 3rd, Hurov KE, Luo J, Bakalarski CE, Zhao Z, Solimini N, Lerenthal Y, Shiloh Y, Gygi SP, Elledge SJ (2007) ATM and ATR substrate analysis reveals extensive protein networks responsive to DNA damage. Science 316:1160–1166

Mauser A, Saito S, Appella E, Anderson CW, Seaman WT, Kenney S (2002) The Epstein-Barr virus immediate-early protein BZLF1 regulates p53 function through multiple mechanisms. J Virol 76:12503–12512

McBride AA, Sakakibara N, Stepp WH, Jang MK (2012) Hitchhiking on host chromatin: how papillomaviruses persist. Biochim Biophys Acta 1819:820–825

Mesri EA, Cesarman E, Boshoff C (2010) Kaposi's sarcoma and its associated herpesvirus. Nat Rev Cancer 10:707–719

Miller CL, Burkhardt AL, Lee JH, Stealey B, Longnecker R, Bolen JB, Kieff E (1995) Integral membrane protein 2 of Epstein-Barr virus regulates reactivation from latency through dominant negative effects on protein- tyrosine kinases. Immunity 2:155–166

Mittnacht S, Boshoff C (2000) Viral cyclins. Rev Med Virol 10:175–184

Moody CA, Laimins LA (2009) Human papillomaviruses activate the ATM DNA damage pathway for viral genome amplification upon differentiation. PLoS Pathog 5:e1000605

Moody CA, Laimins LA (2010) Human papillomavirus oncoproteins: pathways to transformation. Nat Rev Cancer 10:550–560

Moody CA, Fradet-Turcotte A, Archambault J, Laimins LA (2007) Human papillomaviruses activate caspases upon epithelial differentiation to induce viral genome amplification. Proc Natl Acad Sci U S A 104:19541–19546

Munger K, Phelps WC, Bubb V, Howley PM, Schlegel R (1989a) The E6 and E7 genes of the human papillomavirus type 16 together are necessary and sufficient for transformation of primary human keratinocytes. J Virol 63:4417–4421

Munger K, Werness BA, Dyson N, Phelps WC, Harlow E, Howley PM (1989b) Complex formation of human papillomavirus E7 proteins with the retinoblastoma tumor suppressor gene product. EMBO J 8:4099–4105

Nam EA, Cortez D (2011) ATR signalling: more than meeting at the fork. Biochem J 436:527–536

Nikitin PA, Luftig MA (2011) At a crossroads: human DNA tumor viruses and the host DNA damage response. Future Virol 6:813–830

Nikitin PA, Luftig MA (2012) The DNA damage response in viral-induced cellular transformation. Br J Cancer 106:429–435

Nikitin PA, Yan CM, Forte E, Bocedi A, Tourigny JP, White RE, Allday MJ, Patel A, Dave SS, Kim W, Hu K, Guo J, Tainter D, Rusyn E, Luftig MA (2010) An ATM/Chk2-mediated DNA damage-responsive signaling pathway suppresses Epstein-Barr virus transformation of primary human B cells. Cell Host Microbe 8:510–522

Norseen J, Thomae A, Sridharan V, Aiyar A, Schepers A, Lieberman PM (2008) RNA-dependent recruitment of the origin recognition complex. EMBO J 27:3024–3035

Palm W, de Lange T (2008) How shelterin protects mammalian telomeres. Annu Rev Genet 42:301–334

Price AM, Tourigny JP, Forte E, Salinas RE, Dave SS, Luftig MA (2012) Analysis of Epstein-Barr virus-regulated host gene expression changes through primary B-cell outgrowth reveals delayed kinetics of latent membrane protein 1-mediated NF-kappaB activation. J Virol 86:11096–11106

Radkov SA, Kellam P, Boshoff C (2000) The latent nuclear antigen of Kaposi sarcoma-associated herpesvirus targets the retinoblastoma-E2F pathway and with the oncogene hras transforms primary rat cells. Nat Med 6:1121–1127

Rawlins DR, Milman G, Hayward SD, Hayward GS (1985) Sequence-specific DNA binding of the Epstein-Barr virus nuclear antigen (EBNA-1) to clustered sites in the plasmid maintenance region. Cell 42:859–868

Reinson T, Toots M, Kadaja M, Pipitch R, Allik M, Ustav E, Ustav M (2013) Engagement of the ATR-dependent DNA damage response at the human papillomavirus 18 replication centers during the initial amplification. J Virol 87:951–964

Rickinson A, Kieff E (2007) Epstein-Barr virus. In: Knipe DM, Howley PM (eds) Fields virology, 5th edn. Lippincott, Williams, and Wilkins, Philadelphia, pp 2603–2654

Ritzi M, Tillack K, Gerhardt J, Ott E, Humme S, Kremmer E, Hammerschmidt W, Schepers A (2003) Complex protein-DNA dynamics at the latent origin of DNA replication of Epstein-Barr virus. J Cell Sci 116:3971–3984

Rodier F, Munoz DP, Teachenor R, Chu V, Le O, Bhaumik D, Coppe JP, Campeau E, Beausejour CM, Kim SH, Davalos AR, Campisi J (2011) DNA-SCARS: distinct nuclear structures that sustain damage-induced senescence growth arrest and inflammatory cytokine secretion. J Cell Sci 124:68–81

Rosendorff A, Illanes D, David G, Lin J, Kieff E, Johannsen E (2004) EBNA3C coactivation with EBNA2 requires a SUMO homology domain. J Virol 78:367–377

Saha A, Murakami M, Kumar P, Bajaj B, Sims K, Robertson ES (2009) Epstein-Barr virus nuclear antigen 3C augments Mdm2-mediated p53 ubiquitination and degradation by deubiquitinating Mdm2. J Virol 83:4652–4669

Saha A, Bamidele A, Murakami M, Robertson ES (2011) EBNA3C attenuates the function of p53 through interaction with inhibitor of growth family proteins 4 and 5. J Virol 85:2079–2088

Sakakibara N, Mitra R, McBride AA (2011) The papillomavirus E1 helicase activates a cellular DNA damage response in viral replication foci. J Virol 85:8981–8995

Saridakis V, Sheng Y, Sarkari F, Holowaty MN, Shire K, Nguyen T, Zhang RG, Liao J, Lee W, Edwards AM, Arrowsmith CH, Frappier L (2005) Structure of the p53 binding domain of HAUSP/USP7 bound to Epstein-Barr nuclear antigen 1 implications for EBV-mediated immortalization. Mol Cell 18:25–36

Sato Y, Tsurumi T (2010) Noise cancellation: viral fine tuning of the cellular environment for its own genome replication. PLoS Pathog 6:e1001158

Scheffner M, Huibregtse JM, Vierstra RD, Howley PM (1993) The HPV-16 E6 and E6-AP complex functions as a ubiquitin-protein ligase in the ubiquitination of p53. Cell 75:495–505

Schwarz E, Freese UK, Gissmann L, Mayer W, Roggenbuck B, Stremlau A, zur Hausen H (1985) Structure and transcription of human papillomavirus sequences in cervical carcinoma cells. Nature 314:111–114

Seo T, Park J, Lee D, Hwang SG, Choe J (2001) Viral interferon regulatory factor 1 of Kaposi's sarcoma-associated herpesvirus binds to p53 and represses p53-dependent transcription and apoptosis. J Virol 75:6193–6198

Shin YC, Nakamura H, Liang X, Feng P, Chang H, Kowalik TF, Jung JU (2006) Inhibition of the ATM/p53 signal transduction pathway by Kaposi's sarcoma-associated herpesvirus interferon regulatory factor 1. J Virol 80:2257–2266

Shreeram S, Demidov ON, Hee WK, Yamaguchi H, Onishi N, Kek C, Timofeev ON, Dudgeon C, Fornace AJ, Anderson CW, Minami Y, Appella E, Bulavin DV (2006) Wip1 phosphatase modulates ATM-dependent signaling pathways. Mol Cell 23:757–764

Si H, Robertson ES (2006) Kaposi's sarcoma-associated herpesvirus-encoded latency-associated nuclear antigen induces chromosomal instability through inhibition of p53 function. J Virol 80:697–709

Sinclair AJ, Palmero I, Peters G, Farrell PJ (1994) EBNA-2 and EBNA-LP cooperate to cause G0 to G1 transition during immortalization of resting human B lymphocytes by Epstein-Barr virus. EMBO J 13:3321–3328

Skalska L, White RE, Franz M, Ruhmann M, Allday MJ (2010) Epigenetic repression of p16(INK4A) by latent Epstein-Barr virus requires the interaction of EBNA3A and EBNA3C with CtBP. PLoS Pathog 6:e1000951

Smith J, Tho LM, Xu N, Gillespie DA (2010) The ATM-Chk2 and ATR-Chk1 pathways in DNA damage signaling and cancer. Adv Cancer Res 108:73–112

So S, Davis AJ, Chen DJ (2009) Autophosphorylation at serine 1981 stabilizes ATM at DNA damage sites. J Cell Biol 187:977–990

Stewart GS, Wang B, Bignell CR, Taylor AM, Elledge SJ (2003) MDC1 is a mediator of the mammalian DNA damage checkpoint. Nature 421:961–966

Stiff T, Walker SA, Cerosaletti K, Goodarzi AA, Petermann E, Concannon P, O'Driscoll M, Jeggo PA (2006) ATR-dependent phosphorylation and activation of ATM in response to UV treatment or replication fork stalling. EMBO J 25:5775–5782

Stracker TH, Carson CT, Weitzman MD (2002) Adenovirus oncoproteins inactivate the Mre11-Rad50-NBS1 DNA repair complex. Nature 418:348–352

Stracker TH, Usui T, Petrini JH (2009) Taking the time to make important decisions: the checkpoint effector kinases Chk1 and Chk2 and the DNA damage response. DNA Repair (Amst) 8:1047–1054

Subramanian C, Hasan S, Rowe M, Hottiger M, Orre R, Robertson ES (2002) Epstein-Barr virus nuclear antigen 3C and prothymosin alpha interact with the p300 transcriptional coactivator at the CH1 and CH3/HAT domains and cooperate in regulation of transcription and histone acetylation. J Virol 76:4699–4708

Sugden B, Mark W (1977) Clonal transformation of adult human leukocytes by Epstein-Barr virus. J Virol 23:503–508

Sun Y, Jiang X, Chen S, Fernandes N, Price BD (2005) A role for the Tip60 histone acetyltransferase in the acetylation and activation of ATM. Proc Natl Acad Sci U S A 102:13182–13187

Suram A, Kaplunov J, Patel PL, Ruan H, Cerutti A, Boccardi V, Fumagalli M, Di Micco R, Mirani N, Gurung RL, Hande MP, d'Adda di Fagagna F, Herbig U (2012) Oncogene-induced telomere dysfunction enforces cellular senescence in human cancer precursor lesions. EMBO J 31:2839–2851

Tarakanova VL, Leung-Pineda V, Hwang S, Yang CW, Matatall K, Basson M, Sun R, Piwnica-Worms H, Sleckman BP, Virgin HWt (2007) Gamma-herpesvirus kinase actively initiates a DNA damage response by inducing phosphorylation of H2AX to foster viral replication. Cell Host Microbe 1:275–286

Tatsumi Y, Ezura K, Yoshida K, Yugawa T, Narisawa-Saito M, Kiyono T, Ohta S, Obuse C, Fujita M (2008) Involvement of human ORC and TRF2 in pre-replication complex assembly at telomeres. Genes Cells Devoted Mol Cell Mech 13:1045–1059

Thorley-Lawson DA (2001) Epstein-Barr virus: exploiting the immune system. Nat Rev Immunol 1:75–82

Touitou R, O'Nions J, Heaney J, Allday MJ (2005) Epstein-Barr virus EBNA3 proteins bind to the C8/alpha7 subunit of the 20S proteasome and are degraded by 20S proteasomes in vitro, but are very stable in latently infected B cells. J Gen Virol 86:1269–1277

Verma SC, Borah S, Robertson ES (2004) Latency-associated nuclear antigen of Kaposi's sarcoma-associated herpesvirus up-regulates transcription of human telomerase

reverse transcriptase promoter through interaction with transcription factor Sp1. J Virol 78:10348–10359
Verschuren EW, Klefstrom J, Evan GI, Jones N (2002) The oncogenic potential of Kaposi's sarcoma-associated herpesvirus cyclin is exposed by p53 loss in vitro and in vivo. Cancer Cell 2:229–241
Verschuren EW, Hodgson JG, Gray JW, Kogan S, Jones N, Evan GI (2004) The role of p53 in suppression of KSHV cyclin-induced lymphomagenesis. Cancer Res 64:581–589
Walboomers JM, Jacobs MV, Manos MM, Bosch FX, Kummer JA, Shah KV, Snijders PJ, Peto J, Meijer CJ, Munoz N (1999) Human papillomavirus is a necessary cause of invasive cervical cancer worldwide. J Pathol 189:12–19
Wang X, Wang HK, McCoy JP, Banerjee NS, Rader JS, Broker TR, Meyers C, Chow LT, Zheng ZM (2009) Oncogenic HPV infection interrupts the expression of tumor-suppressive miR-34a through viral oncoprotein E6. RNA 15:637–647
Weitzman MD, Lilley CE, Chaurushiya MS (2010) Genomes in conflict: maintaining genome integrity during virus infection. Annu Rev Microbiol 64:61–81
Whitehurst CB, Vaziri C, Shackelford J, Pagano JS (2012) Epstein-Barr virus BPLF1 deubiquitinates PCNA and attenuates polymerase eta recruitment to DNA damage sites. J Virol 86:8097–8106
Yang K, Moldovan GL, D'Andrea AD (2010) RAD18-dependent recruitment of SNM1A to DNA repair complexes by a ubiquitin-binding zinc finger. J Biol Chem 285:19085–19091
Yates JL, Warren N, Sugden B (1985) Stable replication of plasmids derived from Epstein-Barr virus in various mammalian cells. Nature 313:812–815
Young AR, Narita M, Ferreira M, Kirschner K, Sadaie M, Darot JF, Tavare S, Arakawa S, Shimizu S, Watt FM (2009) Autophagy mediates the mitotic senescence transition. Genes Dev 23:798–803
Zhang Q, Gutsch D, Kenney S (1994) Functional and physical interaction between p53 and BZLF1: implications for Epstein-Barr virus latency. Mol Cell Biol 14:1929–1938
Zhao B, Marshall DR, Sample CE (1996) A conserved domain of the Epstein-Barr virus nuclear antigens 3A and 3C binds to a discrete domain of Jkappa. J Virol 70:4228–4236
Zhou J, Snyder AR, Lieberman PM (2009) Epstein-Barr virus episome stability is coupled to a delay in replication timing. J Virol 83:2154–2162
Zhou J, Deng Z, Norseen J, Lieberman PM (2010) Regulation of Epstein-Barr virus origin of plasmid replication (OriP) by the S-phase checkpoint kinase Chk2. J Virol 84:4979–4987
Zou L, Elledge SJ (2003) Sensing DNA damage through ATRIP recognition of RPA-ssDNA complexes. Science 300:1542–1548

Index

B. R. Cullen (ed.), *Intrinsic Immunity*, Current Topics in Microbiology and Immunology 371, DOI: 10.1007/978-3-642-37765-5,

MIX
Papier aus verantwortungsvollen Quellen
Paper from responsible sources
FSC® C105338

If you have any concerns about our products,
you can contact us on
ProductSafety@springernature.com

In case Publisher is established outside the EU,
the EU authorized representative is:
Springer Nature Customer Service Center GmbH
Europaplatz 3, 69115 Heidelberg, Germany

Printed by Libri Plureos GmbH
in Hamburg, Germany